D1068429

THE
ELECTRONIC
EPOCH

Copyright © 1982 by Editions
Hologramme
Library of Congress Catalog Card
Number 82-51015
ISBN 0-442-28254-0

THE ELECTRONIC EPOCH

Elizabeth Antébi

Scientific Advisors
Pierre Aigrain:
Director of the Science and Technical Division
of the Thomson Group.

Pr Philip Warren Anderson:
The Bell Telephone Laboratories. Nobel Prize
Winner 1977.

Pr John Bardeen:
University of Illinois-Champaign. Nobel Prize
Winner 1956 and 1972.

Pr Hiroyasu Funakubo:
University of Tokyo.

Pr William Gosling:
School of Electrical Engineering, University of Bath.

Pr Frank Tetzner:
President of the International Association of
Radiotechnical and Electronic Media.

VNR VAN NOSTRAND REINHOLD COMPANY
NEW YORK   CINCINNATI   TORONTO   LONDON   MELBOURNE

P.W. Anderson: Bell Telephone Laboratories.
P.L. Aryl: Scientific writer.
P. Auger: Member of the French Institute of Sciences.
R. Bernhard: IEEE Spectrum.
G.-A. Boutry: Founder of LEP (Laboratory of Electronics and Applied Physics, in France).
C. Colliex: University of Paris-Sud-Centre d'Orsay.
H. Edgerton: MIT.
Pat Hawker: Independent Broadcasting Authority.
F. Jutant and M. Boyer: Ecole Nationale des Télécommunications.
Pr. Dr. Ing. C. Reuber: Technical University of Berlin.
J. Robieux: Scientific Director of Laboratoire de Marcoussis.
M.-I. Skolnik: Superintendant of Radar Division, Naval Research Laboratory of U.S. Navy.
D. Strelkoff: French television's Pioneer.
B.-J. van Westreenen: Engineer, retired from Philips.

Translation:
Philippa Crutchley and Stephen Wallis Allen Translation Service (Articles)

Art Director:
Tilman Eichhorn.

Drawings:
Bernard Londinsky and Anatole Pasternak: Pages 45, 46, 47, 48, 50, 93, 95, 111, 132, 133, 145, 146, 147, 159, 175, 177, 189, 190, 191, 192, 210, 213.

Photos:
AEG Telefunken and Archives Telefunken, 16 n. 1 - 17 n. 7 - 28 - 70 n. 1 - 81 n. 1 - 116 n. 1 - 126 n. 7 - 127 n. 1 and 2 - 130 n. 3 and 4 - 150 n. 4 - 151 n. 1 and 2 - 152 n. 3 and 5 - 154 n. 1, 5 and 6 - 173 n. 2 and 4 - 203 n. 3 -
Aérospatiale, 38 n. 1
AIP Niels Bohr Library, 166 n. 4 - 221 n. 1 - 232 n. 1 and 2
Ampex, 60 n. 2 - 99 n. 1 - 162 n. 5
BBC, 151 n. 7 - 227 n. 1
Bell Labs 30 n. 1 - 36 n. 2 - 40 n. 1 - 55 n. 2 and 3 - 65 n. 1 and 2 - 80 n. 4 -83 - 98 n. 1 and 3 - 150 n. 5 - 206
Bundespostmuseum, Francfort, 118 - 199 - 229 n. 2 and 3
Cavendish Laboratory, U. of Cambridge, 188 n. 2 - 208 n. 1, 2 and 3
CERN, 183 n. 2 and 3 - 221 n. 3, 4 and 5
Charmet Jean-Loup, 36 n. 1 - 40 n. 2 - 116 n. 4 - 120 n. 1 - 123 - 126 n. 4 and 5 - 150 n. 1
Chrysler, 70 n. 3 and 4
CIT Alcatel, 41 n. 3

and 2 - 44 n. 1 and 2 - 96
CNRS, 143 n. 2, 3 and 4 - 193 - 214
Collection Fritz Trenkle, 166 n. 3 - 168 n. 3 - 169 n. 1 and 2
Collection Mermaz, 25 n. 1 - 126 n. 1
Collection Pat Hawker, 150 n. 3 - 151 n. 8 and 9
Compagnie des Compteurs, 151 n. 3 - 202 n. 1
Deutsches Museum, 24 n. 1 - 54 n. 3 -55 n. 6 and 7 - 63 n. 1 - 80 n. 2 - 112 - 113 -116 n. 2 and 3 - 142 n. 3 - 152 n. 4 - 162 n. 2, 3 and 4 - 166 n. 2 - 184 - 188 n. 3 - 4 and 5 - 196 n. 2 and 3 - 197 n. 1, 2, 3, 4 and 5 - 203 n. 1 and 2 - 208 n. 4 - 229 n. 5
Electronique Marcel Dassault, 30 n. 3 - 31 n. 1 and 4 - 32 n. 3 - 104 n. 1
Fujitsu, 12 - 13 - 52 - 60 n. 3 - 104 n. 2
General Motors, 70 n. 5 and 6
Hewlett-Packard, 117 n. 3
Hitachi, 68 - 70 n. 2 - 78 - 80 n. 3 - 84 n. 1 - 143 n. 1
IBM, 22 - 54 n. 1, 2 and 6 - 56 n. 4 and 5 - 59 n. 3
Imperial War Museum, 120 n. 3 - 166 -167 n. 1, 3, 5 and 6 - 170 n. 3 and 4
Institut de France, 151 n. 5
IRCAM, 19
ITT, 36 n. 3 - 39 n. 1
King's College Library, Cambridge, 55 n. 5 - 229 n. 4
LAAS, 27 n. 5 - 76 - 77
LEP 84 n. 3 - 85 n. 4 - 91 n. 2, 3 and 5 -105 n. 1 - 200 - 205
Magnum (Eric Hartmann), 224 - 235
Manchester University, 57 n. 1 and 2
Marconi, 120 n. 2 and 4 - 229 n. 1
Marcoussis, 84 n. 2 - 90 n. 1 - 91 n. 4
MIT Historical Collection, 59 n. 1 and 2 - 60 n. 1
MIT Radiation Laboratory, 167 n. 2, 4 and 7 - 168 n. 1 - 170 n. 1
Mitsubishi, 16 n. 4 - 164
Mostek, 61 n. 1, 2, 3, 4 and 5 - 109 n. 14 and 16
Motorola, 80 n. 6
Mullard Ltd, 188 n. 1
Musée de l'Holographie, 99 n. 2 and 3
Nasa, 38 n. 2 and 3 - 74 n. 2 and 4 - 75 n. 1, 2, 3, 4, 5 and 6 - 101 - 202 n. 2 - 203 n. 5
Naval Photographic Center, 30 n. 2 and 4 - 33 - 170 n. 2 (U.S.Navy)
NEC, 14 - 16 n. 5 - 41 n. 4 - 59 n. 4, 5 and 6
NHK, 129 n. 1 - 151 n. 4 - 152 n. 6
Pr Okabé, 136 n. 4
Olympus, 39 n. 5
Peugeot, 16 n. 2
Philips, 18 n. 2 and 3 - 129 n. 3 - 163 n. 2
Plessey, 32 n. 1 and 2
P. Pons, 38 n. 2
Public Record, 55 n. 9 and 10
Radioastronomie Nançay, 72 n. 3
Radiotechnique and Radiotechnique

129 n. 4 - 130 n. 1 and 2 - 154 n. 2 and 4 - 203 n. 4
RCA, 142 n. 5 - 152 n. 2
Regis Mc Kenna P.R., 231 n. 1 and 2
Science Museum London, 21 n. 1 - 36 n. 4 - 54 n. 4 and 5 - 56 n. 2 and 3 - 72 n. 1 - 114 - 117 n. 1 and 2 - 136 n. 1 - 137 n. 1 and 3 - 142 n. 1 - 162 n. 1 - 182 n. 1 and 2 - 186 - 194 - 196 n. 1 - 221 n. 2 -226 n. 1
Siemens and Siemens Institut, 25 n. 2 - 81 - 121 n. 1 - 126 n. 3 - 129 n. 2 - 142 n. 2 and 4 - 168 n. 2 - 169 n. 3 - 220 - 221 n. 6
Smithsonian Institution, 55 n. 1, 4 and 8 - 56 n. 1 - 63 n. 2 - 72 n. 2 - 74 n. 1 and 3 -98 n. 2 - 126 n. 2 - 136 n. 2 - 151 n. 6 - 218
Sony, 18 n. 1 - 71 - 80 n. 1 - 89 n. 1 - 154 n. 3 - 155 n. 1 - 162 n. 6 - 163 n.1 and 3
Texas Instruments, 20 - 21 n. 2 - 59 n. 7 - 138
Thomson (P.Y. Dhinaut, G. Perès) and Archives CSF, 16 n. 6 - 17 n. 8 - 24 n. 2 - 27 n. 1, 2, 3, 4 and 7 - 31 n. 2 and 3 - 34 - 37 n. 1 and 2 - 80 n. 5 - 121 n. 2, 3 and 4 - 126 n. 6 - 127 n. 3 - 134 - 139 n. 1 and 2 - 140 - 148 - 155 n. 2 - 160 - 166 n. 5 - 173 n. 1, 3 and 5 - 179 - 180
Total, 17 n. 3
Université de Bruxelles, 226 n. 2
Varian Associates, 137 n. 2 and 4 - 183 n. 1
Western Electric, 39 n. 2 and 3 - 41 n. 1 and 2 - 86 - 88 n. 1, 2, 3 and 4 - 89 n. 2 - 102 - 105 n. 2 - 124
Yagi Antenna, 136 n. 3 - 233

4

#9174905

The author would like to express special thanks to the following people for their patience and encouragement throughout the long process of writing this book:
Pr John Bardeen, Geneviève Doyon, Claude Dugas and Nancy Green.

The book is now being prepared for publication with the collaboration of the following individuals and organizations:

Pr. J. E Baldwin (Cavendish Laboratory)
F. Beck (CERN)
D. Cade (ENT)
Michel Cassé (French Atomic Center)
Pr. Crowley Milling (CERN)
Paul Crozat (Research Director at CNRS)
Charles Dufour
Marcel Giuglaris
Pr. Truman Gray (MIT)
Hiroshi Isobe (Yomiuri Shinbun)
Pr. Tom Kilburn (University of Manchester)
Robert Lattès (Paribas)
Charles Marshall (Parker P.R.)
G. Millault (Société d'Exploitation Marcel Dassault)
Takao Negishi (Electronic Industries Association of Japan.Europe)
Henri Nozières
Takahiro Okabé
Francis Perrin (Member of the French Institute of Sciences)
Maurice Ponte (Member of the French Institute of Sciences)
Mark Popovski (Smithsonian Institution)
Pr. F. H. Raymond (Chaire d'Informatique - Programmation, CNAM)
Pr. Yves Rocard (Former Director of Physics Laboratory of ENS)
Michel Sauzade (University of Paris XI)
Pr. Scaife (Engineering School, Trinity College, Dublin)
Pr. Süsskind (University of California, Berkeley)
Fritz Trenkle
Pr. C. Townes (University of California, Berkeley. Nobel Prize Winner 1964)
H. Vermeij (CERN)
Pr. Karl Wildes (MIT)
Pr. Wilkes (University of Cambridge)

Konrad Zuse

A.E.G. Telefunken (MM. Hahn and Mlitzke)
A.E.I.
Ampex (Derek Ginger)
Bell Laboratories (Robert B. Ford, J. P. Mc Mahon)
C.G.E. (Willy Stricker)
Electronique Marcel Dassault (C. Gerdy, M. C. Cogny, L. Masliah, M. Climaud)
Fairchild (John B. Hatch)
Fujitsu (N. Yamagushi, J. Harigaya)
General Electric (Peter van Every, François D. Martzloff)
G.E.C. Marconi (Betty Hance)
Gründig
Hewlett-Packard (G. Climo)
Hitachi (N. Omiya)
I.B.M. (K. Allen, G. Poetto)
Infomedia (Jacques Vallée)
I.N.R.I.A. (M. Dauzin)
Intel (Chris Butts)
I.T.T. (T. Flynn)
Laboratoire de Marcoussis (M. Muguet)
Matsushita (A. Nagano)
Mitsubishi (T. Tojimbara)
Motorola (I. Carroll)
Nasa (Les Gaver)
N.E.C. (T. Imamiya, T. Toh)
Philips (J. M. D. Brink, H. Bruining, K. Compaan, J. F. Etaix, E. F. de Haan, J.H. Jaegers, D. Latjaden, J. W. Miltenbourg, Ginette Pouvesle, Pr. J. H. Shouten, Pr. Tellegen)
Radiotechnique (J. C. Bonnet, E. Falck, L. Leprince-Ringuet, Y. Salles)
R.C.A. (H. Enders, F. Schubert)
Siemens (Kerr Knapp)
Sony (K. Hiromatsu, S. Fujita)
Tekade (J. Tretter)
Texas Instruments (Jim Muller, Dan Garza)
Thomson (Claude Dugas, Mme Taman)
Toshiba (K. Kosugi, O. Fuji)
Varian Associates
Westinghouse (J. Pope)
Yagi Antenna

Aérospatiale (M. Paulet)
American Institute of Physics
Bundespostmuseum (Herr North)
Cern (Roger Anthoine)
C.N.E.T. (M. Sampeur)
Deutsche Gesellschaft für Ordnung und Navigation

Deutsches Museum
Foothill Electronics Museum
Franklin Institut
French Embassy to Japan (J. F. Mariani)
French Ministry for Posts and Telecommunications (J. Michon)
I.E.E. (S. Deighton, S. Sorensen)
I.E.E.E.
I.E.R.E.
Imperial War Museum
Institut für Informatik (München)
Japanese Electronic Industries Association (Takeshi Takeichi)
King's College Library (M. Halls)
MIT (E. Halligan Jr.)
Museum and Archives of Magnetic Recording (P. Hammar)
N.H.K. (Akio Yamashita)
Royal Air Force Museum (Mr Greenwood)
Science Museum (D. Robinson)
Siemens Institut (Herr von Weiher)
Station de Radioastronomie de Nançay (M. Cordeille)
Technische Hochschule Berlin (H. Müller)
University of Bruxelles, (Archives)
University of Manchester (DR. S.-H. Lavington)
V.D.E. (S. Rögner, Dr. Steinrück)
V.D.I. (W. König)

5

Summary

The origins of electronics may be traced back to many different moments in time. The early Greeks, for example, discovered the electrostatic properties of amber, for which Thales of Miletus (in ancient Ionia) used the Greek word "electron." In the late nineteenth century, Thomas Edison noticed that under certain conditions of vacuum the hot filament of a tube emitted unexplained "electrical charges" that moved in this opposite direction to the main supply current. A little later, O.W. Richardson became interested in this Edison effect and in 1903 he set down his own theory of thermionic emission. The words "electronics" and "electron" cover a multitude of meanings and applications. In 1891 Johnstone Stoney suggested that electron be used to represent the basic electrical particle. Electronic, as a term, first appeared in the title of a confidential report written by John Fleming in 1902. This term received a wider airing in 1904 in the German magazine *Jahrbuch des Radioaktivität und Electronik* (*Radioactivity and Electronics Yearbook*). In 1930 the term was institutionalized when the American editor Ronald Fink used it as the title of his monthly magazine *Electronics*. These different discoveries and developments all stand out as milestones, but they leave the basic question unanswered: what is the real nature of electronics? Is electronics the branch of physics dealing with the behavior of electrons? Or is electronics a branch of technology that is basically concerned with the applications and essential characteristics of the electron: absence of inertia, sensitivity to exterior fields, and ability to amplify?

In fact, both definitions are valid. The history of electronics is the history of a science and its almost immediate technological applications, which in less than fifty years have completely transformed man's traditions, his environment, and his way of thinking.

Electronics, in the modern sense of the word, means utilizing the flow of electrons either within a vacuum or inside matter. This electron flow can be used to transmit, receive, erase, or store information using the techniques of oscillation, modulation, detection, and amplification for the coding or decoding of messages. This information may be transmitted and utilized in the form of electromagnetic waves — ranging from very low frequencies to lightwave frequencies — to produce currents or electrical or magnetic fields, which are all subject to the laws of physics. These different techniques bring into play a number of different disciplines: physics of course, but also mathematics (theories of coding, information, noise, etc.); chemistry (properties of the materials used); and, particularly in the field of computer processing, formal logic, semiotics (semantics, and syntax).

Today electronics occupies a primary role within those sciences and technologies involved in the processing of information, in the broadest possible sense of the term. Electronics was born out of the curiosity of scientists and the ingenuity of engineers, but its evolution can often be linked to the needs of the military establishment and the increasing importance of industrial automation. The directions it has taken and the roles it has played and continues to play are not unconnected to political and social phenomena. Electronics has become a weapon in the fierce competition between companies in the private sector of industry and also in the public arena of national interests. This is something which cannot be ignored.

The history of modern electronics takes root in that great turning point in the thinking of both physicists and philosophers which occurred at the end of the nineteenth century and the dawn of the twentieth century. The Cavendish Laboratory at Cambridge University in England is a symbol of this great divide in thought that provided the base for a reconstitution of twentieth-century physics, through the work of such scientists as James Clerk Maxwell in the nineteenth century, and J. J. Thomson, Lord E. Rutherford, Dutch physicist Balthasar van der Pol, Sir Edward Appleton, and the inventor of the diode John Fleming in the twentieth century. After the formulation of Newton's Laws of Mechanics, scientists attempted to isolate certain natural phenomena in order to study them independently of other phenomena and, thus, to formulate laws of the universe leading to the development of techniques that would become systematically more sensitive, accurate, and reliable. It was at this point that the "mathematical description of nature" proposed by Werner Heisenberg really came under examination; according to the German philosopher Martin Heidegger, it was a continuation of the "mathematical blueprint of nature" first outlined by Galileo in his contention that the most suitable language to explain the universe was the *lingua mathematica,* the language of mathematics. From that point on, political, psychological, and even philosophical language was modeled on mathematical or, more broadly speaking, scientific language. People began to believe that science would eventually be able to explain everything, since everything takes place in a world that is material, tangible, measurable, and capable of being broken down into its constituent elements.

But electronics is also the offspring of electricity and shares the same great ancestors: Faraday, Ampère, Maxwell, Hertz. The concept of the electrical current, which is both immaterial and intangible, had therefore already thrown doubt on this new belief. To avoid dangerous abstraction, scientists very quickly invented the material "ether"; this was supposed to transport the electrical current.

This concept of the "ether" was the subject of much controversy. Heinrich Hertz, for example, was totally convinced of its existence. It was finally proved to be a figment of the scientific imagination through an experiment undertaken by two men, Albert Michelson and Edward Morley, who had actually set out to prove its existence. These first years of the twentieth century saw a whole concept of the universe swing in the balance. Becquerel's work on uranium and the work of Lorentz, Perrin, Wiechert, Kaufmann, Thomson, and many others on the electron attacked the seemingly untouchable edifice of Newtonian mechanics. Work on the electron dealt a blow to the theory which considered the indivisible atom to be the smallest particle of matter. Einstein's restricted theory of relativity and quantum mechanics, which dealt with phenomena of infinitesimal magnitudes, marked a total break with the traditional conception of the physical universe. In the field of the microcosm, quantum theory presented a flagrant contradiction of the old Leibnitz postulation that "nature makes no sudden leaps."

This break with traditional ideas aroused a great deal of controversy. The scientists who made these discoveries had to struggle not only against the violent objections of their contemporaries, but with their own personal and intellectual reticence. In his *Memoirs,* J. J. Thomson discloses his extreme reluctance to announce the discovery of the electron, while the far-from-timid Wilhelm Röntgen, who discovered x-rays, only confirmed his support of the electron hypothesis after a great deal of procrastination. On the philosophical level, the whole concept of determinism seemed to be under attack. This can be shown by Einstein's celebrated opinion: "God doesn't play dice with the world!" and the long correspondence in which he debated this topic with Max Born. The debate still goes on today.

Man is no longer content just to observe matter. We violate it, bombard it, make it explode inside particle accelerators; we are now exploring ideas (anti matter, quarks, etc.) which even more fiercely contradict the systems constructed by physicists. This means that we are slowly getting used to the idea that science does not propose one overall explanation but a number of possible models. Today, scientists do not refer to a particle without also referring to the outside influence; they look at it in terms of the conditions of the experiment. As Werner Heisenberg writes:

When we look at the objects making up our daily environment, the physical process by which this observation is made possible plays only a secondary role. But each and every process of scientific observation causes considerable disturbance to elementary particles of matter. We can no longer continue to talk about the behavior of a particle without taking into account the observation process itself. As a consequence the natural laws, which are formulated mathematically in quantum theory, are no longer concerned with elementary particles themselves so much as the knowledge which we have of them. We can no longer ask whether these particles always exist "in themselves" in time and space in exactly the same way; in fact we can only talk of the events which occur when by means of the reciprocal action of the particle and some other system, such as the measuring instruments used, we attempt to define the particle's behavior. (Heisenberg, 1962: 18).

This reflection has a double significance. On the one hand it takes mathematical theory as enunciated by Descartes, for example, a good deal further, since the scientist can, in the name of technical imperatives, desert the simple study of the nature of the phenomenon and the rules of traditional scientific observation in favor of calculating the interrelated mathematical relationships between these "events which take place."

The same reflection can also open up a totally different philosophical domain. "Natural sciences," as Heisenberg says, "always presuppose the existence of man." Here he leaves the door open for Heidegger's research. To arrive at a way of thinking about nature which escapes science (or, rather, does not come within its jurisdiction) the only method left to us seems to be philosophy or poetry, because science is incapable of reflecting upon itself. It is the impossibility of studying the behavior of the elementary particle in itself, of objectifying it, and the necessity to limit ourselves to the knowledge we have of it, that now preoccupies electronics specialists tempted by the dream of replacing God when they attempt to explain their inventions by analogy to the various parts of the human body; coupling oscillators and the heart, coded modulation and the nervous system, computers and the brain. In fact, we have to put ourselves in the position of consciously controlling a transformation that can no longer be denied; and we must learn to control it carefully, so we are not reduced to mindless cogs in a machine gone haywire. For many years, scientists considered it impossible to bring these questions down to the level of the layman, and this very choice of words signifies the contempt which scientists and technicans feel for anything less than a completely "scientific" explanation for phenomena. But now, more than ever before, these questions have a great bearing on the daily life of mankind. The ongoing fragmentation of techniques affects even the technocrats themselves; they have become links in a long chain, incapable of seeing either its beginning or its end. For many years they have been conscious of possessing a power based on the ignorance of their fellow men; now they themselves feel increasingly isolated and manipulated. They have become victims of an ignorance which they themselves fostered. The end-link of the chain is made up of the end-users of these electronic "miracles" who have "resigned" from the game. Either they have given in to the blind consumerism that sociologists so vociferously denounce, or they have rejected the whole "technological package." In both cases they refuse to even try to understand.

This "generation gap" is already widening into a yawning abyss. A whole generation of people born after 1950 was brought up with television. Another generation, that of the 1980s, is already familiar with the computer. We have all seen those television documentaries in which the interviewer is totally disconcerted by his encounter with young people or even children who speak a very different language from his, but who are perfectly at ease in a dialog with the computer. Some of these adolescents have even succeeded in producing quite extraordinary technical inventions. The old-fashioned question "What kind of a job do you want when you grow up?" has very little meaning for them in the context of the wide range of possibilities from which they will be able to choose.

Nevertheless, we should realize that this is not a total break with the past, but merely a continuation of the "revolution" that started at the beginning of the century. Every generation now alive has grown up and been affected by the radio whose signals reached into even the most distant and wretched corners of our planet. The radio in its various roles — an arm of political propaganda used by dictators, an instrument of advertising suggestion, a mirror of cultural taste, an arbiter of fashion — was the first electronic product to abolish frontiers and transform social behavior. But today, we are still only marginally aware of the important role radio has played in the past and is playing now.

Any technological innovation on this level arouses fear at first, a fear based on ignorance. A science-fiction writer recounts the story of a letter he received after the historic landing on the moon: "So where will our dead go now?" Thus, it is not only timely, but urgent for us to take up the challenge. To orient and control our own lives, which are increasingly affected by developments in electronics, we must come to understand the basic concepts and history of electronics. We must become aware of the importance of science and technology and of the choices they imply. On the political level we must look at implications in terms of power – centralized or decentralized control of telecommunication networks or data banks, the possibility of technologies transfer, increased centralization and bureaucracy. On the social level we have to examine implications in terms of employment and the complementary education of minds and social behavior. On the philosophical level we must look at the effects of electronics on personal freedom, creativity, and thought. On the economic level we must determine the speed of assimilation of technological change and decision making. For an example let us look at one of the most important economic struggles going on in the world today; it centers around a basic component found in almost all electronic components: the integrated circuit.

According to a number of complementary estimates, the world market in integrated circuits which in 1970 was worth under 1 billion dollars, and in 1980 was 9 billion will rise to almost 15 billion dollars by 1985. The world turnover for data processing and peripheral data processing equipment reached almost 55 billion dollars in 1978. In December 1979, *Business Week* published a Dataquest survey that estimated that by the beginning of the 1980s the Japanese would have taken over some 60 percent of the American market for those integrated circuit memories most in demand at the time (64 Kbits ROM). Dataquest showed that NEC, Hitachi, and Fujitsu had, in fact, captured the American market.

The Japanese Ministry of Commerce and Industry immediately earmarked 100 million dollars for a program designed to reduce the size of integrated circuits even more. Most of the leading Japanese electronic firms had already taken part in an earlier joint research project, which cost some 250 million dollars, between 1976 and 1979. New research centers are springing up, particularly in Japan and the United States, within industry and on university campuses. Several recent developments give us food for thought: the defeat of the Swiss watchmaking industry at the hands of Seiko, one of the first companies to understand the importance of a radical technological change in the quartz watch, or the short-sightedness of a number of huge East Coast American companies that did not convert in time to semiconductors and saw the bulk of the electronics industry settle in the West.

We have begun by drawing up a report on electronics through an examination of its applications throughout the world – consumer goods, medicine, defense, telecommunications, industry, and research. Modern electronics is the offspring of what we have somewhat hastily termed "revolutions": the revolution of '48 (the transistor), the revolution of '68 (microelectronics, the laser and quantum electronics). The development of semiconductors, insulating materials which become conductive under certain conditions, has made extreme miniaturization possible. But most of the electronic equipment we use today (radio, television, electron microscope, radar, etc.) was developed before or during World War II, before the completion of basic research on semiconductors, and are in fact based on the technology of the thermionic tube.

If electronics seems at first glance to be a jumble of technical applications, it is nonetheless closely linked to scientific research as carried out by scientists such as J.J. Thomson, J. Perrin, A. Millikan, J.C. Maxwell, and H. Hertz, who had the courage to question the whole traditional view of physics. These were the men who made it possible to define two very important electronic phenomena: the photoelectric effect and the cathode ray oscilloscope, for example. In conclusion, electronics is the history of men who were pioneers and visionaries and of their work in research laboratories all over the world, both on the campuses of large universities and at the giant industrial complexes of the private sector.

利益	先月比	備
246	+6％	
45	−2％	
56	−5％	
445	+12％	
—	—	4月

PART ONE

The true nature of electronics

So I asked Einstein one day, "Do you believe that absolutely everything can be expressed scientifically?" "Yes," he replied, "it would be possible, but it would make no sense. It would be description without meaning – as if you described a Beethoven symphony as a variation of wave pressure."

Hedwig Born, in "Helle Zeit," quoted by Ronald Clark
in *Einstein. The Life and Times.*

Electronics
and the Consumer Market

When people talk about the "electronic revolution," they are referring primarily to the changes electronics has brought about in their daily lives and in the world around them. We tend to talk of a proliferation of "magical" new objects, although in fact what is happening is that objects themselves are disappearing and being replaced by "programs" or "functions." For example, almost identical TV screen data terminals will be used for very many different functions: to control rail and air traffic, to transmit information (ticket reservations, advertisements, televised newspapers, bank statements, cross-sections of the brain, etc.), to reconstitute recorded information (video tape recorders, video discs), for war games, or frequency and intensity analysis of computer-generated sounds.

In any case, the word revolution is misleading. Consider that one of the first electronic "gadgets" to come to the attention of the general public was the electronic flash developed before World War II by MIT researchers Harold Edgerton, Kenneth Germershausen and Herbert E. Grier. And although it has taken ten years for electronic games to become popular, some such as the electronic chess system designed by the Spaniard Torres Quevedo and later completed, with the help of the computer, by Dutch researcher Van der Pol were developed as far back as the 1910s. Along the same lines the video tape recorder, which has taken twenty years to become an everyday consumer item, owes its existence to techniques developed much earlier for television sets and tape recorders; while the principle behind the pocket calculators which are so popular today was discovered during work done much earlier on the development of the computer. Nevertheless, it is true to say that electronics has become increasingly important to our societies since the 1920s.

The key word for all of these inventions is: timesaving. Modern man is woken by his radio or even his television set (the JVC alarm clock connected to a device with a screen hardly bigger than a postage stamp). He can carry his TV set around with him in his pocket (Matsushita's liquid crystal miniaturized TV set). When he is away from home his video tape recorder records his favorite television programs for later viewing. He can find solitude in the midst of a milling crowd with a Walkman headset which he may even use to learn a foreign language. He takes photos without film thanks to Sony's MAVICA (Magnetic Video Camera) which lets him build up video photo albums and transmit them by telephone to friends or relatives who receive the images on their own television sets. We are creating not just new products, but new habits.

If this progress has in fact accelerated since the end of World War II, it is because of the technological development of materials in existence for over a hundred years, but for which production techniques had not been sufficiently mastered. One such example is provided by the semiconductor, which in turn led to the development of the ubiquitous "chip." Chips are fragments cut out from an ultrathin sheet of semiconductor material (usually silicon) from 2 to 5 mm wide. Within this tiny space it is possible to integrate extremely complex circuits containing hundreds or thousands (eventually hundreds of thousands) of transistors. Without these chips it would not be possible to produce microprocessors, the miniaturized integrated circuits for data processing applications, or the high capacity memories that are essential for the storage of data in modern systems. Indeed, chips are everywhere: in induction or microwave ovens, in programmable washing machines, in modulated-power vacuum cleaners, in electronic scales. They may be found in banks for credit card, account control, and management data processing systems; in travel agencies or theatrical agencies where they are used for ticket reservations; and even on the farm in electronic plows. They also play a role in saving energy and avoiding waste by means of electronic power regulation.

One of electronics' star performers is the liquid crystal. Strictly speaking, it belongs to the field of optical electronics, the science of the transmission and reception of data by means of optical techniques and semiconductor technology. We should also emphasize the increasing importance of optoelectronics, which uses infrared diodes to convert electrical signals into light signals (optical measurements, optical data). This technique is used in luminous displays, modulated infrared signals for remote control of television sets and photo-transistors,

Pollution monitoring center, Tokyo Metropolitan Government.

1

2

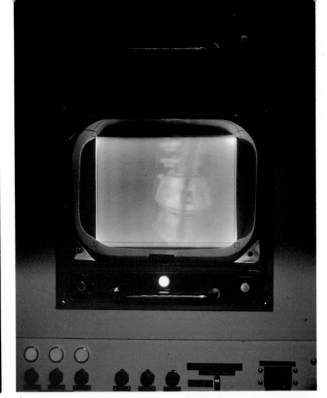

3

1. Car equipped with a radar system which warns the driver if the distance between cars narrows dangerously.
2. This car computer on a Talbot model can hold different sets of data in its memory (mileage, fuel consumption, etc.)
4. Electronic cash registers in a Tokyo department store.
5. Microprocessor-based knitting machine.
6. Flight simulator for the Airbus, used to train flight crews. A color television screen presents the pilot with detailed information concerning aircraft performance and navigation through all stages of the flight. The displayed data is simplified as much as possible to avoid misinterpretation.
3. On board the "Pelican" : a B.O.P. or Blow Out Preventer, a safety system designed to prevent sudden explosions of hydrocarbons.
7. Coding systems like this one have been developed to protect data transmissions: *left,* the text in its uncoded form; *right,* a text whose code is modified each time a transmission is made.
8. Head up display system used on board Air Inter's Mercury aircraft. This landing aid consists of a transparent panel which the pilot swings into position during the final phases of a landing. A computer calculates the required angle of approach and superimposes the theoretical contour of the approaching runway onto the transparent panel. The pilot then modifies his approach angle until the real runway contour coincides with the projected contour.

4

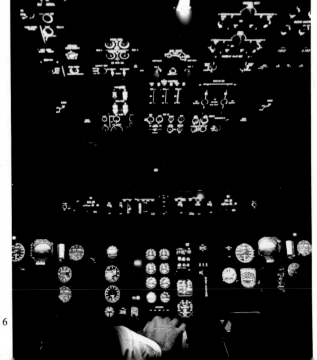

5

6

which receive and amplify light energy.

One of the most interesting developments in this field is the flat, wafer-thin TV screen used for pocket TV sets. The liquid crystals are crossed by electrical wires in the form of a mesh, and picture definition capability has been shown to be comparable to that obtained by a conventional 625 television receiver. When current is applied the liquid crystals become alternately opaque or transparent, and in this way a picture can be built up on the flat screen. So far the technique is capable of displaying only black and white pictures.

The conventional picture tube uses cathode rays to build up the picture point by point while liquid crystals build up a succession of black and white spaces rather similar to the grid pattern of a cross-word puzzle. Seven or eight years from now, these small flat screens will almost certainly transform family attitudes to television. The TV set will become more and more like the transistor radio and like it, will be carried around freely from one room to another.

Within the next ten years electronics will also lead to important modifications in the automobile industry. Since the beginning of the oil crisis, automobile industry technicians have been studying various ways of reducing gas consumption and increasing safety. Electronics will play a major role in this effort, particularly in terms of combustion control. An electronic drive system, the automatic speed regulator, is already fitted in 35 percent of American cars and is expected to become common throughout Europe. But the main area of concern for automobile manufacturers is fuel injection and ignition control systems. Bosch was the first company to develop an experimental ignition and fuel injection technique. It is estimated that by 1985, built-in electronic systems will account for 10 percent of overall vehicle cost. By the year 2000 the car may be fitted out like an airplane, with a computer display screen and an automatic pilot. It may even be able to speak, thanks to a synthesizer unit!

Today three sectors symbolize the transformation taking place in the world: video tape recorders and video discs; the arts, in particular electronic music; and electronic games.

Electronics and the Consumer Market

* The various units used and what they represent:
Hertz (Hz): unit of frequency
Watt (W): unit of power
Ampere (A): unit of electrical current flow
Volt (V): the potential difference between the ends of any given conductor will be 1 V when the current flow is equal to 1 ampere and the power dissipated or absorbed is equal to 1 watt.
Kilo-(k): 1,000 times (1 kHz = 1000 Hertz, 1 kW = 1000 Watts
Mega-(M): 1 million times (1 MHz = 1 million Hertz etc)
Giga-(G): 1 billion times (1 GHz = 1 billion Hertz etc)
Tera-(T): 1,000 billion times (1 THz = 1000 billion Hertz etc)
An inverse relationship exists between frequency and wavelength. A frequency of 1 MHz = a wavelength of 300 meters so:
10 MHz = 30 meters
100 MHz = 3 meters
1,000 MHz = 30 centimeters

Video Tape Recorders and Video Discs

In 1951, Charles Ginsburg developed the video tape recorder for the Ampex company. The video tape recorder uses magnetic tape to record and reproduce not only audio signals but video signals also. The combined signal can be reproduced on a conventional television receiver or, thanks to a recent development, can be projected onto a large screen. In the future it may also be used to record television transmissions that are frequency modulated onto a 5 MHz wide carrier wave.* The recording process had to be adapted to very wide frequencies; the width of sound channels does not exceed 20 kHz, whereas a television signal can extend to 10 MHz. In 1956, the American firm CBS gave a demonstration of the VR 1000, the first video tape recorder ever developed for the general public. However, the video tape recorder did not appear on the market for another 15 years, and when the Philips VCR (video cassette recorder) did appear in 1972, it had nothing like the success predicted. The Japanese products made their appearance in 1974 and proceeded thereafter to dominate the field. Japan now supplies 85 percent of the world market.

Three different systems are in competition: the Sony Betamax, the VHS (developed jointly by Victor Japan, Mitsubishi, Hitachi, and Akaï) and last but not least the Philips Video 2000, which has two advantages over the other two: recording time of twice four hours (reversible cassette) and slaving of the reading head to the track.

The video disc has not met with the same initial success. For a long time it was the flagship of the electronics fleet — but a flagship which was very rarely seen. The video disc is based on a very simple principle: all the information (both pictures and sound) is recorded on a disc according to a predetermined coding standard. The disc is played back using a rather unusual pick-up system. There are three types presently in competition:
— the mechanical pick-up system (R.C.A., J.V.C.) with a reading head guided either by the groove or electronically (the second option is still at the laboratory stage). In this case an electrical current is modulated by the presence or absence of microcapacities on the surface of the disc. The disadvantage of this system is that the disc eventually becomes worn owing to friction from the reading head.
— optical reading by laser reflection (Philips). The laser scans the disc in the form of a beam of light which is reflected with a more or less intensity according to whether or not a micropit passes through it. This disc is designed like a sandwich with a layer that stores the information (analog storage), another layer

obtained by evaporation of a very thin film of aluminum serving as a mirror, and a transparent protective layer that keeps dust from the disc. This Philips video disc provides stereo sound with two sound tracks and the possibility of connection to a Hi-fi system.
— optical laser reading by transparency (Thomson). The laser scans the disc's microcells with a beam of light which passes through the disc and is picked up by the reading head located on the other side. This type of disc has no protective layers and is very thin and fragile.

Video tape recorder and video disc could complement each other. The video tape recorder allows the user to record television broadcasts or to reconstitute the image of a film which he may even have produced himself. The cost of a pre-recorded cassette is at the moment higher than a video disc.

The video disc is always pre-recorded. Its main function is to store information which, for the optical videodisc, may be "frozen" at any point without the disadvantage of scraping or scratching the reading head and wearing out the tape, as is the case with the video tape recorder. It will probably be used with personal microcomputers allowing user participation (optical videodisc). It may also be used for frame-by-frame readout in such uses as recipes, educational programs for children, professional and adult training etc.

Still in the field of recording and reproduction, Philips has developed the Compact audio digital disc, 12 centimeters in diameter and 1.2 millimeters thick, with a single recorded face allowing one hour's stereo listening.

The system uses a mini-laser to optically read the disc, from which the incident light ray is reflected by a metallic film containing all the numerical information. A transparent coating is applied to the disc to prevent scratching and wear. The programming facility allows the user to select the particular part of the program he wishes to hear. This process may well be adopted as the international standard for all future digitally recorded audio discs and is due to go on sale some time in 1983.

1. Inside view of Betamax videocassette recorder.
2. Philips videodisc.
3. Compact Disc: its diameter 12 cm, one hour's playing time on one side.

1

2

3

Electronic Music

The invasion of electronics into the world of music has completely transformed our ideas about musical instruments and musical composition. Electronic music was born in 1951 with the creation of the Cologne Electronic Music Studio under the auspices of Nordwestdeutscher Rundfunk and in collaboration with composer Karlheinz Stockhausen. At the Cologne studio technicians worked with magnetic tapes, a technique that was far from new, but with a difference — they used synthetic rather than natural sounds.

Another device often used by a composer of electronic music is the frequency analyzer; this presents a visual display of the intensity and frequency of sounds (this frequency is generally between 0 and 5,000 Hz; i.e. 5,000 vibrations per second). In music we distinguish between three sorts of waves: the fundamental wave, the harmonic wave (which governs the instrument's timbre), and the transient wave (which occurs when the sound begins to vibrate; this wave lasts only a few milliseconds and, like the harmonic wave, is different for each instrument). To understand what goes on in the listener's ear, we should call to mind some elements of acoustics: sound waves, i.e. those under a frequency of 10,000 Hz, cause the tympanic membrane to vibrate; these vibrations are then transmitted by the otic bones (hammer, anvil, and ossicle). If the sound wave frequency is slightly higher than 10,000 Hz it is transmitted directly to the "outer ear" and is felt as a pressure in the Eustachian tube. The sound wave is transformed into an electrical signal in the "inner ear" by means of thousands of small nerve cells: the pulses penetrate inside the brain through the "positive nerve." The brain is able to make distinctions that the computer is incapable of making; it can distinguish between the different sounds, it can recognize the piano, the violin, the violoncello, the guitar, and the drum and then separate them out from each other. It is also capable of associating these sounds with ideas: sadness with the minor keys, for example, joy with a brisk allegro, and so on. The computer is incapable of this type of appreciation, although it can, like the human ear, measure the pitch of sounds and their intensity.

The two basic units of electronic music are the synthesizer and the computer. The first music synthesizer was developed at RCA in 1955 by Harry Olson and Herbert Belar. An improved version, the Mark II, was installed in New York four years later. The synthesizer is an instrument capable of producing on request sounds of the most diverse frequencies, intensities, duration, and timbre. It can "imagine" speeds or rhythms which the most gifted player would find impossible to reproduce. It has only two limitations: the operator's degree of knowledge of the code and the threshold of sound audible without pain. It is a polyphonic instrument with several different voices all independent of each other, and whose note sequence is controlled independently in the same way as separate instruments playing together in an orchestra.

The fundamental wave is produced by pressing a button; the wave will continue to be generated as long as the button is pressed. The harmonic waves, and finally, the transient waves are then added. Synthesizers possess several memory banks capable of producing at will a timbre resembling that of the violin, a particular sound effect, or any other kind of sound. The computer makes this musical synthesis possible and allows more complicated research into the organization of the sounds. In 1957 the "Illiac" computer at the University of Illinois composed the "Illiac Suite for String Quartet." The computer was, of course, programmed in line with certain laws of musical composition.

In 1963 the Bell Laboratory team led by M.V. Mathews developed the first direct synthesis of sound by computer using digital coding methods. The computer works with figures rather than directly dealing with the vibrating waves themselves: the wave's progress is divided by cutting the sound up at regular intervals to measure the pitch at each point. These pitch levels are then transcribed into a series of perforations on a program card which the computer reads to create the desired timbre (See "Telecommunications" for an extended discussion of this "division process").

The electric organ is designed according to the same principle: by changing the program the operator can change the timbre, so that different types of organ, classical, romantic, etc., are available in the same instrument. The computer makes it possible to produce all the different special effects that were formerly achieved by manipulation of magnetic tapes mixing, alteration, change of speed, reverberation, and so on.

The synthesizer and the computer are highly flexible instruments with an almost unlimited "imagination" and are capable of changing the composition while it is being "created." The words "imagination" and "creation" are in quotation marks because the machine of course only possesses these qualities to the extent that they exist in the programmer. This type of instrument has been created to respond to the needs of musicians who were obliged to use the "old" instruments to create "new" sounds. The instruments of the seventeenth century were unable to meet the needs of a very different public approach to music. Today, musicians are looking for new instruments to create new works, rather as architects are looking for new building materials. Musical composition is becoming a collective work in which the artist carries on a dialog with the technician, as in architecture, when the architect collaborates with the engineer.

Another explosion in our concept of art: the application of the computer's power to the video image. Such famous film producers as Francis Ford Coppola or George Lucas have opened ultra-modern studios to explore all the possibilities of this new field.

The computer allows us to escape from the restraint of observed reality and to create imaginary structures. It makes it possible to calculate individually, within the parameters provided, each one of the 250,000 to one million points making up the video image and thus to create the imaginary shapes or colors desired.

Research in this field began towards the middle of the sixties when artists like Lilian Schwartz or Stan van der Veek started to work with data processors. The army very quickly saw the advantage it could draw from these image syntheses by applying the principle to such devices and systems as flight simulators.

Philippe Quéau's survey reports that I. Sutherland designed a headset equipped with two ultrasonic sensors and two mini TV receivers fed by computer. As the wearer of the headset moved about, the movements of his body were transmitted to the computer which calculated the images he would see if, for example, he was moving around inside the brain, a molecule or some imaginary architecture. (Quéau, 1981: XIII, XV).

Pierre Boulez (left) was born in 1925 and studied with Messiaen and Leibowitz. He is permanent leader of the BBC Symphony Orchestra and the New York Philharmonic. He has been the head of IRCAM (Institute for Acoustical and Musical research and Coordination) since 1971.

Electronic Games

The idea of play, already present in electronic music and in the inventiveness and special effects associated with television screen drawing and design has not escaped observers of the electronics scene. We need only look at how laboratory scientists all over the world, in the most varied environments, are "playing" with computers. As early as the nineteenth century, one of the most famous pioneers in the field of calculators, the Englishman Charles Babbage, invented an electronic game that was produced in the United States almost a century later. From now on sheepish adults no longer have to hide in a corner to play with their electric train sets; bars and other public places have been invaded by electronic games specially designed for adults.

Electronic games underwent particular growth after 1951 thanks to the invention of the chip. The first of these games used the principle of probability calculation now used in electronic chess, Mastermind, and electronic naval battles. It did not take long for this principle to be applied to such educational games for children as "Little Professor," "Speak and Spell" (a game produced by Texas Instruments) or "Dataman" (a game used for teaching children mathematics). Magnavox was the first to market a video game, "Odyssey."

1975 saw the appearance of Atari's "Hockey Pong," developed in close collaboration with American Microsystems, an integrated circuits manufacturer. But the real expansion of electronic games dates from March 1976 when General Instruments announced the production of integrated circuits for television games (AY 3 8500). This opened the way for games simulating tennis, automobile driving and football on television screens. Finally came the development of keyboard control units accompanied by cassettes or memory cards with game rules, which could be used to create new games. This is the "home-computer" generation, with its digital keyboard allowing dialogue with the machine; a system that was developed in the United States in 1977 and 1978. It uses R.A.M.'s (Random Access Memories) or active memories that allow the operator access to any of the information it contains at any point, and R.O.M.'s (Read Only Memories) or dead memories that allow the operator to follow the recorded program only.

This period of the 1970s saw the appearance of a whole spate of firms specializing in electronic games: Commodore, Plustronic, Casio, Rexton, and others.

Probably the most famous electronic game is "Star Trek," which works in the following way: the computer gives the enemy's location, the energy needed to trap him, speed and remaining fuel indication, firing strength, and various other parameters. The player uses these data to attack other spaceships, to make certain that he does not run out of fuel and to avoid collision with meteors. Another popular game is "Hammourabi," the "businessman's wargame." The player becomes Hammourabi, chief lawgiver of a primitive kingdom. He must make decisions concerning the distribution of resources, agricultural planning, and the conduct of war and peace. He also has to deal with the unexpected in the shape of epidemics, invasions, and any other "plague" that threatens to complicate the implementation of his strategy. Other companies such as Ohio Scientific and Micro Biz have invented games based on juggling with the principles of economy and trigonometry. Some Atari games deal with various other disciplines such as history, physics, and psychology.

In this way everyone will shortly be able to recreate the universe in his own 10 by 12 studio apartment. You will be playing golf on the most prestigious golf courses in the world without leaving your own home, balancing the family budget, codifying the universe and waging your own personal war against the adversary of your choice all without having to get up from the armchair to open the door for the dog. There will be a voice (or bark) analyzer to do that.

Some people are alarmed at the possibility of increased isolation of the individual and the atrophying of creative faculties. But electronic games have already made their appearance in public places and have encouraged players excited by the growing complexity of problems to join together in clubs to play against each other or even collaborate against the machine. Another positive development has been the invention of specific modules such as the Atari that allow the player to invent games.

Manufacturers are already thinking along the lines of exploiting all the computer's resources and turning it into an "intelligent" machine, by programming it with so wide a range of rules that it will be able to play any game and even teach the beginner.

Texas Instruments' "Speak and Spell" was the first electronic game to enable the general public to hear the synthesizer "voice." From the first hesitant research carried out at the beginning of the 1960s to the invasion of speaking games in 1980, chips have allowed unbelievable miniaturization; the synthesizer industry created by small manufacturers has become the domain of the great integrated circuit manufacturers. This market, which was worth several hundred million dollars in 1980 is expected to reach almost 1 billion by 1985 and 5 billion by 1990*. One reason for this expansion is that the word synthesizer has hundreds of applications in telecommunications (information, talking clocks), in the public sector (aids for handicapped persons), and in industry (remote monitoring systems, etc.).

The principle behind the word synthesizer is fairly simple: it consists of a power source capable of transforming the basic frequency of the human voice (from 60 to 400 Hz) into pulses; a filter that mimics the vocal process, converts these pulses into audio-signals, causing the loudspeaker membrane to vibrate. The main problem is one of software and the fact that the memory is still limited: the machine's vocabulary has to be prepared and stored; the sound signal, as well as being divided into phonemes, the basic units of the spoken word. Spoken dialog between man and machine is not yet possible,

* These figures advanced by *La Recherche* magazine (January 1981).

Speak and Spell, one of the most famous educational games ever created.

not sufficiently advanced. However the Americans, the Japanese, and even the Europeans are carrying out systematic research in this field, and it seems probable that this man/machine or user/programmer dialog will be available in the near future.

The electronic chip can be seen either as a miraculous or a monstrous invention; as a new tool for well being and creativity or a highly manipulated and controlled universe of leisure and games. In fact, electronic games do to a certain extent symbolize this ambivalence. The games do play an educational role, but they may also be used to encourage the child's instinct for competition and domination (war games, for example) or direct him towards the solutions authorized by the authorities in power (presenting him with programmed solutions in games dealing with energy choices, for example). Such apocalyptic warnings often reflect a very reasonable and responsible desire for caution. But on another and less accessible level, they may be the reflection of unconscious and irrational fears aroused by the imagined enslavement of the world to fantastic robots − a scenario that owes more to such films as "Metropolis" and "Goldorak" than to any of the present generation of electronic machines.

Nevertheless, electronics is disturbing; it forces people to take stock of their everyday customs and habits, their ideas on education and creativity. It calls for a new way of listening to and looking at the universe that surrounds us. Our task is to enter fully into this universe which we have brought to life and helped to create. More than ever, it is important for us to understand the advantages, the limitations, even the dangers of these "strange machines" that man has called into being.

1

1. "The Rainbow's Egg", by C. Emmet and A. Kitching. An example of Computer Art, employing techniques originally developed for cartoon film animation.

2. Personal computers for the "microkids" generation.

2

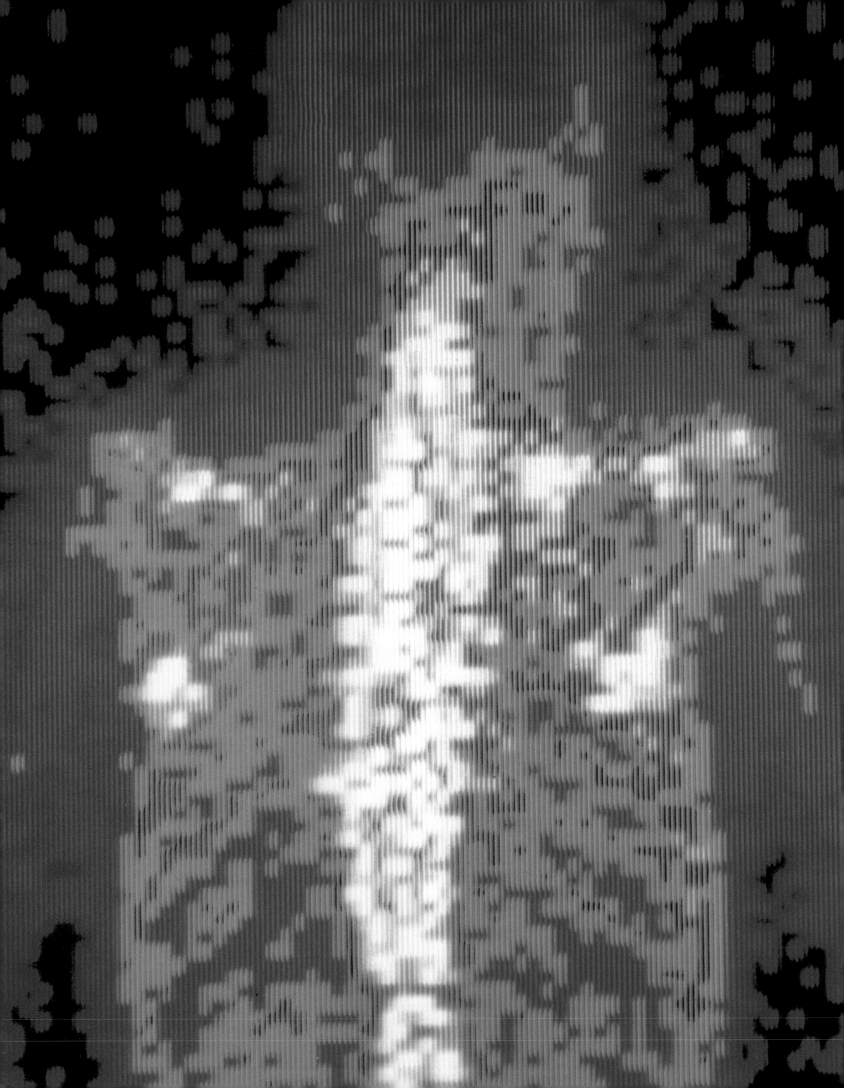

Electronics and Medicine

Electronics and medicine first joined forces after the discovery and almost immediate application of x-rays. The second major step was the discovery of amplifiers and the development of the first measuring oscilloscopes. But the therapeutic effects of electricity had been appreciated for over 2,000 years, thanks to the since-explained phenomenon of the electric ray. In the eighteenth century the French clergyman Jean-Antoine Nollet and a number of other scientifically minded hobbyists gave their celebrated demonstrations in the various salons of the day and at the French court. People flocked to be "electrified," to receive that bizarre and delicious shock; some even attached electrodes to their bodies while embracing to experience truly electrifying body contact. In 1789 Jean-Paul Marat published a *Note on Medical Electricity*. But right up until the beginning of the twentieth century research was ill-defined and rather sporadic in nature, with the possible exception of Jacques d'Arsonval's work on the use of high-frequency waves to stimulate the muscles.

Electronics was first used to carry out comprehensive examination of the patient's body in order to give the doctor an accurate overall idea of his state of health. Shortly afterwards, it was used to actually treat some diseases or to compensate for natural deficiencies. Today there are three main fields of application: diagnosis, therapy and computer monitoring of seriously ill patients receiving intensive care or undergoing surgery. We could mention a fourth field of application: simulation and prevention. Surgeons are now even beginning to have recourse to electronics to assist in the conduct of the operation itself. Behind its current use is a long history of experimentation on animals dating from the end of the 1930s. For the most part this work was carried out by physiologists such as Americans Joseph Erlanger and Herbert Spencer Gasser, pioneers in the field, who carried out experiments on nerve conduction. Their efforts were rewarded by the Nobel Prize for Physiology and Medicine in 1944.

The electroencephalograph was invented in 1929 by Hans Berger, a German psychiatrist. At first nobody had any faith in the new machine (the same thing happened with the photo scanner, which was more or less ignored until an insurance company came up with a way of using it to the advantage of its clients). In the 1930s it was through its championship by the great British physiologist Edgar Adrian (Nobel Prize, 1932) rather than any sales initiatives that this attitude of reluctance changed. In the meantime the electrocardiograph had become electronic, and string galvanometers had been partially replaced by tube galvanometers, although both systems were in parallel use until 1940.

The basic theory of the functioning of the nerve impulse, which was of capital importance to the development of medical electronics and medical research in general, was constructed by two English scientists, Alan Lloyd Hodgkin and Andrew Fielding Huxley as a result of their fundamental research on the giant axon of the squid (the giant axon is the long part of the nerve cell that conducts an impulse away from the cell body). The two Englishmen received the Nobel Prize for their work in 1952.

Diagnosis

The electroencephalogram, the first medical application found for electronics, very quickly became a clinical tool; by 1935 – 36 it was being used in the treatment of epilepsy and brain tumors. Almost all the current methods for detecting physiological signals (electric signals, temperature, pressure, etc.) use electronics. The basic principle of this detection is that each signal must be converted into electrical energy, the only form of signal which can be easily amplified and processed. Almost all physiological examinations are thus dependent on electronics, including radioactive tracer examinations in which all transmission systems are electronic, and ultrasonic methods.

The photoscanner was invented around 1955 and developed by the English firm EMI. It inaugurated a new generation of equipment in which the computer was to play an integral role. The photo scanner scans a particular region of the body – line by line – by means of a very fine beam of x-rays whose absorption is continuously measured about a 180 degree scanning path. The computer processes the results obtained and builds up a series of cross sectional images ("image reconstruction") showing the slightest differences in absorption. In this

IBM computer-controlled scanning and display equipment used in cancer research at a Paris clinic.

which often have harmful effects, in measuring differential absorption and thus detecting the presence of organ degeneration, tumors, infarctions, or edema.

The Mayo Clinic in Rochester has just developed a special x-ray scanner with a moving section weighing 17 tons; this arm moves around the patient and produces three-dimensional images of the vital

1. Electrotherapy: the patient is "electrified" by the doctor using a cylinder-type generating machine. Vases filled with water act as a resistance.
2. Manufacture of x-ray tubes.

end of 1980. The x-rays are produced by means of a scintillation camera; the impact of photons (elementary light particles) on a large sodium iodide crystal produces local scintillation that is picked up by a series of photomultipliers. An auxiliary device, usually connected to a computer, locates the point on the crystal where the scintillation originates and draws up a map.

Scintography is used in the study of opaque bodies. A scintillation counter follows the path of an active gamma ray emitting isotope through an organism. If, for example, we inject iodine into a normally functioning thyroid gland the gland will appear on a screen. If it is not functioning we see nothing at all. If it is partially functioning we see both a black area and a visible area. Possibly cancerous nodules appear as black dots.

The ultrasonic technique exploits the echo phenomenon and the acoustical properties of the human body. It gives an image of internal organs and body tissue and records any anomalies. This method uses high-frequency waves and works by reflecting ultrasonic pulses for anatomical and physiological examinations. Thanks to the extremely short wavelengths, the beams may be focused on precise points and penetrate very deeply. This technique is particularly suitable for the examination of organs not easily accessible to x-rays, the liver, the spleen, the pancreas, the heart, and the breasts. The field in which it is most often used is gynecology. In the field of neurology, low-power electrical pulses of only a few microvolts may be detected on the surface by means of electrodes applied to the head and linked to amplifiers. The difference between the radioactive tracer method and the ultrasonic method can be seen in terms of their relative results: the radioactive tracer

data; the ultrasonic method cannot. In contrast the ultrasonic method can provide images of the heart valves. A deviation of the intermediate cerebral septum can be detected by the ultrasonic method, but it can be exactly located only by use of the radioactive tracer method, that is by using a photoscanner.

Another category of functional examination is the impedance method; this may be used, for example, to separate the operation of the left lung from that of the right. A very weak electrical current (0.5 milliamps) is sent at a relatively high frequency (between 2,000 and 4,000 Hz). A measurement of deep tissue resistance is obtained; when the lung fills up with air, this air acts as an insulator and the blood as a conductor. In this way the increase of resistance at the moment of respiration can be noted.

Blood circulation may be observed in the same way. When blood is driven out of the heart during the systolic (contraction) phase, tissue resistance decreases. This method has also been used to examine the main arteries in the limbs. Various other

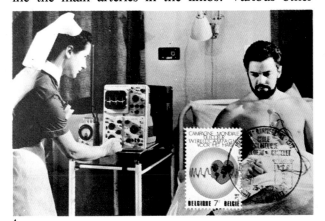
1

or the surgeon during an operation. Doctors have access to thousands of medical descriptions stored in a control unit and x-ray reports giving all diagnostic possibilities. In many clinical cases, physiopathological procedures are known and laboratory data are measurable. The diagnoses thus presented are usually based on probability calculations and determination of probable classification. In the same way television pictures stored during the course of an operation may later be analyzed, and x-ray images may be used as an adjunct in postoperation analysis.

Therapeutic Applications

The therapeutic applications of medical electronics are much more recent than the diagnostic applications. Some of them – pacemakers, heart stimulators, devices that destroy certain tissues by high-energy electron bombardment, devices that destroy cancerous tymors by irradiation – are very well-known. Others are hardly out of the experimental stage. One example is the defibrillator; this machine produces electrical pulses that paralyze the heart when an uncoordinated movement of the ventricular muscle cells occurs. Other new methods include diathermy (heating of internal tissues by 10 kHz waves), ultrasonic treatment of the ears or gallstones, and the use of the laser in cytology (affecting cell mytosis) or genetics (destruction of chromosomes in a single cell).

The introduction of microprocessors in medical data processing has led to the development of analysis and self-monitoring machines (pregnancy tests, weekly insulin schedule for diabetics, etc).

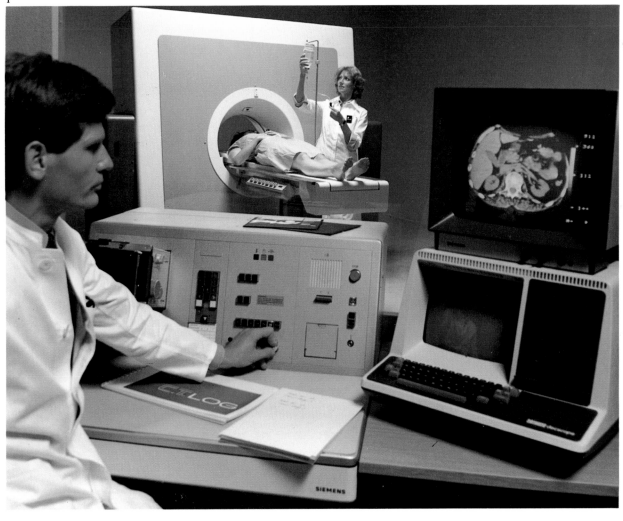
2

1. The electrocardiogram provides a chart recording of currents produced by the contraction of the heart muscle and can detect abnormal rhythms or heart disease.
2. The German Somatom, the first scanner capable of "scanning" the patient's whole body.

tronics is the monitoring of patients receiving intensive care or undergoing an operation. Such physiological measurements as pressure (intra-arterial, intravenous, and intraventricular), electrical activity of the heart, temperature, blood density after injection of an indicator dye, urinary flow, extent of drainage from the thoracic cavity during heart surgery, etc. may be continuously monitored using electronic devices.

A computer then analyzes the different curves and processes the data, giving immediate warning of the slightest anomaly. This warning is automatic once the upper or lower alarm limits set by doctors or surgeons are exceeded. Such anomalies as tachycardia (an abnormally fast heart rate) and its opposite, bradycardia, are detected by continual monitoring of the heartbeat and comparison with preprogrammed limits. For each incident the machine memorizes the date and time at which the attack begins, the duration of the attack, and the abnormal heart rates occurring during the attack. Doctors also use probability data in a statistical analysis to follow significant changes in a patient's condition.

Since 1972 various experiments have been made using both closed circuit and satellite television links to monitor the condition of patients. Canada was the first country to install this system to monitor patients in a small Ontario clinic from the central hospital. In 1964 a satellite system linking the University of Nebraska, the Omaha Psychiatric Institution, and Norfolk Hospital was set up in the United States. This was the beginning of what is now known as diagnostic screening facilities. In 1967 a television hookup was installed between Massachusetts General Hospital and Boston International Airport. It was the first such hookup to be used for diagnosis and clinical treatment; the 1964 hookup was used only for psychiatric consultation and had a more administrative bias.

Surgery

Surgeons have recently begun to use an increasing number of simultaneous measurement devices such as the laser blood cell counter, to follow the progress of the patient on the operating table. But as far as electronic surgery is concerned one of the major areas of progress has been simplification of hemostasis; faster coagulation is achieved by applying clamps to small blood vessels while needle or flat electrodes are used to stop hemorrhages and encourage volatilization of the tissue sections before cutting. In addition the laser can be now used to operate on detached retinas without hospitalization and even without anesthetic. The laser beam passes through an endoscope (an optical instrument used for internal examination of the eye) that is hooked up to a computer; the laser has well- established antihemorrhaging powers and leaves much cleaner scars. The laser knife was used for the first time at New York's Montefiore Hospital in 1964. In 1978 it was used for the first time in acupuncture and also in microsurgery of the living cell by French professor Maurice Bessis. It may eventually be used in place of the dentist's drill and for vascular surgery, brain surgery, and dermatology.

Simulation and Prevention

Data processing also has certain applications in terms of simulating vital processes and studying their functions. The simulation model is designed to imitate a given physiological state as accurately as possible, providing theoretical curves of physiological

allows an operator to measure the effect of variations in one or several parameters on the whole system. It may be used as a diagnostic aid and also to measure the consequences of new drugs on the organism. Thanks to data processing, doctors now have access to a data based system upon which they can establish long-term medical therapy; these systems can also provide back-up for preventive medicine or control of such chronic disorders as hypertension. A program of this type has been in service in Tokyo and Osaka for a number of years and has so far shown interesting results.

The computerization of patient information and diagnostic screening and treatment facilities have given rise to some controversy over the dehumanization of medicine and the Kafkaesque metamorphosis of man into a machine. But technicians are nowhere near the stage of being able to completely understand the "human machine"; neither can they reproduce every element of it. What is more, they are unlikely ever to possess these means, because the real barrier here is one of cost. A man himself is priceless, but the medical equipment for observing him, caring for him, saving his life, or simulating his body functions is subject to very definite cost limitations. The competition of a number of increasingly sophisticated systems creates a crucial overinflation of the market and imposes a choice of priorities. The introduction of such systems has often been regarded with suspicion in medical circles, partly as a kind of conservative reflex of course, but also partly through the justifiable fear of losing the warm, spontaneous, and trusting relationship between doctor and patient.

Electronics can be a valuable tool in diagnosis, detection of anomalies, clinical care and postoperative observation, but it is only a tool and should not eliminate necessarily the human rapport. When interwiewed on French television, Professor Pierre Rabishong of the French National Institute for Medical Research in Montpellier, France, said:

If we asked a technician to produce an optical system which with the help of, say, two cameras, could reproduce stereoscopic vision, in color, with the definition which we find in the human retina, with its 130,000 pick-ups per square millimeter; to retain a certain convergence in lateral movements and to analyze all the images and store them in order to finally reconstitute them in a machine weighing no more than 300 grams, he would tell us that the task was completely impossible. Guirardoni, 1981.

Electronic man will not replace his more fleshly "ancestor" tomorrow, nor in any foreseeable future.

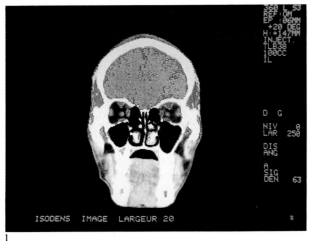

1

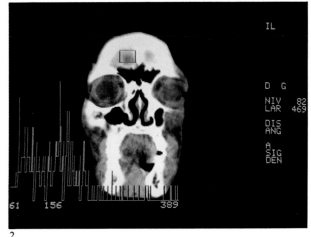

2

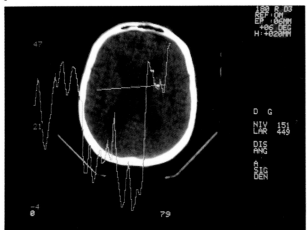

3

4

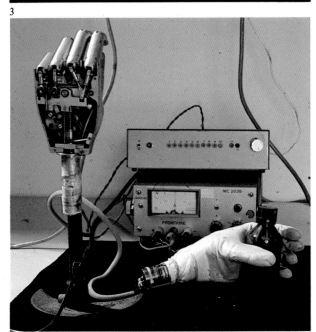

5

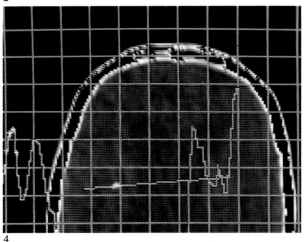

6

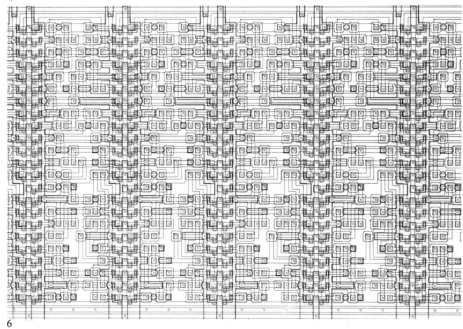

7

1, 2, 3, 4. This scanner continuously scans the brain and records its electromagnetic radiations. These recordings are then analysed by a computer which allows high definition cross sections of the brain to be displayed on a T.V. monitor.

5. Artificial skin was first tested by Mr. Clot at LAAS (Laboratory for System Analysis and Automation). This unique system appeared in 1978 and is used for automatic shape recognition, the study of friction-free beds, analysis of tactile functions and balance, automatic safety and management systems and to measure athletic effort.

6. Prototype of the artifical throat for word synthesizers.

7. Linear accelerator for medical use, used in radiotherapy treatment, particularly for cancer.

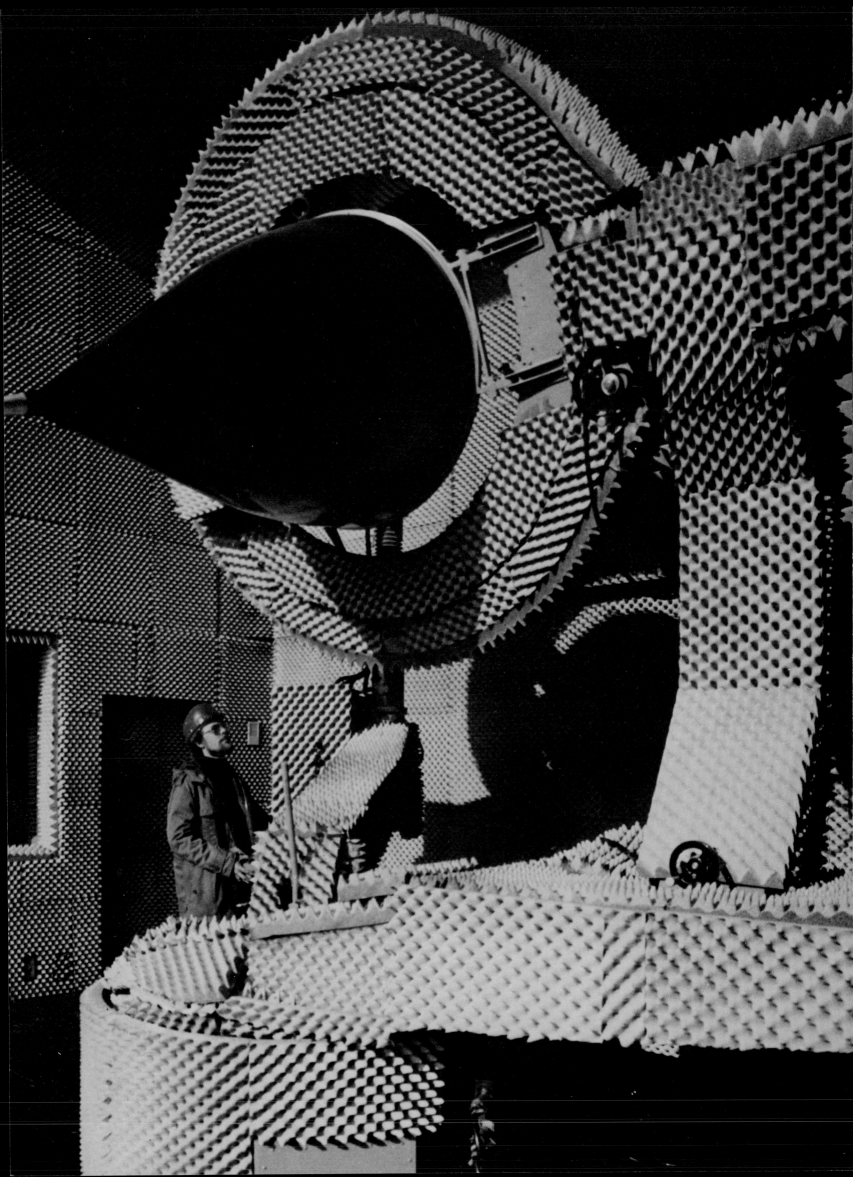

Electronic Warfare

Popular fancy and science-fiction writers have together created a vision of the electronic war, fought by fleets of pilotless aircraft or "drones," Night hawk helicopters equipped with computers, teleguided missiles, Nike-Zeus missile interceptors (1957 style), M.I.R.V. s (Multiple Independently targeted Reentry Vehicles consisting of several individual guided missiles), and the infamous death rays. In reality lasers have not yet obtained the great power and flexibility of these sci-fi death rays; the electronic war is above all a defensive war and its battlefield an electromagnetic one.

Electronic warfare is not a recent invention; during World War I the Germans were able to jam the radio transmissions sent out by the British cruiser "Gloucester" as it pursued two German cruisers. In World War II, the British developed "Bromide," a device that sent out radio beams to disturb the aim of German aircraft using radioelectric signals to guide their night bombers. The Allies also used "windows" (aluminum strips) to confuse German Würzburg radars. During the Korean War, the U.S. Air Force equipped its Tb-25 J's with jammers against antiaircraft guns using radar fire-control. The United States and the USSR have long been interested in the potential of electronic warfare, but there has also been a great rush on the part of other countries – Europe and the Middle East at first, but also Japan, India, South Africa, Brazil, Argentina, and some Southeast Asian countries – to purchase this highly sophisticated equipment. Their share of the market at the beginning of the 1980s exceeds 1 billion dollars. To what do we owe this sudden interest?

Three events in the last ten years have been mainly responsible for arousing the interest of various military and defense research organizations throughout the world. The first was in 1968 when thanks to the jamming and interference devices that had been installed along the Czech border Warsaw Pact forces invaded Czechoslovakia without immediately triggering the NATO warning systems.

The next event occurred during the closing stages of the Vietnam war when the Americans staged their successful B-52 bombing raids on Hanoi and Haiphong in 1971 despite the fact that the North Vietnamese were using one of the most powerful antiaircraft defense systems ever developed. To overcome this system the Americans used the first completely integrated airborne jamming system, the EA-6B Powler designed by the U.S. Navy and Grumman (the system is also known by the initials T.J.S. – Tactical Jamming System). The T.J.S. consists of ten very powerful jammers installed in five pods located around the fuselage. The EF 111A escort aircraft has now been fitted out with this electronic countermeasure equipment.

But it was really the Yom Kippur War (October 1973) that revealed the overriding importance of very rapid response time in any electronic war: in two weeks the Israelis lost 105 aircraft because they had not designed an effective countermeasure system against a completely new Soviet ground-to-air missile system that proved effective at low altitudes. This was the SAM6 (a medium- and long-range heat seeking missile that required no further assistance from the ground after firing and that used infrared sensors to track down attacking aircraft) and the short-range version, the SAM7. For the first time ever, Soviet antiaircraft systems could cover the whole defense spectrum.

Paradoxically, however, the Yom Kippur War was also a triumph for the American Navy in terms of electronic warfare: the Styx missiles fired by the Arab Navy were neutralized. And yet the Navy had woken up to electronic warfare long after the Air Force did. It also received three warnings: the destruction of the "Eilat" and an Israeli merchant ship in 1967 by six radar-guided missiles; the losses sustained in 1971 during the India-Pakistan dispute when the Indian Navy launched thirteen Styx missiles; and the capture by the North Koreans of the American electronic intelligence-gathering ship "Pueblo," which had no long distance detection systems to avoid capture or provide any type of defensive action.

Electronic warfare research can be broken down into three main areas. First is *electronic intelligence-gathering*, which brings together information on enemy objectives and methods. The characteristics of each radar (carrier wave frequency, mode of transmission and power, pulse length, repetition fre-

Radar in an anechoïc chamber, a "free space" room devoid of echoes and reverberations. Reflection-free sound conditions are vital for testing equipment that produces and receives sound, or for studies of speech and hearing.

Electronic Warfare

1. In April 1952, a new era of technology dawned when the NIKE Ajax system developed by Bell Laboratories destroyed a B-17 drone in flight above the White Sands, New Mexico, testing ground.
2. The EA 6 B Prowler, an all-weather electronic countermeasure aircraft on patrol.
3. Firing a Matra Super 530 air to air missile.
4. The EA-6 A Intruder, a special version of the A 6 A attack aircraft, awaits final clearance before a test launch from a catapult at the naval air test center at Patuxent River.

1

2

3

4

quency) are analyzed, fed into computers and stored. As the parameters and type of information received are numerous, the analysis is done by very fast computer; the histogram method which analyzes the frequency of appearance of a particular piece of information (amplitude, repetition frequency, etc.) has become increasingly popular; so has digital coding of the various parameters. These different electronic intelligence-gathering methods are used to deduce the strategic or political objectives of a potential enemy.

In the second area, that of *countermeasures*, "passive" or "confusion" methods may be used: these include limited or intelligent jamming, for which it is necessary to know the exact frequency of the enemy radar. In contrast, barrage jamming, a method that uses the whole frequency band, requires very high power levels. Western scientists are working on improving intelligent jamming methods and Western nations have therefore concentrated on miniaturization, rapidity, and sophistication while the Eastern bloc relies on large concentrations of equipment to compensate for a certain technological lag (for example in semiconductor technology). This choice also corresponds to a basic difference in approach. The Eastern bloc countries operate a large number of radars across several different frequency bands; this forces an airborne adversary to counter jam several frequencies at the same time. The more frequencies an electronic countermeasure system is required to jam, the less power it has available to jam any one frequency. This, in turn, has led to the radiated power management system, based on the principle of concentrating a jammer's available transmission power on whichever signal is computed to represent the major threat. This system brings into play computerized assessment of spacing, timing and frequency classification and is a high-activity field of research at the present time.

Harry Eustace, editor of *Electronic Warfare,* estimates that in the case of a threat from Eastern bloc countries (where the probable signal density would be around 2 million pulses per second at 20 GHz), the survival chances of Allied aircraft would essentially depend upon their skill and capacity to interfere with or neutralize enemy fire-control radar. The receiver must be capable of determining jamming priorities, measuring the frequency of the radar selected, and then locking onto the jamming frequency, analyzing the repetition frequency to allow synchronization of jamming signals, and measuring the scanning frequency. This new method of radiated power management makes it possible to concentrate a greater number of watts onto each frequency jammed.

At the same time, the interval of time during which full jamming power is applied is calculated at the exact moment at which the echo returns towards the radar. In this way intermittent jamming is just as effective as continuous jamming, and by juggling the time factor as well as the power factor, one unit can jam several radars at the same time. Eustace points out that by applying these principles we can multiply by 100 the effectiveness of countermeasures in terms of radiated power. In other words, to obtain such an improvement without digital radiated power management, one would have to multiply the transmitter's size, weight, and energy consumption by 100, which would obviously be prohibitive.

To confuse enemy radar screens, the West still uses decoy methods such as "chaffs" (plastic or glass strips covered with a layer of metal or metal needles and strips). Heat-producing decoys are also used to send approaching radar-guided missiles off track.

"Active" or "deceptive" methods designed to mislead the enemy are also used in the field of countermeasures. False messages may be introduced into the enemy transmission system, for example, and he may thus be prevented from receiving certain information signals, or find his processing capacity overloaded. In this case it is necessary to know not only the frequency of the carrier wave, but all the other parameters of the particular radar installation concerned.

In the case of electronic warfare on the ground, the main objective is the disruption of radio transmissions through circular networks of high-frequency passive antennas or fixed and mobile station networks controlled by a computer that automatically intercepts, analyzes, and identifies enemy transmissions. Miniaturized radar warning systems, such as the one produced by the British firm Decca Radar, are now being considered. A series of research efforts in the field of personnel countermeasures have already begun. These include methods of transmitting light pulses concentrated in beams by means of parabolic wave guides. This serves to disorient the visual sense and disturb the sense of balance; a countermeasure of this type is particularly effective against antitank missile crews, for example.

Since 1975, the USSR has been working on a high-power laser program designed to neutralize space satellites; the Americans have responded by developing systems such as the LAHAWS (Laser Homing and Warning System) with a charge transfer device that detects the energy generated by the laser transmitter's optical system and diffracted by the atmosphere by using a pair of PIN diodes and a fast high-capacity memory.

Finally, in an attempt to prevent enemy jamming of radio or radar signals and allow continued trans-

1

2

1. Aida II: Automatic fire control radar for guns and infrared missiles.
2 and 3. Anti-aircraft and anti-tank missiles.
4. The Matra 30 missile's auto-guided AD 26 system.

3

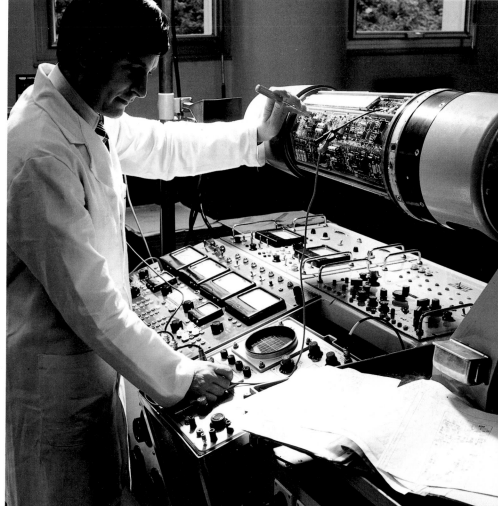

4

1

1 and 2. The Plessey Digital Message Terminal (DMT) in use with a battlefield manpack radio enables the soldier to remain in contact in difficult terrain where normal methods of radio communication have proved ineffective.
3. Rasura: Battlefield close range surveillance radar.

2

3

mission despite enemy measures, counter countermeasures have been developed; one of the most important is frequency evasion in which the receiver automatically changes frequency to find a section of the spectrum free of any jamming interference, or alternatively, the enemy is forced to dissipate his jamming power over a wide range of frequencies thus obliging him to employ a smaller watt-MHz ratio.

One of the greatest problems in the field of electronic warfare is reaction speed. This problem has been tackled by the development of totally computer controlled systems with real time processing units allowing immediate modification of a program in response to an unexpected change. Jammers can thus adapt automatically to detection systems and determine jamming parameters; one example is the D.T.P.E.W.S. (Design to Price Electronic Warfare Suite), a modular system designed by Raytheon for the American Navy. Other developments have been made in the field of ultrasophisticated signal processing techniques whereby a message can only be received by an operator with access to a code (semiconductor); without the semiconductors, all that can be heard is a series of meaningless bleeps. More advanced research concerns transmission systems reaching up into super-high frequency bands and the design of very highly directive antennas. The basic aim behind all this research is to considerably reduce background noise and pick up signals too weak for present rapid-detection radar systems. The principles have been known for at least twenty years but the techniques necessary have only really been developed over the last few years; they include cross field amplifiers and traveling wave tubes through which the wave is propagated in the same direction as the movement of the electron flow.

The installation of electronic warfare devices in aircraft, helicopters, and ships at the construction stage is a recent development. During the Korean War, the first remote radar detection sets appeared along with the Whiff radar, an antiaircraft fire control system for ground artillery. During the course of this conflict the Americans found it necessary to build jamming transmitters into their aircraft; they began by modifying older bombers such as the TB-25J. From 1958 onward they began to produce countermeasure equipment pods; the U.S. Marine Corps EF 10B may be considered the first in a series of tactical electronic warfare-equipped aircraft. Now they are producing systems that are built into the aircraft during manufacture such as those installed in the N-1 or the F-15. However, a less expensive alternative is to have the air or naval fleet accompanied by an escort ship or aircraft specially deployed for electronic warfare. On these aircraft the space previously taken up by weapon systems, bomb bays, etc., is used to house the electronic jamming equipment and the comparatively light weight of electronic equipment enables an effective number of jamming systems to be carried.

Chiefs of Staff all over the world have long dreamed of the "invisible" aircraft. On August 22, 1980, American Defense Secretary Harold Brown announced that the Americans had succeeded in developing such an aircraft, the "Stealth", that could not be detected by any known electromagnetic, electro-optic or infrared system. This aircraft would be constructed using materials that absorb radar waves, and would carry devices to suppress noise and heat radiation plus a complete electronic countermeasure system. The aircraft could only be detected when it was very close to enemy radar, by which time it would be far too late to intercept it.

This, of course, brings to mind the famous story of the U2 and its pilot Gary Powers; this American reconnaissance plane went undetected until it was actually sighted and shot down by the Soviet Air Force. The military establishment in every country is working on this type of countermeasure, although so far no one has come up with an aircraft that can escape detection by any imaginable type of radar. The situation is bound to become even more complicated as the Soviets, now forewarned, react to this new illusionist's challenge by trying to develop techniques to turn the invisible aircraft into a highly visible one.

The winner of the electromagnetic war will no doubt be whichever group first achieves mastery over the electromagnetic spectrum. Aerial warfare was the first to be modified by the electronics "explosion."

Naval warfare is now following suit with the introduction of fast patrol boats equipped with antiship missiles; defense systems will eventually be perfected and installed on large combat vessels. Electronic warfare can also be effective in peacetime. Used as an element of dissuasion, for example, decoying an enemy reconnaissance aircraft rather than shooting it down, could avoid triggering off a whole series of reprisals and counterreprisals. The very nature of warfare may change as a result. Major General E.S. Fris of the U.S. Marine Corps recently told the Senate that with the Soviets clearly disposed to increase the military capabilities of Third World countries, it was highly likely that we would shortly see the appearance of very sophisticated forms of warfare in regions that, up until now, we have only envisaged fighting on the tribal level.

A 6 A attack aircraft with hardware

Telecommunications

The term "telecommunications" stands for a number of different techniques of widely varying complexity; some we are very familiar with, others are new to us. The most common of these techniques is the telephone; this is based on a particularly elaborate infrastructure and represents the major part of the world's current inventory of telecommunications networks. To understand the operation of this service, we could begin by tracing the normal path of a telephone conversation. As we already know, each subscriber has a telephone that is linked to a telephone exchange by a two-conductor cable. The built-in microphone in the handset transforms the spoken word into alternating electric current and the built-in earpiece retransforms this current into audible sounds on reception. What happens when a subscriber wants to call another subscriber who is linked to a different exchange? First he takes the receiver off the hook; this opens a multicontact switch inside the set. The switch enables a direct current to flow through the line, sending a signal to the automatic line connecting equipment at the exchange and indicating that the subscriber wishes to dial a number. A patch recorder then connects itself to the line requesting the connection and sends back a dial tone indicating that the number required may now be dialed. The first part of the number designates the exchange requested and a charge meter control system selects the appropriate charge rate while a marker hunts for an available circuit. Interexchange connections can be achieved in a number of different ways. Several calls may be carried by a single metal conductor (this is known as multiplexing). Multiplexing provides good line and financial economy. When these circuits are overloaded or when the exchanges are located long distances from each other (intercity or intercontinental traffic) exchanges may be linked by means of a transit center. As soon as the automatic line-connecting equipment receives the last figures of the number requested — unique to that subscriber — it sends an alternating current along the requested party's line. At the receiving telephone, the current passes through a condenser placed between the two wires of the line and makes the bell ring. At the same time a ringing tone is sent back to the caller indicating that the line is free. When the person picks up the receiver, the tone generators stop immediately and the two lines are both free to begin conversation.

This, then, is the normal path of a telephone call from one subscriber to another. It introduces the concepts of "patching" the call through automatic line-switching equipment at telephone exchanges and the way in which trunk lines interconnect these exchanges.

Transmission networks

We should emphasize that all transmission networks rely on an already-existing infrastructure and that new networks are rarely created, except, for example, in the case of high-speed data transmission. Because of faster response times and the absence of moving parts, the telephone industries have opted for electronic digital transmission and temporal switching, which we shall examine in detail. The introduction of electronic equipment into transmission networks has made it possible to increase capacity, reduce size and use increasingly higher frequencies. These improvements concern equipment as affected by technical progress. But, at the beginning of the 1970s, the evolution of technology imposed a radical change in the whole concept of telephone transmission. The capacity of traditional networks and their band width (minimum signal width necessary to let frequencies pass without weakening them), have evolved considerably. In particular the new optical fiber has opened up a whole new field of possibilities because of the large number of signals that it can carry, its small cross-section, and its insensitivity to electrical interference.

Cables

The first cable was inaugurated in 1850. It was a telegraph cable (the first telephone cable did not appear for another forty years). It was not until 1943 that technicians succeeded in installing the first underwater cable electronic tube repeaters (amplifiers which boost electronic signals along telephone cables). The first cables were balanced thin lines (two parallel insulated copper wires). The first steps toward coaxial cables (with an insulated copper central conductor and a copper shield enclosing it like a tube) were taken in 1929 thanks to Lloyd Espenchied

A telecommunication's tower

1. The automatic telephone, first shown at the Paris Fair in 1927. It caused a sensation.
2. In San Francisco for the opening of the first transcontinental line on January 25, 1915. Thomas B. Doolittle perfected a process for making hard drawn copper wire which speeded early development of long distance telephone services.
3. Neptune and the Naiads disturbed by a diving-suit-clad worker laying the transatlantic cable.
4. Sosthene Behn, founder of ITT.

* A bit (**b**inary dig**it**) represents the basic quantity of information supplied by the knowledge of a binary number. (See chapter on data processing.)

kept within the circumference of the cable to minimize energy losses. Larger band widths can be fed along coaxial cables than along balanced twin cables. Progress in the field of telephone cables can be divided into three main areas: first, insulator quality, thanks to the discovery of polyethylene; second, repeater performance with improvement in amplification factors and band width; and third, in the case of underwater cables, the laying of signal equalizers that can be remotely adjusted to compensate for the disadvantages and drawbacks in telephone communications caused by the marine environment.

Long distance land cables were developed through the use of high frequencies that made it possible to simultaneously transmit 10,800 channels per pair of coaxial cables. These cables are used in television, radar, cable television transmission, and so on. The life cycle of an underwater cable is twenty-five years, as opposed to seven years for a satellite. Underwater cables are often used in tandem with satellites. The satellite covers a much larger area, but cables are often more competitive for large capacity arteries joining two heavy traffic zones. The two techniques are combined to connect two points situated at opposite ends of the earth; underwater cables are used for the Atlantic portion; the Pacific section of the path is covered by satellite. The Pacific is inadequately supplied with underwater cables; however a complete satellite connection involves a longer signal transit line and therefore poor retransmission quality. Cables and satellites are used in parallel in high-security areas; if one of the two systems is out of commission following a technical incident, the other mode can still guarantee connections. Although satellites can handle data signals at a much higher speed than cables (1,300 bits* per second for space link, 160 bits per second in translatlantic cables), the cable connection can correct an error, once detected, ten times faster than could be done over a satellite link due to the differences in lengths of the signal paths. We should also point out that underwater cables have lower capacities than earth cables (at present, 5,520 channels although technicians believe that it will be possible to obtain 15,000 channels in the years to come).

Radio Links

Microwave links consist of a series of relay stations that can provide radio links over very long distances. However, owing to the curvature of the earth's surface, the maximum distance between each relay station is limited to about 100 km (60 miles). These relay stations operate at wavelengths lower than 30 cm (12 inches). The first radio links were established at the end of World War II for military purposes. They have since been used in various specific cases in remote and inaccessible areas or political trouble spots, where one end of the link (or both) is mobile or where the high number of signals to transmit demands a very large band width. The first radio links all suffered from the same defects: the equipment was not very reliable, it was difficult to power directly from the main supply and there were frequency stability problems. But as time went on, remote controlled equipment was increasingly used in switching units in cable and radio link networks for such purposes as automatic channel selection and emergency frequency operations. In 1953 the International Radiocommunications Consulting Committee decided to standardize the main characteristics of radio signals for international connections. 1970 marked the beginning of the modern era of radio communications, with the transistorization of equipment leading to improved performances, increased reliability, better price ratios, and simplified operation and maintenance conditions. The next step occurred around 1975 with the advent of digital wide band radio links (140 Mbits, 1,920 channels). At the present moment we have obtained a capacity of 650 Mbits, which represents 7,680 channels.

Satellites

Artificial earth satellites for telecommunications purposes rely upon radio links from the ground to control and maintain communication. But we should emphasize that, unlike ground networks, the satellite can simultaneously amplify several signals by means of a single repeater. Satellite repeaters work on a principle of multiple access frequency distribution using analog carriers. They both amplify and change the frequency of signals received from ground-located transmission stations. These "first generation" satellites, which normally occupied a band width approaching 500 MHz in the 6 and 4 GHz bands, may be described as simple signal repeaters whose main advantages include the simplicity of their modulation and filtering equipment, but that have major drawbacks including intermodulation problems and the large band width required.

The adventure began in 1958 with the first telecommunications satellite, SCORE (Signal Communication by Orbiting Relay Equipment), a time-delayed relay system that recorded and later retransmitted the stored message by remote control. It was used for collecting or transmitting data. In October 1960 Courier I B, constructed in the United States, featured an experimental process which ensured privacy and transmission secrecy, but it had only a four-minute memory. Later, the active Telstar and the passive Echo satellites made their appearance. But the real birthdate of satellite telecommunications is July 20, 1962, when the first television picture was retransmitted by Telstar between Andover in the United States and the French station at Pleumeur-Bodou. Shortly afterwards, the USSR succeeded in establishing the first interspace link between two manned satellites. Over the last few years, geostationary communication satellites have been launched into orbit 36,000 km above the earth. These satellites have an equatorial and circular orbital path so that they revolve at the same rate as the earth and therefore appear to be stationary. In 1965 the United States "Early Bird" (Intelsat 1) became the first of a number of world coverage TV satellites. The Intelsat system, which links 101 countries with the notable exception of the USSR, has an almost complete monopoly over intercontinental satellite communications. In 1969 the Intelsat III series opened up the era of multiple access links from ground-located satellite stations. The Russians launched "Molnya" in 1965, and in 1967 developed the Orbita network. In 1971 it collaborated with the Eastern bloc countries to build the Interspoutnik satellite. Since 1962 most of the improvements made to these first generation satellites have concerned the questions of channel quality in terms of power, characteristics stability and reliability, and access mode. But the principle has stayed the same — they are primarily relay stations.

Scientists are now conducting research into a new generation of satellites that — as well as their relay functions — will perform electronic switching functions thanks to two essential and complementary techniques: multiple access by temporal distribution and digital concentration of conversations. As Didier Lombard, head of France's CNET (National Center for Telecommunications Research) Space and Transmission Division has noted: *"To organize the use of the radioelectric spectrum to obtain the greatest possible capacity per hertz, and to avoid inter-system interference, will be the basic precept of radioelectric transmission development in the coming years."* We have already seen this in basic radio

1

2

1. Army Radio links in France.
2. Satellite telecommunication ground station in Zaire.

Telecommunications

1. The most powerful of today's civil telecommunications satellites, Intelsat V, with its 12,000 duplex telephone circuits and its two television channels, has a capacity twice that of Intelsat IV.

2. Solar system view of the flight of an unmanned Mariner spacecraft as it passes Saturn. The craft is dominated by the large dish antenna needed for long distance radio contact with Earth. The three circular canisters at the bottom are nuclear-electric power generators. The boom at the top holds TV cameras and other science instruments. Sensitive magnetic field detectors are on the large boom at the bottom. The shorter booms are radio antennas.

3. An Atlas-Centaur rocket carrying the Mariner 6 spacecraft lifted off from Cape Kennedy's Launch Complex, February 24, 1969. The spacecraft's journey to Mars will span five months, bringing it to whithin 2,000 miles to help determine whether that planet can support life.

communication technology. Mr Lombard adds: *In the field of space communications, apart from the problems of the allocation of frequencies, there are those concerning the allocation of geostationary orbits for telecommunications satellites.* In fact in 1984 an international conference will discuss the task of guaranteeing quite concretely to each and every country equal access to geostationary satellite orbits and the frequency bands allocated to space communication services. The multiple access mode using temporal distribution techniques eliminates all the problems of intermodulation as only one signal is selected for processing in each short time unit by the space repeater. The digital concentration of conversations makes it possible to obtain additional savings. This technique consists of introducing into the gaps between the words of one telephone conversation (gaps represent around 50 percent of total call time) additional words from another conversation. The actual conversation between the two subscribers will, of course, sound quite normal while this concentration system is operating. Second-generation satellites will use a new 500 MHz band at 11 or 14 GHz.

The first satellites lost radioelectric energy when covering "useless" areas like deserts or oceans. The

new generation satellites like Intelsat IVA, for example, will be equipped with zone coverage directors that will only serve a specific predetermined area and so for the same repeater power, the down link can be more effective while facilitating higher transmission densities or the use of more effective modulation systems. The new generation of Intelsat V satellites – unveiled by the United States in 1980 – can handle 13,500 simultaneous conversations, against 6,000 at the present time, and only 340 for the first generation of Intelsat satellites. Some ten of these satellites are expected to be ready for service sometime between 1981 and 1983. The first videoconference and data transmission tests (IBM's Business Satellite System) will use these satellites; they will also serve as a means of communication for press agencies (UPI and AP already communicate with each other by satellite) and daily newspapers (the *Wall Street Journal* and the *New York Times* publish simultaneous editions linked by satellite). Another possible use would be to supply Third World countries with educational programs and medical assistance programs.

1

2

3

Optical Fibers

From the first days of research, pioneers in the use of the laser envisaged the possibility of developing a coherent controllable light source that could be used to replace the radiowave in radio or radar applications with optical waves. The theoretical communications capacity of a single laser beam is enormous because of the very high frequency of the carrier. However, there are a number of physical limitations, such as the fact that the atmosphere absorbs optical waves; we must either limit ourselves to space links or protect the beam emitted by the laser or light-emitting diodes with tubes' similar to wave guides. The first demonstration of the principle of guiding light through water was given a century ago. In 1910, Peter Debye proposed a type of light wave guide made of multiple layers of transparent material, and established that a material's ability to guide light was heavily dependent on its refraction index. The first industrial research was undertaken in 1966 by Charles Kao and George Hockham (ITT).

One of the most difficult problems to solve was that of attenuation varying according to the wavelength. The light beam was weakened by absorption and diffracted by impurities or tiny fractures. In 1970 the American firm Corning Glass began to market its own silica optical fibers. The first experimental links were set up in 1976 in Canada, the United States, Japan, Holland and France. Japan has even begun to export optical connectors. Current research predicts that the lifetime of laser diodes would be around 1 million hours (120 years) and that attenuation could be as low as 0.2 dB/km, which means that we may envisage systems in which the signal would only need to be reamplified every 50 km using a wavelength of 1.3 micron and gallium arsenide laser diodes. It is now estimated that at 140 Mbits per second, the cost per kilometer would be less than metal cable transmission. Optical fibers offer two main advantages: the band width and thinness (the width of a human hair) enable us to envisage its future use for subscriber networks carrying both telephone and video pulses. In addition, optical links do not use electromagnetic waves, and it is thus impossible to jam them if trying to locate a ship or alter the course of a laser-guided missile. They ensure perfect discretion and secrecy and for this reason are already used on board some military ships, such as the US cruiser "Little Rock."

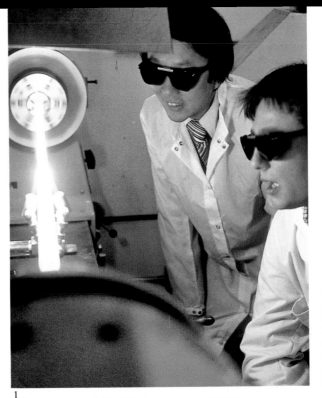

1

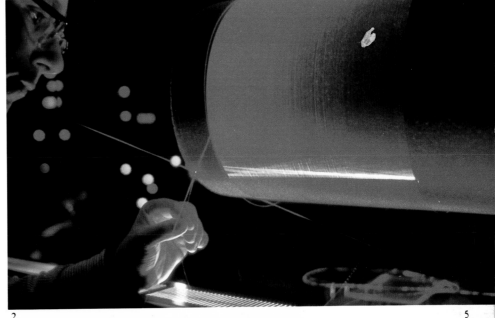

1. Charles Kao (left), one of the inventors of optical fibers
2, 3 and 4. Optical fibers.
5. One specific application of optical fibers: these probes are used to observe, photograph and even operate inside the human body.

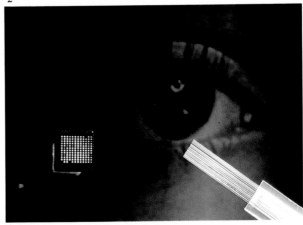

2

5

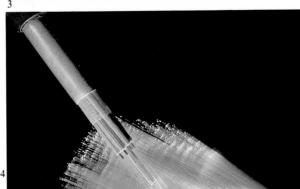

3

4

1

1. In 1948, Claude Shannon, then 32 years old, published a complete mathematical model of the communication process and the laws that govern it. In a later experiment (shown here) he used switching relays like those used in a dial telephone system for logic and memory experiments.
2. "A telephone exchange in Berlin in 1950" by N. Zemme.

2

Electronic Switching

When talking about switching it is essential to recognize the difference between electromechanical and electronic switching systems. Electromechanical systems include a certain number of electromagnetically operated relays and solenoids that operate in a sequential order based on cable logic design techniques. Electromechanical systems do contain a certain number of purely electronic devices; but in a purely "electronic" switching system, a computer controls certain functions of the circuit and processes data concerned with the "space" requirements of an electronic signal or the "time" requirements of several electronic signals. This "time" processing involves mixing together a number of separate signals which are later reconstituted in their original form (the principle of the temporal technique).

Whichever of these techniques is used, electromechanical, spatial or temporal, switching is an essential element of telephone communication. We have already seen that by picking up the receiver, an electrical signal establishes a liaison between the user and the receiver's exchange by means of the first three numbers dialed. In the days of telephone exchange girls and manually operated equipment, the operator connected one exchange and another by placing a plug in a hole on the switchboard to complete the electrical link to the required exchange. But one day in the 1920s, an American funeral parlor director invented an automatic switching system which dispensed with the services of an operator. He was inspired by the fact that his rivals in the funeral business were bribing telephone operators to send his "clients" to them ! From then on, switching was handled by electromagnetic relays connected to a whole system of selectors and recorders. Then, in the 1950s, the Crossbar system was invented; it enabled accelerated connection time thanks to a horizontal and vertical grid of crossed lines; it was a great deal more effective and much faster than the rotating drum system.

The spatial technique, introduced at the end of the 1960s, was an electronic adaptation of the classical technique used in electromechanical automatic exchanges. It is called analog because the voice is retransmitted in the form of an electric signal that faithfully reproduces its initial form. By means of a physical connection network, individual and permanent contact is established between two subscribers throughout the duration of a communication; the computer makes it possible to increase speed and reliability in all areas including maintenance. But a single electrical junction mounted in the machine can only handle a single conversation. The temporal technique, so described because it uses division and processing of the word on a time allocation basis, allows several communications to be handled on one junction. The signal is replaced by series of numbers that are translated into digital language. In temporal type electronic exchanges there are usually three successive word processing operations.

Sampling: Temporal communication is based on a property of the human ear which, like the eye, can reconstitute all the information from very brief pulses taken at regular intervals from the electrical current and which thus reproduce the acoustical variations of the voice. In sampling the spoken word is thus broken down into a series of brief electrical pulses at a frequency which is generally of the order of 8,000 samples per second or one every 125 microseconds. These figures are not just guesswork; like the figures which follow they are taken from very accurate studies such as those done by Claude Shannon at Bell in 1948.

Coding: The amplitude of the sampled signal is quantified and each sample value is given a binary number made up of a series of eight elements. Thus, to reproduce a signal there is a series of binary elements at a rate of 8×8000 which equals 64 Kbits/second. Pulse amplitude must be reproduced to an accuracy of 0.25 percent to ensure high-transmission quality. So, instead of sending the pulse itself, which lasts only about four-thousandths of a second, the number representing the electrical value of the pulse is transmitted. Computers transcribe it in the following manner: pulse equals 1, no pulse equals 0.

Multiplexing: As transmission circuits can handle traffic much higher than 64 Kbits/second, digital information from other channels may be interwoven in the dead time. This makes it possible to handle thirty conversations simultaneously in one 2048 Kbits multiplexer. To explain this series of operations by using a more familiar image, let us imagine that every second we are dispatching a train of 8000 passenger cars each containing thirty numbered seats. In each car, passenger number 1 represents one-eight thousandth of the conversation transmitted by correspondent 1. The subscriber must receive the 8,000 passengers destined for him at the final destination. As we have already pointed out, there are thirty different subscribers. We can thus imagine that the station contains thirty loudspeakers, each of which communicates only to passenger 1, 2, or 3 (up to 30) of each car. In the following second another train of 8000 cars arrives and the operation is repeated. The conversation is thus reconstituted upon arrival, with each train setting down its passengers in the space of one second as the cars reach the platform, to be met by their thirty recipients.

In January 1970 the world's first temporal electronic switching exchange was introduced into a public network in Guingamp in Brittany (France). This was PLATON (Lannion prototype digital organization temporal switching system). It was in nearby Lannion that the CNET was set up after World War II. France was one of the first countries to opt for the challenge of temporal switching, which involves the two basic principles of digital modulation by pulse and coding (practically insensitive to distortion and electrical interference) plus the separation of switching and management functions into separate processes by means of a central computer. The E 10 temporal switching system, which was subsequently marketed by CIT-Alcatel, a subsidiary of the CGE (French General Electricity Company), is a pre-programmed electronic system. Microprocessors, Reprom memories, and hybrid circuits (electronic components), have taken the place of relays in the switching equipment. Only the ends of the subscriber lines are equipped with minirelays that are still indispensable for the telephone standards now in use. In the beginning, temporal switching aroused more doubt than enthusiasm; but in less than ten years, CIT-Alcatel succeeded in proving that temporal switching could be integrated into the existing network. By October 1977, the battle had largely been won. During the Intelcom '77 international telecommunications conference held in Atlanta, Georgia, the breakthrough of temporal switching into the market could no longer be ignored. We should also point out that Bell had been conducting research into electronic switching since 1955, and that it developed the Electronic Switching system whose first model, the ESS 1 (spatial switching mode) used sealed magnetic relays for connection points. The system is controlled by two interdependent calculators that operate in unison and control each other. The ESS 4 (temporal switching mode) has a solid state central section, digital computers and a huge memory and programming section. It was introduced into service in Chicago in 1976.

Between them, CIT-Alcatel and Western Electric have installed more miles of public telephone lines (the E 10 and the ESS 4) than any other company in the world (including systems on order). These systems are compatible with decentralized networks, transit centers or heavy traffic urban zones. Thanks to modular decentralized design, in most cases they coexist with analog spatial systems already in service (superimposed networks). The recent development of temporal switching is now permitting the design of "integrated service networks." This will make it possible to considerably increase the number of communications exchanged and to integrate a certain number of new services. At the same time it will reduce the cost of exchanges. Indeed, other networks matched to the evolution of user needs (telex, data transmission by Tymnet in the United States or by Transpac in France) have been installed alongside traditional telephone networks. They use specialized switching systems to process data in the form of packets or complete messages.

The move towards temporal techniques and integrated digital networks does pose some social and economic problems of which the industry is becoming aware. These problems call for overall reflection on the future evolution of the zone to be equipped. Some theorists have even begun to talk of a "new philosophy" of telephone networks. In any case, temporal switching poses strategic "social choice" problems. Temporal systems offer the public a range of highly-advanced terminals and over the next few years, we will undoubtedly see an evolution of the communications modes available within our societies. The problem can be looked at on three different levels:

Political: Are we moving toward increased centralization and state or multinational monopoly of dissemination of information, or toward decentralization and power sharing?

Social: Do we not risk accentuating the social cleavage between individuals and also, more importantly, between nations, by operating a kind of economic segregation?

Economics: Temporal techniques need three times

less operating personnel than spatial techniques and could lead to an increase in unemployment. However, redeployment of personnel is now starting to be a distinct possibility because of an increasing trend towards telematics.

1. The Crossbar switching system, invented in the 1950's.
2. The 4 ESS, an American temporal switching system.
3. A printed circuit board for the E 10 French temporal switching system.
4. The NEAX 61, the Japanese digital switching system.

Telematics

Since the French first invented the term "telematics" it has passed into common usage, although the technique it refers to still provides heated debate for symposiums, newspaper articles and books. The new trinity of telephone-computer-television receiver will undoubtedly lead to social, ethical, and political upheavals in the field of telecommunications, and we are only now beginning to get some idea of how far-reaching these may be. There are two main divisions in industrial policy. In countries like the United States, private companies manufacture the equipment, components, and terminals; and they control their own telecommunications networks. In 1981 IBM, in association with Comsat and the Aetna insurance company, launched its own telematics satellite, the SBS. In other countries, such as France, the two functions are completely separate. Companies manufacture the equipment and undertake advanced technological research (CIT-Alcatel, for example), but the networks are owned and managed by the state. This separation of functions means that industrial companies in the second category must create a domestic market to be able to undertake mass production of equipment at low prices. This leads to a number of different problems: evaluation of the demand and the market to be created, the necessity of taking into account the reduced purchasing power of potential customers, and the fear of the threat posed either in the medium or long term to employment.

Peripheral Equipment

In the field of peripheral equipment, an electronic system that leaves the hands free has been developed. The receiver is eliminated and the listening unit replaced by a loudspeaker that amplifies the conversation. This system provides abbreviated dialing, individual call notification and automatic redialing, conference facilities, storage of frequently dialed numbers on a microcircuit memory, storage of urgent numbers (police, fire department, etc.). A display screen shows timing of calls, verification of numbers in the memory and the like. A visual facility with microphone, loudspeaker, camera and television screen may be installed in the same terminal, thus allowing the dialer to see a picture of his correspondent. Processing of the spoken word (automatic recognition of individual words pronounced that belong to a preprogrammed vocabulary of some 100 words, synthesis of certain phrases) will allow a subscriber to announce the number he requires, thereby eliminating dialing by ordinary rotary dials or touch-tone units; control of subscriber's installation (connections, disconnections, transfers, etc.); consulting services by telephone, centralized multimode answering-recording machine capable of providing all kinds of functions (start, stop, recording, playback, erasing) from any point in the network.

Videotex

Videotex is in a sense the favorite child of telematics. It consists of displaying alphanumeric or graphic information on the screen of a large dissemination terminal (an extensive public network terminal). It may be divided into two main systems: The first system is transmitted to the consumer by radio waves that incorporate a digital keyboard allowing the user to select a predetermined "page" of information. The second system is an interactive system allowing dialogue between the user's terminal and the information center. In the first case (Antiope in France and Viewdata in England), information is transmitted by a television channel and displayed on a standard television receiver. In the second (Teletel in France and Prestel in Great Britain), information is sent by telephone channel. Antiope (digital acquisition and television display of images organized in written pages) and Viewdata share the problem of providing data transmission without additional consumption of the frequency spectrum. When a new service is created, it is usually necessary to liberate a frequency band. Technicians have long known that there are sections of a television signal that are not used. This is the so-called "dead time" used for transmitting synchronization and color reference information (Dead time consists of a blank period, invisible to the eye, that occurs while a picture is being built up on a screen). These periods represent free spectrum space that can be used for other purposes. This phenomenon gave technicians the idea of using the dozens of "lost" lines for transmitting other completely different messages.

In 1967—68 the BBC considered using these unused sections to broadcast subtitles for the hard of hearing. The idea was later taken up again by the BBC and also by the IBA; each developed their own videotex systems, CEEFAX and ORACLE. In 1972—73 the French decided to develop an even more advanced system. In the British system, owing to British synchronization standards, it was impossible to fit in a method of individual packet transmission.

A general view of Antiope.

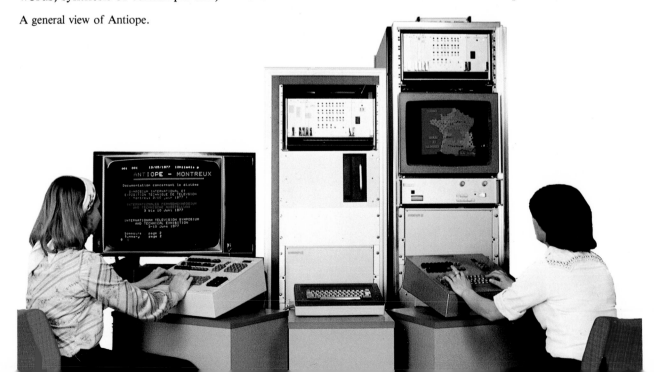

Therefore the English system can only transmit a fixed (100 word) "page," whereas the French system can send one word at a time, placing it in the exact spot desired and operating a kind of "page-setting" system.

One of the most spectacular applications of this research is the electronic telephone directory which went into experimental operation in 1981 for telephone subscribers in the Ile-et-Vilaine region in France. The desired number is dialed on the telephone, then the message is typed on an alphanumeric keyboard linked to a data bank and the requested information, for example, a page of the directory, is displayed on a screen. The whole page is displayed so that the subscriber may make the right choice between several subscribers with the same name. Names of new subscribers may be introduced immediately, outdated subscriber information erased and new numbers added as soon as they are allocated. Employment in the subcontracting industries should greatly increase as these new techniques are commercialized. The subscriber will also have access to other services: programmed recording of a transmission at a predetermined day or hour in the user's absence, special programs reserved for particular socio-professional categories, such as doctors, or for general subscriber use. These special services will be accessible only to those possessing an electronic control key.

Some countries, such as England, have chosen to sell fully-equipped terminals that do not bring the family television receiver into play for receiving these special video transmissions. Others, such as France, have decided to equip the television 'receiver with a special connector to which may be added new access channels as services become available. From March 21, 1980, all new television sets sold in France are mandatorily equipped with the new Scart peripheral television socket. Only the electronic directory is available as a separate unit.

In Japan, research on the Captain project (Character and Pattern Telephone Access Information Network System) began in December 1979, with a test group of 100 Tokyo households. It uses the normal telephone network, the family television set, and a computer equipped with a 63 Kbits memory and a selection of 4000 Kanji characters. The CATV (Community Antenna Television) system was tested in January 1976 at Tama New Town near Tokyo, on a sample of 500 households, using a single 27-channel coaxial cable to retransmit all kinds of information: weather forecasts, food prices, educational programs for children, etc. But the Japanese have so far disseminated little information on their results.

The advent of telematics will also provide teleconference facilities either via a sound-only link with speaker identity signalling (audioconference) or via audiovisual links where the image of the speaker will also be transmitted. Many of these new services are already out of the research stage and are beginning to be widely used. The evolution of telematics is no doubt irreversible and many countries are now, somewhat belatedly, beginning to consider the sociological changes and upheavals that it will almost certainly produce.

Although industrial strategy was quickly formulated, political and sociological adaptation has lagged behind; and we risk finding ourselves facing the same problem as occurred earlier with the explosive arrival of television in everyday life. People were bewitched by these "strange new windows" opening up before them, but no one was very sure what to look at through them. In other words, we are still in the process of asking ourselves what a television program should really be. We are being confronted with one new invention after another, before there is any real consensus on how it should be used. Imagine the extraordinary changes that will occur when telephone operators can communicate directly with subscribers from their own homes, or when television correspondence sale replaces the door-to-door salesman. Will these changes improve the quality of life, or will they isolate people from each other? What will happen to the written media with the development of immediate advertising and sale by television? Are we heading toward technological unemployment, or are we beginning to prepare for a redistribution of work? Will we have to completely change our mentality and our mode of working?

1. With the videoconference system, the user can carry on a discussion with colleagues in other countries without leaving his armchair. He may choose to view them as in this photo or may use a computer with text transferred to memory.
2. Visiophone allows you to see your correspondent while making a telephone call.

1

2

Telecommunications

Electronic Office Systems

In the field of electronic office services, experimentation has already been undertaken into "electronic mail services." They may eventually be extended to the general public, although these services will never eliminate the conventional methods of sending personal messages or packages. Industrial concerns already use a familiar electronics communication service, the telex. The market is now moving toward reasonably-priced telecopying machines capable of transmitting letters and documents by telephone. In 1980, 240,000 telecopiers were already in use in the United States (a growth rate of 20 percent) and 150,000 in Japan (a growth rate of 30 percent). The second basic tool of electronic mail systems is the teletext or remote typing device which provides a number of different document authentification services (date, times of communication, identification of message limits, etc.). Once this system is connected to the telephone network it allows modification, addition, and rearrangement or correction of words, phrases, and paragraphs before the printer types out the page in thirty seconds. Thus a text may be transmitted at the rate of 240 characters per second, by telephone line, optical scanner and computer. Another important element in electronic office services is "word processing." The message is typed on a keyboard and displayed on a screen where it may be corrected, arranged in columns or tables and then transferred to the memory. Once the text has been processed, the printer types it out at a rate of ninety characters per second. It is then transmitted to the correspondent by telephone line at the rate of 6.6 characters per second. The machine can undertake call procedures, dialing, dialing code control, line clear, and composition and message transmission while the printer is available to copy any messages that may come through when other preparation work is being done. In certain cases (discussions between architects and scientists, for example) plans and drawings may be transmitted by telewriting, using a television screen, a telephone, a drawing board with a sheet of paper, and an electronic pen. The drawing board contains a printed circuit, a microprocessor and a REPROM memory. It also contains a black box destined to encode the graphic signal and mix it with the telephone signal on transmission, with the opposite process taking place on reception. The world market for electronic office systems should be very near 40 billion dollars by 1985. The figure is a little controversial as the term "electronic office systems" covers a wide range of material and takes in some fairly vaguely-defined notions. The companies best placed to take advantage of this new market will no doubt be data processing firms, office supply companies that already have solid sales networks, telecommunications companies and a few other challengers who managed to foresee the way the wind was changing.

Data Banks

Today, one of the fiercest battles in the field of electronics centers around the data bank; the largest in existence is the ISI bank in New York which adds 1 million new scientific references each year. We could recall the case of an American company, Mead Data Central, that decided to store on memory the whole Federal criminal code and all civil and criminal legislation for more than thirty states. Lawyers paid little attention to this development until they realized that a subscription to Mead Data was a valuable weapon in the fight against a better informed adversary. A subscription is now considered indispensable; this marks the acceptance of an important new challenge. Mead Data then decided to store on memory the last five years of the *New York Times*, the *Washington Post* and the weekly magazines *Time* and *Newsweek*. Here it found itself in competition with the *New York Times,* which was creating its own memory bank. Speed won out. Mead Data provided the raw material (such and such an article appeared on a particular date in a particular year in a particular newspaper) while the *Times* invested time and gray matter in supplying syntheses and resumes. Even with this extra information, it appears that the *New York Times* is losing the battle.

But this little anectote is reassuring. It demonstrates that the person who contacts a data bank wants to obtain just that data. Machine memories cannot replace intelligence. Rather, they store and regurgitate a large range of information, and it is the work of an intelligent person to make a selection between the data proposed. Software (manufacture of programs) is important, and the task of making software belongs to the human operator. A kind of fantasy spawned from ignorance and wrong information has sometimes taken root in the public mind. But unless techniques change so drastically that they develop into something completely beyond our current conceptions, the machine can only divulge what has been programmed into it. The fantasy of programmed access to the individual thought process is merely a science fiction nightmare. Nevertheless, this does lead us to the true problem, that of power. Who programs the machine? Who controls it? Who has access to the data bank? The real political debate will center around these questions.

1. Tele-writing is used to transmit drawing, sketches and graphs over a distance.
2. Word processors are used to compose, correct and transfer texts to the memory and also to transmit them to other word processors.

1

2

Electronics and telecom- munications

by F. Jutant, Assistant Lecturer at the French National School for Telecommunications, and M. Boyer, Chief Engineer at C.I.T. Alcatel.

In this article, F. Jutant and M. Boyer point out the importance of the introduction of electronics and microprocessors in telecommunications systems. They explain the basic principles of telecommunications and analyze the differences between analog transmission and digital transmission.

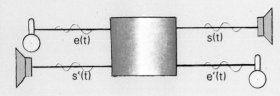

Communication between people utilizes the five senses and the facilities of interpretation that are the fruit of the experience of prior communications. The physical presence necessary for this communication constituted the first obstacle to be overcome in order to develop it.

The use of concise or patterned, that is, symbolized carriers was the first stage in the development of communication between correspondents removed in space and in time. As this development progressed, there appeared the necessity of overcoming the limitations at the level of the time of transport, of the limited delivery and of the lack of interaction to establish a real long-distance communication or telecommunication. Means making use of some optical device were the first step in the development of communication over a distance. The discovery of electricity as a means of transport made it possible to overcome the barrier of sight communication by furnishing the means of communicating rudimentary signals over long distances permitting the transmission of text by coding: this was the Morse telegraph.

It was the invention of the telephone which permitted the first true person-to-person long distance communication. This restored, under conditions of listening that were not very comfortable, a portion of the richness of face-to-face communication.

The telephone was a crucial invention, one that has remained to this day the principal base of interpersonal telecommunication. The development of electronics has made it possible to make of it a means of mass communication by its technical and economic assets. At the same time as another element in the world of communication came into being: television.

Parallel with the development of data processing a new need for telecommunication arose to do with data of the numerical type.

At present, the leap forward realized in the integration of ever more complex electronic circuits and the emergence of new transmission carriers such as optical fibers make it possible to envisage an unprecedented growth of the flow of telecommunicated data. This growth poses the problem of moving beyond the sole telephonic communication carrier toward a multiservice network.

In order to come to grips with this problem, it is necessary to know the technical data that condition any information transport. The telephone network, which to this day remains the dominant carrier, provides one of the best illustrations of the existing constraints for the transmission of data.

The Information-Carrying Electrical Signal

A microphone furnishes, on the basis of pressure variations, an electromotive force, which literally means a force capable of moving electric charges in a conductive circuit. It constitutes what is known as a *transducer*.

The movement of charges taking place in the circuit can be characterized by the number of charges passing through a circuit element per unit time: electric current. The variations of the electric field corresponding to the flow of the charge along the circuit define *differences of potential,* also called by improper extension *electric tension* (voltage).

The variations of the electromotive force supplied by the transducer produces variations of the charge flow in the circuit and provides the possibility at any point of the circuit of recovering the initial information from the current or voltage and with the aid of an inverse transducer (the receiver), to reconstitute the initial voice signal.

Such was the principle of the first telephone communication.

If we put aside the problem of the transducers at each end of the connection, the question arises of understanding how the information-carrying electric signal is carried and perturbed by the circuit.

Analog Transmission

The electric signal furnished by the transducer carries the voice information in its loudness (amplitude) and its variations in time.

The transmission circuit known as channel must furnish, at the other end of the connection, an image $s(t)$ of the input signal $e(t)$; it must also realize a transmission circuit in the opposite direction in order to establish a bilateral communication.

In order to preserve the information in its entirety it is necesary that $s(t)$ be analogous to $e(t)$; that is to say, the signal furnished by the channel must be identical in form to that of the signal emitted. Any deformation would run the risk of destroying a portion of the information.

The transmission channel must, therefore, be such that the signal it furnishes is proportional, in constant conformity in the course of time, with the original signal.

It is said that the transformation realized by the transmission channel must be linear and stationary. What does this mean in practice? The channel spends a certain time propagating the electric signal and imparts to it an attenuation and deformations.

Is it, under these conditions, possible to preserve the intelligibility of the conversation to be transmitted?

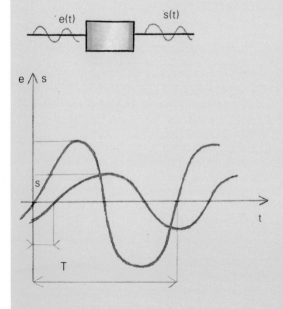

ω = pulsation
f = frequency
T = periode
ωt = phase

$$f = \frac{1}{T} = \frac{\omega}{2\pi}$$

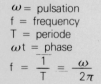

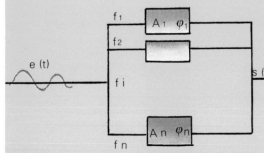

What, then, is the range of the communication?

In order to answer all these questions, we must analyze the delays, attenuations, and deformations imparted by the channel. This analysis calls for the construction of a model of the transmission channel that makes it possible to predict, for a given input signal, the shape of the corresponding output signal. This model will then permit the characterization of the transmission channel and consequently enable us to realize the operations necessary in order to ensure a good transmission of the voice signal and to bring about an increase of the range.

Linear and Stationary Systems

The transmission channel for furnishing a signal analogous to the input signal must, as we have shown above, be *linear* in order to preserve the shape and *stationary* so that the proportionality ratio between the two signals does not vary in the course of time.

The first difficulty in characterizing a system resides in the great diversity of the possible shapes of the input signal. This diversity can be reduced in the case of a linear and stationary system. There exists, in fact, an elementary signal shape: the sinuosity for which the passage through such a system can be defined simply by the ratio of the amplitudes of the output and the input signal *(attenuation)* and by the divergence in time of the zero-passages of the two sinusoids *(delay)* or, in other words, of the corresponding angular displacement of the two sinusoids *(phase shift)*. The attenuation and the phase shift can depend only on the frequency of the input sinusoid.

These sinusoidal functions have, on the other hand, the property of constituting a base of decomposition of any physical function; that is to say, any signal to be transmitted can be broken down into a sum of sinusoidal signals; the whole of the amplitudes and of the phases associated with each sinusoidal signal of the breakdown constitute the spectrum of the signal.

It appears, in that case, possible to analyze the passage of a signal through a linear and stationary system in the following manner:

Note that e (t) breaks down into a sum of sinusoids, the system attenuates and dephases each sinusoid, the sum of the attenuated and dephased sinusoids forms s (t).

The modeling of the linear and stationary system can then be expressed by two functions: the attenuation of a sinusoid and its phase shift as a function of its frequency. A knowledge of these two functions then makes it possible to predict the shape of s (t) for a given signal e (t).

What, then, is the condition for a linear and stationary system to produce an output signal of the same shape as the input signal? It is necessary that each component sinusoid be attenuated and dephased in an identical manner. In order to prevent a linear and stationary system from deforming the input signal, regardless of what this input signal may be, the following two conditions must be met: a constant attenuation, that is, an attenuation which is independent of the frequency and a phase shift proportional to the frequency, or zero.

The Linearly Modeled Transmission Channel

A real channel will show deviations with regard to these two laws; it will, therefore, lead to deformations of the signal known as *linear distortions*. In order to characterize these distortions, the attenuation and the time of group propagation of the channel are plotted as a function of the frequency (the propagation time of the group being defined as the derivative of the phase shift with respect to the time).

The *passband* of the system is then defined as the frequency region in which the attenuation is confined between a maximum value AM and a minimum value Am.

The transmission channel transmits, therefore, only a portion of the components of the spectrum of a signal, in the sense that it strongly attenuates the components outside the passband. It also imparts to the signal "within the band" certain deformations that are due to the variations of the attenuation and of the time of group propagation.

Nonlinearity of the Transmission Channel

The linear modeling of the transmission channel is, in fact, only approximate; the departure from linearity of the real system produces a deformation of the signal as a function of its amplitude. In order to understand this deformation, let us assume that the system is acted upon by a sinusoidal input; because of its nonlinearity, the output will be periodic but not sinusoidal. This signal may break down into a sum of sinusoids, some at the frequency of the input sinusoid — the *fundamental frequency* — others at multiple frequencies — *harmonics*.

The channel can then be characterized by the ratio between the power of the harmonics and the power of the fundamental; this ratio is a function of the amplitude of the input signal and is called the *nonlinear distortion rate*.

If we now introduce at the input a composite signal, the nonlinear system causes the appearance, over and above the responses expected for each component, of sinusoids at frequencies corresponding to combinations of the fundamental frequencies figuring in the input signal and of their harmonics.

This results in a deformation of the signal, which can ally itself with an interference if the signal has a dense spectrum and if the nonlinearity is marked; one speaks in that case of intermodulation noise affecting the quality of the transmission. From this it must be concluded that the nonlinearities of the channel should be very slight.

Noise of a Transmission Channel

In any electric circuit, signals of an uncertain character — called *noise* or *parasitic signals* — develop and attach themselves to the useful signal. The channel is characterized by the mean power of the noise it produces at the output; this mean power increases with the passband of the channel.

Dynamic of the Channel

The *dynamic* of a channel is defined by the ratio of the maximum power of the electric signal, which it is able to supply at the output without exceeding a fixed rate of nonlinear distortion and the power of the noise it produces.

If we call S the power of the useful signal and N the power of the noise and if we know that the power of signal and noise is additive, then:

$$\text{Dynamic: } d = \frac{S + N}{N} = 1 + \frac{S}{N}, \text{ this}$$

expressed in decibels, $D = 10 \log d$s.

Establishment of a Long-distance Connection

The transmission channel imparts to the signal which it transmits an attenuation that increases with its length. Since there is a limit to its dynamic, it is impossible to increase the power of the emitted signal beyond a certain limit. And, consequently, after a certain distance, the attenuation becomes too great and the signal "drowns in the noise." This means that the dynamic of the signal or the signal-to-noise ratio has become insufficient.

This would, for a given carrier, result in a maximum range of communication. This limitation can be overcome if the signal in the channel is picked up before its signal-to-noise ratio becomes insufficient and if energy is supplied to it by an amplifying operation. The apparatus producing this amplification is known as a *repeater*.

However, each amplifier in its turn contributes distortions and noise and does so in a cumulative manner at each repetition step, leading to very severe constraints with regard to the linearity and noise properties of each repeater in

order to realize the large number of repetitions necessary for a long-distance communication. There are, in this case, great technical difficulties which vary according to the carrier (cable, radio frequency, Hertzian beams) and have affected the development of the telephone network and the improvement of its transmission properties.

The criterion of a good transmission is the maintenance of the intelligibility of the communication, a criterion that is affected by the subjectivity of the receiver (the human being, not the telephone). This criterion expresses itself by a minimum level of the signal to be recognized in the noise, by a certain ease of listening, by a deformation sufficiently small for the speaker to be recognized, etc. Standards have been defined which guarantee "good intelligibility." They impose, more specifically, a passband of 300–3400 Hz and a minimal dynamic for the signal transmitted.

Telecommunication Network

The problem consists in putting into potential communication a large number of interlocutors (subscribers). If each person were connected, point by point, with all the others, it goes without saying that the number of connections to be established would be very large and would become economically prohibitive, all the more so if the rate of utilization per subscriber is low. It appears, therefore, necessary in order to obtain an economically viable network to concentrate the means in order to limit the number and the length of the connection, and to improve the rate of occupancy. This has led to the idea of constructing a network organized according to the following scheme:

The exchanges are connected with each other by channels capable of carrying a large number of communications in parallel for multiplexing.

We shall leave aside the problem of control signal switching in order to turn our attention to multiplexing.

Frequency Multiplexing

The voice signal can, as we have seen, be limited to the 300-3400 Hz band without destroying its intelligibility. The transmission carriers (cables, radio waves, Hertzian beams) permit much larger bands. It is, in this case, sufficient to shift the frequency of the spectrum of the signal to the limited band by a modulation operation and to juxtapose the signals thus obtained, in order to form a composite signal whose band width will be *n* times the width of the elementary channel, if *n* is the number of lines to be multiplexed.

Upon its reception, the signal is demultiplexed with the aid of $f'_1...f'n$

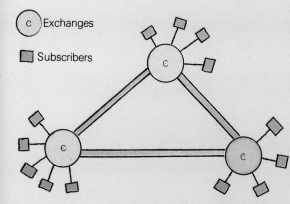

C Exchanges

■ Subscribers

47

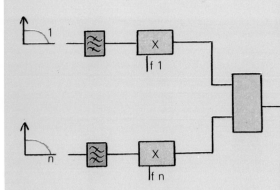

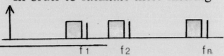

frequencies, which are as close as possible to the multiplexing frequencies and filtered so as to recover the channel signal in its baseband.

In order to facilitate these filtering operations, Δ kHz is set aside for each line, and multiplexing of up to 10,800 channels is thus realized.

The limitations of the number of lines of the multiplex and of their quantity stem from the limitations with regard to the passband of the carriers and to the linearity and noise of the repeaters.

Let us note that the nonlinearities affect the transmission with a new defect known as *cross talk* which expresses the interferences between different lines of a multiplex due to the nonlinear behavior of the amplifiers in particular. Frequency multiplexing and spatial commutation have been at the base of the development of the present telecommunication network.

From Analog to Digital Transmission

As we have just seen, a pure analog transmission system is afflicted with cumulative defects introduced by circuits (repeaters, filters, lines, etc.) that limit its range and quality. Since World War II remarkable advances have been made that have given the telephone network an altogether suitable quality.

In the meantime, the awareness of new technologies, of the cost reductions resulting from them, and the necessity of integrating new services have led to a new orientation: the introduction of digital systems.

The first step in this direction was taken at the time the pulse techniques were developed: amplitude-modulated pulses (PAM) and numbers-modulated pulses (PNM). The grand perceptions resulted from the fact that one could thus get rid of the nonlinearities and intermodulations introduced by the amplifiers during multiplexing and increase the power emitted and reemitted by *sampling* the signal.

The decisive step was taken by Alec Reeves who suggested that the value of the sample taken from the signal be replaced by an integer of elementary voltage values (quantification) and that this number be transmitted in binary code. The code signal is then no more than a sequence of pulses representing the binary 0's and 1's.

The idea was to arrive by this relatively simple transformation at a transmission system that was almost insensitive to the perturbations introduced by the transmission channel since at the reception it was only necessary to recognize the 1's and 0's.

Such is the basic idea of digital transmission whose importance and also limitations we shall try to demonstrate.

Digital Transmission

In order to define what is meant by digital transmission we must first present a diagram of the system (below):

Decoding

A source emits a message consisting of symbols taken from a set constituting the alphabet of the source. This message is realized by a digital signal of characteristics proper to the source.

This digital signal is transcoded by an operation known as *in-line coding,* which furnishes a digital signal that can be identified by the receiver.

This transmission scheme calls forth several comments: the source itself may comprise several subunits for processing the original message for the purpose of imparting the right characteristics for a better transmission; in that case the receiver comprises the reverse-process functions.

We shall not go into these various processing steps known as *coding.* They are the subject of a special field of study going beyond the scope of the present article.

We shall however try to show their full importance for the improvement of digital transmission.

Having said this, we must now give a better definition of what a digital (electric) signal really is. It is *a signal to which are assigned values belonging to a finite number of unconnected fields.* These values can be electrical levels, frequencies, and phase shifts. The simplest example is that of a signal with binary levels.

Since a physical signal has no discontinuity, this means that the signal carries information only during disjoined intervals of time.

The information carried by a digital signal only at particular instants is called *discrete,* in contrast to the information carried continuously in time by an analog signal.

When this digital signal is now transmitted it leads, as in the case of any physical signal, to a weakening by deformations and to the addition of noise; but since the signal received at the moment when it is sampled (that is, when it is picked up) is in the validity domains of the information, the latter is preserved.

The use of a digital signal as information carrier confers upon the latter a protection against noise and deformations (one speaks of noise immunity).

| Source | In-line Coding | Canal | In-line Decoding | Receiver |

Therefore, since the information is preserved, it is possible to restore it to its initial property. The signal can in some way be *regenerated*. This feature is the major attraction of digital transmission, that is, the possibility of recognizing the information in the signal transmitted and of reemitting it, thus forming a repeater without modification of the information of the original signal. We can therefore see here the means for remedying the major shortcomings of an analog transmission system.

However, in order to transmit a large number of data per unit time (corresponding, for example, to a large number of lines in parallel) the electric signal must be able to pass rapidly from a value pertaining to one domain to a value pertaining to another. This poses for the channel the problem of the speed of variation of the signal, which gives rise to a concept, that can be linked to that of the passband of a linear system.

This brings us to the problem of the information flow allowed by a transmission channel.

Information Capacity of a Channel

The maximum capacity C of information which a channel is able to supply has been defined by Hartley as

$$C = \lim_{T\infty} \frac{\log (N(T))}{T}$$

where $N(T)$ is the number of different signals of duration T allowed by the channel. Claude Shannon has established that the maximum flow C is equal to

$$C = B \log_2 \left(1 + \frac{S}{N}\right)$$

where B is the passband of the channel in Hertz and $1 + \frac{S}{M}$ is its dynamic.

The choice of the base 2 for the logarithm corresponds to the bits/sec as the unit for the information flow.

Without going into the details of the proof, we state that quite naturally this capacity of the channel to furnish different signals is linked to the number of different states that can, on the average, be recognized in these signals: *dynamic* and, at the rate of change of the signal it allows, *passband*.

It should also be noted that dynamic and band are not independent factors since the power of the noise increases with the passband of the channel. In practice, this maximum information flow is never attained. There are several reasons for this.

First of all, obtaining this maximum flow presupposes that the signal presented at the input of the channel uses to the fullest all the transmission possibilities of the channel.

In practice, the adaptation of the source to the channel, called *in-line coding*, leads to circuits of increasing complexity and cost if one tries to approach the optimum.

Moreover, when in order to transmit rapidly one tries to make fuller use of the possibilities of the channel, the risk of error increases. This fuller use presupposes, in effect, the search for a larger number of value-domains, thus diminishing their size and their span and the diminution of the size and of the spans between the intervals of time when the signal carries the information. In doing so one increases the probability that the noise and the deformations place the signal in an invalid zone or that the uncertainties at the reception at the instants of validity of the information place the observation outside these instants of validity, thus leading to errors in transmission.

Because of this, a digital transmission is characterized by its flow (bit/sec) and by its rate of error.

It should, however, be noted that there is a possibility of obtaining between the source and the receiver an extremely small, theoretically, even zero, rate of error by using a conditioning of the signal prior to in-line coding that permits an error correction (repetition, range control, error-correcting codes).

However, this conditioning (which calls upon the science of coding) is paid for by a lengthening of the message and, therefore, by a diminution of the useful flow of the transmission. Even if this lengthening can theoretically be kept quite small, the processing operations for approaching this goal again involve circuits of increasing complexity and cost.

We realized, therefore, that the complexity and the cost will, whether at the level of coding or of in-line coding, be the higher the harder we try to approach the maximum flow a channel is capable of delivering.

The Various Systems for the Transmission of Digital Signals

In its initial form, the information signal presents itself in a binary form such that it can be transmitted in line, but the effectiveness of the transmission is limited because of the very large spectrum it occupies. Harry Nyquist's criterion shows us in effect that in order to transmit such a signal, the channel must have a frequency band equal to at least R/2 where R is the velocity of the signal in bit/sec., that is, the inverse of the period between two bits and, more specifically, under the following conditions: infinite attenuation outside the band and phase shift proportional to the frequency or zero.

This criterion by Nyquist can be illustrated in a simple manner. If a

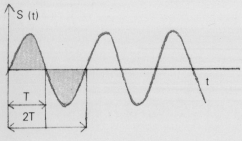

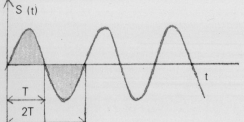

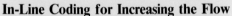

system is able to pass a frequency $\frac{F}{2}$ of period 2 T, in one 2 T period, it is able to transmit the equivalent of two binary symbols.

Moreover, for a physical system, the conditions of infinite attenuation outside the band and of linear phase cannot be rigorously realized. The distortions of the attenuations within and without the band and of the time of propagation lead to a deformation of the pulses emitted and to an overlapping of the successive pulses at the reception, a phenomenon known as *intersymbol interference*.

One has, therefore, the choice between limiting the speed of transmission, that is, to space out the pulses in order to reduce the interference, and realizing the phase and amplitude correction systems known as *equalizers*.

In-Line Coding for Increasing the Flow

The goal is to modify the emitted in-line spectrum for the purpose of utilizing the frequency band and the dynamic of the transmission carrier more effectively. At this level, we distinguish between two types of in-line coding.

Baseband coding. With this type of coding, the information is coded in levels (bipolar, biphase, ternary codes). One of the prime concerns of this type of coding is to get rid of the passage to the zero frequency and of the very low frequencies and to reduce the width of the spectrum to be transmitted by a different distribution of the signal in the band (one will also endeavor to place it where the distortion of the channel is at a minimum). Bipolar coding can, for example, lower the maximum frequency of the spectrum by one-half and do away with the necessity of transmitting the zero frequency. As has been pointed out above, for a given channel these improvements are realized at the cost of an increase of the rate of error.

The apparatus for realizing for a transmission line the baseband in-line coding in one direction and the decoding in the other direction is known as a *baseband emitter receiver*.

The second type of in-line coding utilizes, as in the case of frequency multiplexing, the possibility of transposing the spectrum of the original digital signal with the aid of a sinusoidal carrier. The carrier can be modulated by the digital signal with respect to amplitude, phase, and frequency in order to obtain a new in-line digital signal better adapted to the characteristics of the channel (for example, minimum amplitude and phase distortion). Thus, if it is desired to transmit digital information over the analog network, one uses a transposition

designed to position the spectrum of the signal to be transmitted in the band of a telephone line or in the band of a group of several lines (primary or secondary group) depending on the flow required by the transmission.

The apparatus performing the function of modulation at the emission and of demodulation at the reception is known as a MODEM.

With this type of apparatus it is necessary to distinguish between the speed of modulation expressed in baud and the speed of transmission in bit/sec. For example, a MODEM performing a phase modulation at B states of the signal at a speed of 1,600 times per second or 1,600 baud with a flow of 4,800 bits per second can, with B states, makes it in fact possible to code the information to be carried by three successive bits of the input signal.

In order to assess the inputs, it should be noted that in the telephone band one currently uses *MODEMS* with a flow ranging between 50 and 9,600 bits/sec. for a 4-wire line (the flows obtained over 2-wire lines being smaller mainly because of the echo problems) and for primary groups, speeds ranging between 42 and 72 Kbits/sec.

Digitalization of the Telephone Signal

In the perspective of the progressive changeover to the digital in the telecommunication network motivated by the intrinsic advantages of digital transmission and of the objective of a multiservice network integrating the transmission of data, there arises the problem of the digitalization of the telephone signal. Two approaches are possible. The first approach aims at trying to preserve the shape of the signal through sampling and analog-to-digital conversion in one direction and digital-to-analog conversion in the other.

In the case of a signal that can be restricted to a band of 300 to 3,400 Hz while preserving its intelligibility, this presupposes a sampling frequency of at least 3,400 Hz (Shannon's theorem). In practice, a frequency of 8,000 Hz is used, which corresponds to a sample taken every 125 μ s.

This sample is then quantified and converted digitally, which gives rise to a quantification error.

For the dynamic retained for the telephone signal, 256 levels of quantification are sufficient. These can be coded with 8 bits. It must be pointed out that the coding of the 256 levels is not done linearly but according to a logarithmic law (compression) so as to maintain a ratio of signal to noise due to the conversion of about 33 decibels in the entire band of the levels of the average speaker. The required flow is, therefore, of the order of 8000 x 8 = 64,000 bits/sec. It is interesting to

50

compare this flow with that which can currently be transmitted over a standard analog telephone line with the aid of a MODEM. The flow required is much greater. The digital transmission of the telephone signal over a single line of the analog network is, therefore, not possible in this manner.

However, the sampling of the telephone signal allows us to envisage another type of multiplexing, called *temporal* that consists in intercalating on one scanning field the signals coming from several lines.

For example, a 30-line (PCM) multiplex signal can be realized by using 32 intervals of time, one being reserved for the synchronization, the others for signaling. This implies a scanning flow of $64,000 \times 32 = 2,048$ million bits/sec (Mbits/sec).

Also for the connections comprising a large number of lines (interexchange connections) typically digital feeder lines have been constructed. At present, there exist connections with 140 Mbits/sec corresponding to a 1920-line multiplex.

The digitalization of the transmission between exchanges is already well under way.

The digitalization of a portion of the network poses the problem of partly analog, partly digital interfaces.

The system that performs the analog-to-digital conversion (or coding) in one direction and the inverse conversion in the other is known as CODEC. It should be noted in this connection that the reconstitution of the original analog signal and the filtering are possible only if the original signal itself has been filtered prior to the analog-to-digital conversion.

We shall conclude with the possibility of arriving at a digital continuity in the networks thanks to the development of temporal commutators that realize the switching of the communications (commutation) starting from the samples of the telephone signal and from their address, a technique concordant with temporal multiplexing in transmission.

The second approach to the digitalization of the telephone signal consists in analyzing the analog signal to be transmitted in order to extract from it its information content, then transmitting the corresponding data that at the reception will make it possible to synthesize an analog signal as close as possible to that of the original signal.

This approach is linked to the problems of speech analysis and synthesis, a field presently in full development. By sacrificing some of the fidelity, it is even now possible to reduce greatly the flow required for the transmission of the voice signal. The apparatus for performing this operation is known as the *vocoder*.

At the emission, the coding of the spoken word is done by a brief spectral analysis that determines with the aid of a bank of filters the existence or nonexistence of a fundamental frequency and its value as well as the characteristics of the resonant cavities that constitute the sound-producing apparatus.

At the reception, a synthesizer comprising a frequency generator, a noise source, and an adjustable filter reconstitutes an electric signal close to the signal to be transmitted.

At present, flow reductions by a factor of 20 are being realized, but not without detriment to the fidelity of the transmission.

Conclusion

Electronics has played and continues to play a basic role in the development of telecommunications. It has made possible the realization and the improvement of the properties of the passband and of the linearity of the circuits allowing high-level multiplexing at costs low enough to permit the mass development of telecommunications.

The advent of large-scale integrated circuits has permitted the actual development of digital systems that must satisfy not only the needs of the telephone services but also those of the new services.

The increase in the volume and the power consumption brought about by integrated circuits has allowed the realization of increasingly complex digital systems — high-flow systems, electronic commutators, concentration systems — that utilize the moments of silence in the conversation.

The technical possibilities of electronics linked to the reductions in cost resulting from mass production make it possible to envisage the putting at the disposal of the subscriber of a multiservice terminal allowing access to digital data banks. This means "telematics" at the subscriber's disposal, the present goal of development in telecommunications.

5

Data processing and peripheral data processing

The computer generation is the latest manifestation of a historical process made up of a number of very diverse currents. The first, the logical current, goes back as far as Aristotle and his principles of identity, contradiction and the excluded middle. These principles, as elaborated in the middle of the nineteenth century by Gottfried Leibnitz, who invented the binary number system, and English mathematician George Boole, led to the so-called "digital" system, based on binary logic: a statement is either true or false, yes or no, "1" or "0".* Another current, the arithmetical current, had its beginning in the development of the first methods of calculation; this current inspired the invention of the slide rule in the 17th century (the work done during the same period by Wilhelm Schickard at the University of Tübingen on calculating machines), as well as Pascal's work on addition, Leibnitz's research into four-function machines, and today's analog computers.

The third current, the concept of programming, dates back to the ancient Greeks, more specifically to the Alexandrian mechanics and their use of the cam. A more recent pioneer was the Englishman Charles Babbage, a friend of both Darwin and Dickens.

Babbage wrote a description of his Analytical Engine which included the concept that the "program" could be used to solve not only mathematical problems but also to perform intellectual operations. However it was an American, Herman Hollerith, who in 1890 was the first to use an electromechanical system to transpose census data into code. He used punch cards to program his machines, following the principle first enunciated by Falcon and J.M. Jacquard, two French inventors who used it respectively in automatic calculation and weaving looms.

The final current, the memory, provided the technological foundation of the modern computer; this development first appeared in 1919 with the flip-flop system designed by W.H. Eccles and F.W. Jordan on the basis of the oscillator invented by French scientists Henri Abraham and Eugène Bloch. A series of control pulses were fed into a circuit made up of two triode tubes (or, from the 1950s onwards, two transistors) and each successive control pulse would stop one tube conducting and turn on the second tube, thus producing a see-saw type action. However the basic principle of the memory (prophetically described by Babbage and employed by Konrad Zuse) was defined by the American J. von Neumann at the end of World War II. Memories made it possible to store and process comparatively simple information, an the combination of calculator and memory heralded the beginning of the computer age.

Computers operate in either of two modes: the analog mode, mainly used for calculation, or the digital mode used to process data and to control the operation of one computer or group of computers by another computer or group of computers. The term "analog" was originally taken from the field of hydraulics, and refers to the process whereby variations in fluid densities are calculated by noting the corresponding (analagous) variations of the temperature and pressure of a given gas. Analog mode is used to measure variable factors. The slide rule allows addition or subtraction of numbers represented on its surface by predetermined lengths, and was the first example of analog calculation. In a calculator, the numeric value of a number is represented by a certain given voltage level. Thus, analog calculators are used to evaluate lengths or distances, automatic airborne flight systems, bombing systems, navigational aids, flight simulation, product development schedules, high voltage transportation, and distribution networks. In 1925, an MIT research worker, Vannevar Bush, began work on a mechanical device that later evolved into the differential analyzer, the first analog calculator capable of integrating ordinary differential equations.** He achieved this result in 1930, but soon afterwards began to work on the possibility of developing a thermionic tube calculator; he published a number of papers on this theme between January 1937 and November 1938. Around the same time one of his assistants, H.L. Hazen, advanced a theory of servo-mechanisms; that is, machines that control other machines, a field which we now refer to as *automation*.

A digital mode computer analyzes the various combinations of electrical impulses rather than measuring the size of signals received. In order to simplify these operations, we replace the decimal system, which would require 10 different states, by the binary

Kanazawa Institute of Technology: the Japanese FACOM TSS-CAI system is designed to help students teach themselves.

* The binary system was first used by Francis Bacon (1561 – 1626) to transpose secret messages into code. Jacquard's looms were inspired by this system as was Emile Baudot's telegraphic code, which is known as the Gray system in the United States. It is based on the two symbols 0 and 1 and operates as follows:

0:0	4:100	8:1000	12:1100
1:1	5:101	9:1001	13:1101
2:10	6:110	10:1010	14:1110
3:11	7:111	11:1011	15:1111
			16:10000

** for some researchers, as for W. Giloi, V. Bush's name is connected with mechanical integrators, and the first electronic analog computer was built by Helmut Hoelzer at Peenemunde (around 1941).

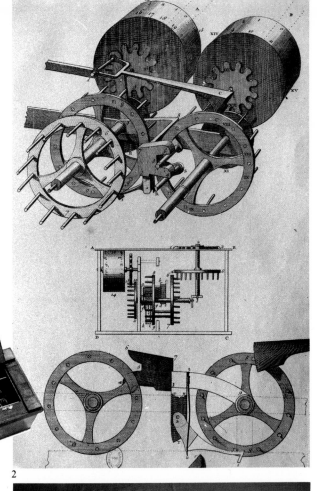

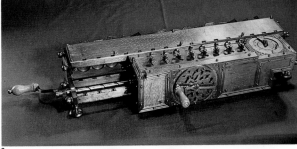

1 and 2. Blaise Pascal (1623-
1662), French mathematician,
physicist and philosopher,
built his arithmetical machine
in 1652. This adding machine
was based on tiny wheels
engraved with numbers, with
a gearwheel system for
carrying units over.
3. Gottfried Leibnitz
(1646-1716), German
philosopher and
mathematician, invented
differential calculus in 1676.
4. Up until his death in 1871,
Charles Babbage had spent
many years on the design of
his "analytical engine" for
automatically solving
mathematical equations. The
basic principles are similar to
those underlying the first
electronic computers.
5. The Countess of Lovelace
(1815-1852), daughter of Lord
Byron and a mathematician,
was history's first computer
programmer. She participated
in Babbage's work and lost
an enormous sum on her idea
for an "infallible" betting
system. She died, despairing
and financially ruined, at the
age of 36. Her name has been
given to the computer
language "Ada".
6. Falcon's machine
introduced perforated cards
for the first time.

system which only uses two symbols, 0 and 1. We
use the same system when we compose a number of
a value higher than 9 in the decimal system. In the
same way, after the number 99 we create yet another
space to the right of the 10 and start over again at 0,
which gives us the number 100.

According to the binary system, $1 = 1$, $2 =
1 - 0$, $3 = 1 - 1$, and so on, because we are no longer
working with ten figures but with two. In the computer, flip-flops are programmed to represent these
binary numbers, and a "1" or "0" value is transposed to "on" or "off" or a magnetized or nonmagnetized state of the flip-flop circuitry. The first
binary circuits used for addition, subtraction, multiplication, and division, binary-decimal conversion
and vice-versa, were designed by George R. Stibitz
for the Bell Laboratories and were used in the development of telephone exchange switching equipment
(relay computers).

The first high-powered digital calculator which
still used electromagnetic relays was the Mark 1,
developed in 1944 by a Harvard physicist, H. Aiken.
A German researcher, Konrad Zuse, designed the
similar Z4 around the same time, but he never succeeded in convincing the German army of its potential. However he completed the first *operational*
digital computer in the world.

All these different developments were to give
birth to the *first generation of computers.* The first
model, the ENIAC (Electronic Numerical Integrator
and Computer) apppeared in 1945; it was a monstrous machine weighing 30 tonnes and housing some
18,000 electronic tubes which gave off enough heat
to boil an egg on its surface! Its constructors, John
W. Mauchly and J.P. Eckert, decided to replace the
electromagnetic relays by electronic flip-flop circuits
whose electrical outputs were used to either illuminate or extinguish panel lamps. When they began
working in 1943 aerial combat was intensifying and
their immediate goal was to improve the method of
ballistic equation calculation in order to produce new
firing tables, a job which up until then had been
done by differential analyzers.

ENIAC's computation time for addition was 0.2
milliseconds and 20 milliseconds for division. It was
not only the world's first electronic computer but
also the fastest: 1,000 times faster than the Mark 1.
Another electronic machine, the Colossus, designed
by T.H. Flowers and W.W. Chandler, was probably
already in operation during the war years at Bletchley Park, the British secret research center. Its existence was not made known until much later but its
capacity was probably rather less than that of the
Mark 1.

It was at this point that a certain character who
was to become a legendary figure in both scientific
and political circles made his appearance on the
scene. This was John Von Neumann, the man chiefly
responsible for "liberating" the computer from its
role as an instrument of mathematical operations
and making it a machine capable of processing information. Von Neumann had been influenced by the
ideas of a young Englishman named Alan Mathison
Turing, who in 1936 had advanced the thesis of the
"universal machine" capable of programming other
machines. Flowers, one of the inventors of the
Colossus, later suggested that during the war Turing's
team working at Bletchley Park had in fact developed electronic calculators with electromechanical
inputs and outputs, but even now the matter is far
from being clarified as most of the documentation
concerning Turing's activities at the secret research
center is still classified Top Secret. To what extent
did Turing influence Von Neumann and vice versa?

1. Vannevar Bush (left) at MIT.
2. Computer designed by G.R. Stibitz.
3. The mathematician G.R. Stibitz.
4. Mark I, built by Aiken.
5. Alan Turing.
6 and 7. Konrad Zuse in 1949 and the Z 3 which was rebuilt after the war.
8. J. von Neumann.
9 and 10. The Colossus.

1

2

3

1. Eckert and Mauchly in
front of their machine.
2 and 3. Professor Wilkes
and EDSAC in Cambridge.
4. IBM's 704 electronic tube
computer.
5. IBM's 7090 transistor
computer.

4

5

This is a very difficult question to answer.

In 1943, Von Neumann was appointed consultant to the Los Alamos Laboratory and became associated with the Manhattan Project. He worked in close liaison with the Göttingen physicist Oppenheimer with whom he had been friendly for many years. Von Neumann was exasperated by the lengthy business of calculation. Two incidents were to push him into taking an interest in electronic calculators. First, during a trip to Europe he became interested in designing a program for an accounting machine. A little later, on a railway platform in Aberdeen (he was also a consultant at the nearby firing center) he was waiting for a train with the mathematician H. H. Goldstine whom he knew by sight; Goldstine mentioned the ENIAC electronic calculator that had been built under his direction at Moore School.

At this point Von Neumann became quite enthusiastic and a few days later went to have a look at the machine himself. The research workers at the school were impressed by the rapidity of his calculations, his almost photographic memory, and his very solid grasp of the theoretical underpinnings of their work. The resulting collaboration led to the publication of a report signed by Arthur W. Burcks, von Neumann and Goldstine, "A preliminary study of Electronic Design," which contained an extremely important section dealing with the concept of the prerecorded program. Von Neumann introduced the idea of the necessity for a code and the concept that the machine itself would read, interpret and carry out this code.

Thus Von Neumann made Turing's 1936 thesis, the "program" is itself the "information," a reality, to the extent that the system memory contains data to be processed and instructions, thereby making it possible to modify the program.

Von Neumann was not the only scientist working on the idea of computer memories (Babbage and later Zuse proposed mechanical storage for data but were unable to develop it further). By the end of 1944, Vladimir Zworykin (the father of electronic television), Mauchly and Eckert had all proposed their own solutions. Eckert was particularly drawn toward the idea of the ultrasonic mercury delay system that had already been developed by radar experts for the measurement of range by comparing the reflected signal with a locally delayed signal. This idea was later taken up by the English, who were the first to apply Von Neumann's ideas (the American EDVAC computer was only finished at the end of 1951). During the summer of 1946, the forerunners in the field of digital calculation − Eckert, Mauchly, von Neumann, Goldstine, and Burcks − gave a number of lectures for European scientists. The first man to apply all he had learned in the United States was a Cambridge research scientist, M.V. Wilkes, who in 1949 completed the EDSAC (Electronic Delay Storage Automatic Computer), the first computer equipped with internal programming facilities. This development signaled the birth of the *second generation of computers*.

Wilkes himself notes that the summer of 1946 marked a turning point in his career. As a student at the Cavendish Laboratory, his main field of interest had been the ionosphere; in the course of his work he became interested in a differential analyzer (constructed by J.B. Bratt) based on ideas put forward by J. E. Lennard-Jones. Wilkes used the analyzer to solve a differential equation preoccupying him at the time. During this period, Lennard-Jones was toying with the idea of setting up a laboratory for the study of calculators. He actually founded the laboratory in October 1937 and appointed Wilkes as Director of Research. When war broke out Wilkes took part in

work on radar development, but in May 1946 L.J. Comrie, on his return from the United States, gave him a copy of von Neumann's EDVAC theory. Wilkes then decided to apply Von Neumann's ideas and submitted his project to J.R. Womersley, director of the National Physical Laboratory. Womersley invited Wilkes to attend a series of lectures which Turing was giving in London. However, the two men did not establish any basis for cooperation; Wilkes disagreed with Turing's ideas and Turing was completely uninterested in Wilkes's project. In 1977, Wilkes wrote: "Turing seemed to me to have blinkers on and I believed that he was generally outside the main development stream of the computer." In a NPL internal memo from Turing to Womersley in which Turing commented on proposals put forward by Wilkes for ACE systems, he wrote: "The code which he (Wilkes) suggests is in contradiction to our whole line of development and much more in line with the American tradition of resolving difficulties by adding more equipment rather than through logical development." In 1949, Wilkes produced the EDSAC, a binary memory calculator with a bank of mercury reservoirs; at the end of 1950, he published *Preparation of Programs for an Electronic Digital Computer*, the first book on computer programming.

But another computer may have beaten Wilkes's EDSAC to first place. This was the Mark 1 prototype built by F.C. Williams and Tom Kilburn at

1. A revised version of the first computer program (Manchester University's Mark 1). Taken from a contemporary notebook kept by G.T. Totill.
2. Tom Kilburn (left) and F.C. Williams shown at the control panel of the Mark 1. The prototype ran its first program on June 21, 1948.

the University of Manchester on June 21, 1948. The battle lines over the different computer memory design methods were now drawn up, and from 1946 onwards two opposing approaches were in existence: the EDSAC system, based on acoustic waves propagated by a mercury delay line, and the Mark 1 system, based on electrostatic charges stored on the luminescent screen of a cathode ray tube.

In the fall of 1946, Williams came up with a method for memorizing digital data on the cathode ray screen of a specially designed television set. He patented his invention on December 11, 1946. When he was appointed Professor at Manchester University he invited Kilburn, with whom he had previously worked on radar, to join him. One by one the old members of the Bletchley team, some of whom had worked on Colossus, joined the new Manchester team. In October 1945, Turing joined them, although his role was hardly more than a secondary one. His ideas, however, were developed between 1948 and 1951 by a National Physical Laboratory team headed by J.H. Wilkinson and were the basis for a third machine, the ACE. The three teams worked independently of each other; their ideas and their aims were very different, as were their ideas of what a computer should be. In Cambridge, Wilkes paid more attention to programming than to technique. The Williams-Kilburn team in Manchester put the emphasis on technological improvements, and Turing, as we have seen, was more concerned with the theoretical and philosophical aspect of research.

The 1950s proved to be the first great turning point for the computer. People began to look at machines in a different way. In 1948, Norbert Wiener published his book, *Cybernetics, or Control of Communication in Animal and Machine.* Scientists began to compare the mechanism of the computer with the functioning of the human brain, and the idea was expanded further by Von Neumann in his theory of automatons.

It was also an important period in the commercial development of the computer. Up until then computers had only been used for scientific research or military purposes. In 1950 the ENIAC team in the United States set up the UNIVAC sales company, and in 1953 the EDSAC group launched their LEO on the market. Some several thousand models of the IBM 650 electronic tube computer, with its associated monoprogramming system, were sold. To give an idea of the technological progress made since the development of the first computer, the machine weighed 3 tons, used 20 kW and cost 1,000 times more than a modern industrial calculator using a few milliwatts.

Some writers use the road analogy to describe digital machines; according to this description the bits (units of information) are the cars and the programmers who organize and regulate the passage of information may be compared to the highway police. The memories and the terminals for data input and output operations are the ramps and exits, while the various storage systems for unrequired data may be thought of as parking lots.

The machine is organized around five units: (1) the input unit (definition of the problem to be solved) which converts the request into coded data and instructions for the machine; (2) the control unit (where the data is collected by the technicians) which is linked to (3) a logic unit whose function is to make the calculations and take a decision in response to a signal sent by the control unit; and (4) a memory unit where instructions and data are stored. These four units make up the processor part of the machine. The fifth unit is the output which converts the results from machine code into normal language.

One of the most common coding methods chosen is undoubtedly pulse code modulation. As early as 1938, Alec Reeves, an English scientist working in France for ITT, took out a patent for his pulse code modulation system; this system would only appear on the market following the development of semiconductor circuitry and the invention of the transistor.

Pulse code modulation (PCM) includes three successive stages:
– The word is broken down to a reasonable speed, and its value is measured at each time interval.
– These values are quantified to the nearest whole number.
– These numbers are transposed to ordinary telegraph code.
The criterion used is the presence or absence of a pulse. Alec Reeves wrote:

Pulse code modulation, or coded step modulation (which I think would have been an apter name), is a good example of an invention that came too early... When PCM was patented in 1938 and in 1942, I knew that no tools then existed that could make it economic for general civilian use. Only in the last few years, in this semiconductor age, has its commercial value been felt... Having had it patented, for understandable reasons I then let the invention slip from my mind until the end of the war. It was in the United States during World War II that the next step in PCM's progress was made, by Bell Telephone Laboratories. In this important stage, a team under Harold S. Black designed a practical PCM system later produced in quantity for the U.S. Army Signal Corps. Research was also done under Ralph Bowen (Reeves and Deloraine, 1945:2).

By December 1948, almost all the major elements for effective long-distance pulse code modulation had been invented and a number of countries began to study its potential. In November 1945, a Philips team consisting of J. F. Schouten, E. de Jager, and J. A. Greefkes published the results of its work on Delta modulation which also broke down a word for the purposes of binary mode transmission. In 1951 the Japanese electronic communication laboratories (Nippon Public Telegraph and Telephone Corporation) began their own research, and Ki Yasu invented the reflected binary code, known as the Gray code in the United States.

In 1960 Alec Reeves concluded: *Pulse code modulation has been a child with a long infancy; except for certain military uses not described here, in application it is still only in the adolescent stage.*

The year 1950 also saw the development of the transistor. This invention, the vanguard of the approaching armies of integrated circuits, made it possible to reduce size and cost and to construct increasingly rapid and reliable memories. The first transistorized computers (IBM's 7090 and Bull's Gamma 60) appeared around 1955. Between 1959 and 1964, IBM invested billions of dollars to develop the 360 series which was responsible for the company's subsequent success and its domination of the world market. At the end of the 1960s and the beginning of the 1970s another important turning point was reached with the development of LSI, or large-scale integrated circuits; the infant computer, confounding everyone by growing smaller rather than larger with each new day, had been endowed with a new quality, the gift of intelligence. This was the *third generation of computers*, the generation of multiprogrammed machines with shared time and remote access using integrated circuits and mass memories. Over the last twenty years, computation

Specialist Wiener, engaged in trajectory research.

1 and 2. Norbert Wiener (1894-1964) was the founder of cybernetics, the science of communication and control in machines and living beings. A system evolves to a new state previously defined by two mechanisms: feedback and memory.
3. The IBM integrated circuit and ferrite computer with central and peripheral memories.
4. Optical character reader.
5. Line printer.
6. Spinwriter.
7. Standard calculator on an MOS chip.

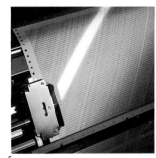

Data Processing and Peripheral Data Processing

speed has increased about 1 million times, and costs have fallen despite a dizzying increase in sophistication. Microprocessors brought about a revolution in the computer's whole mode of "thinking". Information is no longer treated only in the central unit, but processing is distributed among the different system units, each of which has its own particular function. This makes for increased operation speed, with several elements, each tackling a different part of the same job. Reliability is increased even further because in the event of breakdown the task normally alloted to the malfunctioning unit can be transferred to another.

Modern microprocessors are made up of multiple and complex systems and are referred to in terms of integrated systems rather than integrated circuits. By the end of the 1980s, it will no doubt be possible to produce chips containing millions of transistors, and elements and interconnections will be smaller than the wavelength of visible light. These extraordinary developments have already tended to revolutionize the approach to and construction of computers; computer designers who up until now have devoted themselves to producing circuits and logic in close collaboration with semiconductor manufacturers, seldom participating in the specification and design of circuits, will no longer be confined to working on standard integrated circuit systems and will instead be thinking in terms of integrated architecture and design.

Since the 1950s there has been much improvement in memory design and function: research work has been oriented toward storing the maximum amount of information in the smallest possible space. The first memory to allow passage from the analog to the digital was the "ferrite core" type. Theoretical and practical progress was made in the field of magnetism – a phenomenon that had been known even to the early Greeks – through the work of Langevin and Weiss, and ferro-magnetic materials played a basic role in electricity and the beginnings of electronics. In the 1930 s the Philips company and a number of university teams began to produce new compound materials called *ferrites*, ceramics made from an iron oxide base. The theoretical work on this material was carried out in France by Louis Neel in the 1940s; the French scientist was inspired by earlier research work done by the German scientist Hiltelt and the Dutch scientist Snoek. Before the tiny, ring-shaped toroids were developed, data was recorded onto cylindrical layers of magnetic material in the form of drums. At the beginning of the 1950s, ferrite cores began to be used in memory systems. A feature of their natural properties enables them to be easily magnetized or demagnetized, and their state at any given moment may be defined as "1" or "0." A control current can therefore program each core with a "1" or "0" state, and another signal is able to read the core: the information is erased and must be restored by rewriting it into the core. This type of memory was developed as the result of research work done by a number of American scientists, in particular Jay Forrester (MIT) and J. Rajchman (RCA). Although very rarely used today, it was an important milestone in the development of the modern computer.

There are three main types of memory: 1. random access, which gives direct access to information; 2. sequential access, in which the whole tape has to be read to obtain the desired information; and 3. "associative memories," where the information is located by analogy with its content. Modern memories are either of the magnetic memory type, which has lower storage costs than electronic memories and is often used to store files in data processing systems, or completely electronic memories, the most common of which are semiconductor memories. A number of other promising developments are beginning to appear: magnetic bubbles, Josephson superconductive diodes, and coupled charge devices (CCD's), for example. Magnetic memories give block access, that is access time depends on the exact spot where the information is located on a thin magnetic film with a nonmagnetic backing (ribbon, plastic disc, aluminum drum, etc.); this spot is determined by mutual displacement of the magnetic film and the reading/writing head. C.A.M.'s (Content

1. Jay W. Forrester, who developed the ferrite bead memory and directed the Whirlwing 1 air defense system project for MIT.
2. Ferrite, no longer used for memories, today constitutes a component used to control the self-induction of high-frequency coils.
3. The world's first 64 Kbits MOS RAM.

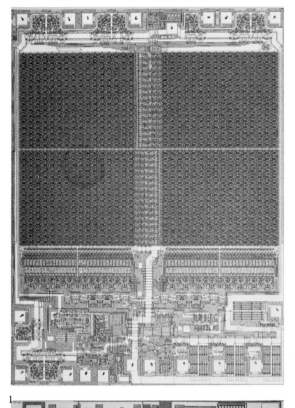

1

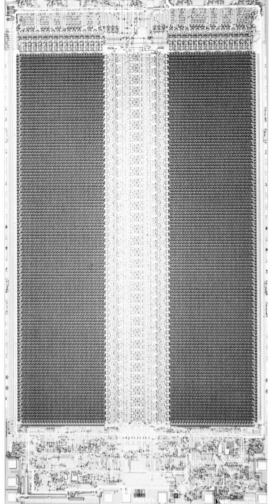

2

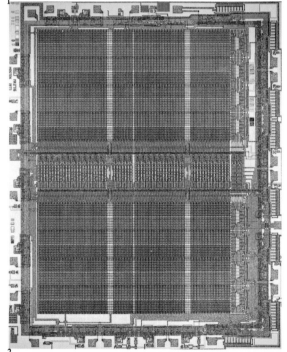

3

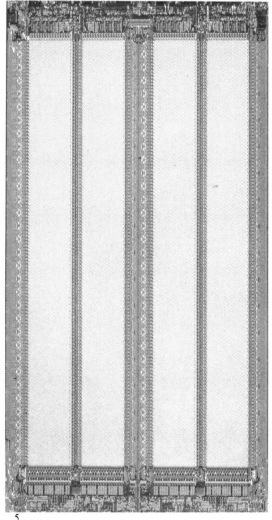

4

5

1. 4 Kbits *Static* RAM: Flip-
flop memory system,
presently available only at a
maximum capacity of 16
Kbits (64 Kbits in the
foreseable future).
2. 16 Kbits *Dynamic* RAM:
In a dynamic RAM, a bit is
represented by an electric
charge. Maximum capacity of
64 Kbits or (in the near
future) 256 Kbits.
3. 16 Kbits EPROM.
4. 64 kbits ROM.
5. 64 Kbits RAM.

Adressed or associative type Memories) are memories in which the content or partial knowledge of the word serves as a reference. The per bit cost is still very high, which means that they are used mostly for reservation of places or file research and for "random access" memories where access time is independent of the information's position. The most well-known memory is the random access memory (RAM). These are live memories capable of storing up to 16,384 bits (the next generation is already in production and will have a capacity of 64,536 bits. 256 Kbits chips will be available in a couple of years from now. These electronic memory devices offer the lowest cost per bit.), read out/writing type with flip-flop memory system. Access time to information varies between 0.05 and 1 microsecond. R.O.M. (read only memories) are dead memories whose recorded programs are very rarely modified.

Various different versions of these memories also exist. The P.R.O.M. for example (Programmable R.O.M.) is programmed by the user and not by the manufacturer, and the R.E.P.R.O.M. (Reprogrammable R.O.M.) is a memory that may be erased by optical means. These memories are made from M.O.S. transistors (see the chapter on microelectronics for a description of this technique) with two superimposed grids. To erase and updata data the circuit is exposed to an ultraviolet light source that makes the insulator (silicon oxide) partially conductive and discharges the lower (or floating) grid where the information is stored. However, because the elements are extremely small, the operator cannot select what he wants to erase and the whole memory disappears. The use of E.A.R.O.M. memories is now under consideration; this version enables certain information to be changed without erasing the whole memory, but it is little used at present. Coupled charge devices and magnetic bubble memories are both new developments. They are series access memories with access time dependent on the chosen access path. They are expected in the medium or long term to replace tape or disc memories, as their storage surface is smaller and, consequently, cheaper. In 1975 the first CCD (coupled charge device) had a capacity of 9,000 bits; in 1976 its capacity had already reached 16,000 bits; and by 1977 the figure had climbed to 64,000 bits.

Magnetic bubble memories, developed in 1966 by Andrew Bobeck at the Bell Laboratories, have been on the market since 1977; they can store even more information than other types of memory. Magnetic bubbles are tiny magnetic areas less than 1/16th the diameter of a hair built into a magnetic support which itself is oppositely charged. The bubbles move with great speed in precisely controlled directions and retain their polarity when power is removed from the memory. Although CCD memories are highly volatile, they lose their information if the memory power supply is interrupted, but they are ten to a hundred times faster than bubble memories. In addition, CCD memory systems can be integrated on a single chip whereas bubble memories require separate components and auxiliary circuits as well as wire reels to generate magnetic fields. Now CCD's seam to be dead; and it is still an open question whether the bubble memory will make it in the face of competition from the VLSI semiconductor memory.

Current research is investigating the possibility of using Josephson superconductive diodes for memories (see cryoelectronics). These allow even greater data density since any heat losses would be negligible and the speed ten to a hundred times faster. The main difficulties involved in the production of this type of memory concern amorphous oxide technology which must be done in situ and whose extreme thinness leads to mechanical constraints that have not yet been overcome.

But despite all these encouraging new developments, we are still a long way from the magician's dream of recreating the human brain with its 100 billion neuron information system capable of processing 100 million billion events per second; and even further from duplicating that aspect of mental activity that does not depend on mechanical processes and that we call the soul or the imagination.

The data processing market wasted no time in creating another parallel commercial network, peripheral data processing, that is all the materials and equipment associated with the computer (terminals, visual display units, etc.), along with the different services that data processing has inspired, such as computer assisted design.

Although the Americans are still the market leaders, the Japanese have begun to conquer an increasingly large share of the market, to the dismay of the United States. In 1979 Hitachi produced the biggest and fastest general-purpose computer in the world, the HITACHI M-200 H. One of the major problems facing the Japanese has always been the difficulty of transposing the Japanese alphabet, which consists of kanji, or Chinese characters, and two phonetic alphabets, hiragana and katakana, into a suitable code form. Toshiba was the first to develop computer technology for reading kanji script, but Hitachi, Fujitsu, and IBM Japan were not far behind, and in 1978 IBM developed an alphanumeric and katakana keyboard. Japan has also produced a laser printer (the Oki) with 4,000 kanji characters. J.E.F. has produced a kanji-kana system with a phonetic (kana) input and a conversion key that when activated, causes the simple or compound identical-sounding kanji character selected by the kanji memory to appear on screen. The many homonyms in the language are memorized in order of frequency of use. The system also contains the Roman alphabet and the Japanese syllabary. This work has made it possible to automate newspapers by using computer networks in systems such as the Nelson (New Editing Layout System of Newspapers), experimental televised newspaper systems, such as the one installed at the Hotel Imperial in Tokyo, and systems that provide photocopies of a given page on request (Tama trials). These are the kind of developments that have made it possible for the Asahi newspaper, with its six morning editions and its three evening editions, to be five years ahead of the *New York Times* in terms of technological progress.

Computer-assisted design is still in its infancy. Its basic function is to examine all types of design hypotheses within the shortest possible time, to materialize the imaginary and to determine any application drawbacks, to exhaust the whole range of technical and architectural design possibilities in the most varied areas – construction, mechanics, yacht-design, footwear, textiles, furnishing and so on. Using the most varied and accurate data supplied by the designer the computer draws up scaled plans and drawings, figured indications, starts a dialog between the designer and his own proposals, and allows him to visualize projects for the client; this last property makes it an effective decision-making and sales tool.

The computer can even be ordered to modify a design, to show it in perspective, or to make it move across the computer screen at various speeds like a cartoon. A recent development has dramatically reduced the time factor involved; the moving pen system is replaced by an electrostatic unit loaded with specially treated presensitized paper onto which the computer reproduces the computer design in the form of electrostatic charges. The picture becomes visible after the paper is passed through a carbon-charged development bath. The computer can now be used to design the cabling for the next generation of computers. In the United States some companies are already using computer-assisted design and production systems, whose digital control programs make it possible not only to design but also to produce such industrial parts as tire moldings.

One of the most interesting branches of computer-assisted design is the area called *calculation of structures by finite elements*. By using this method it is possible to considerably improve the efficiency of material durability research. The part to be manufactured is divided into small, simple, geometric elements that are studied separately to determine the distribution of forces and the possible warping or buckling involved; this technique makes it possible to reinforce the critical points and to save on excess material. This method, used not only in the fields of aviation, oil transportation, nuclear power stations and public works, but also in boiler works and other more traditional activities, reduces costs and increases safety.

The computer is also used in "value analysis," in line with a principle discovered in 1947 by L. D. Miles at General Electric. In 1955 U.S. Secretary of Defense MacNamara insisted that this system be adopted by all large suppliers to the Defense Department. In 1977 the use of value analysis saved $100 million for the U.S. Army, $34.5 million for the Navy, and $16 million for Chrysler. Product functions are established by a multidisciplinary group that determines the product's uses, the constraints to which it will be subjected, the portion of total cost attributable to each function, and the best methods of design and modular production. Increasingly powerful computers are being designed to handle these problems; the present "champion" is the Cyber 205 (Control Data) which can carry out 800 million operations a second, although it is so astronomically expensive (between 10 and 12 million dollars) that only about 30 models have been produced.

Computer-assisted design and all the other systems described above have encouraged related development in the field of peripheral data processing. A peripheral system usually consists of an external memory; an input and output system (with ultra-fast printing models using a helium-neon laser); terminals (data bank terminals or specific terminals such as the automatic design machine which transmits the computer's final draft onto paper); data systems with access or transfer procedures; and increasingly complex and sophisticated control systems made up, for example, of a message chain, an input/output chain, and a modulation/demodulation chain. These control systems are intended to provide centralized monitoring for thermal energy stations, nuclear power stations, or highway networks.

Finally, software is becoming increasingly important, particularly in terms of flexibility and end-user applications. Software research has come up with concepts such as program and application modules. MITI (the Japanese Ministry of Industry) will, between 1981 and 1983 devote some $1 billion to software research in an attempt to oust France from its position as the world's second largest software producer. Peripheral equipment is accounting for an increasing proportion of total investment in the computer industry (50 to 60 percent at the present time). Further research is oriented towards increasing

computer memory speed and capacity. IBM is studying ways of increasing the density of memorized information by adopting a design based on holographic principles in association with the surface of a magnetic disc; this technique could theoretically make it possible to obtain 50 million bits per square inch. In the field of computer memories (but also in that of medical applications and various other fields), researchers are working on a phenomenon they have only recently succeeded in explaining: superconductivity, which has itself given birth to a whole new technique, that of cryoelectronics. The phenomenon of superconductivity was detected at the beginning of the twentieth century thanks to the work of Dutch scientist Heike Kamerlingh-Onnes. Kamerlingh-Onnes liquefied helium gas at 4.2°K. He then tried to solidify it by lowering the temperature and reducing the vapor pressure above the helium bath by means of pumping; he managed to obtain a temperature of 1°K. But the result was the opposite of what he expected: the helium became more liquid. In 1911, after continuing his research, he established the theory of the superconductivity of mercury, that mercury heated to the temperature of liquid helium has no resistance. It is known that the resultant effect of free electrons on impurities and thermic vibrations exhibits a resistance which diminishes as the temperature of the metal is reduced towards 0° absolute. The most important discovery in the field of superconductivity was when it was shown that for metals such as lead, tin, or niobium, resistance vanishes completely when a certain temperature close to zero is reached (the exact temperature is dependent on the different characteristics of each metal) and the metal thus becomes superconductive.

In 1935 another physicist, F. London, discovered that the phenomenon of zero resistance was associated with magnetic expulsion i.e. that a large superconductor did not register the presence of a magnetic field. If there was no resistance there was obviously nothing to restrict the circulation of current; therefore it was quite unsurprising that a magnetic field could not be introduced. However, if the superconductor was heated above critical temperature and a magnetic field introduced, a rather strange phenomenon occurred: as the superconductor cooled, it lost its magnetic field. Given that there was no current in the initial stage it appeared strange that a current could be created.

In 1940 the Russian physicist Piotr Kapitza established the theory of superfluidity, a complement of superconductivity, where the viscosity of helium becames nil. Another Soviet physicist, Lev Landau, then took this theory further, giving a better understanding of the various properties of superfluid helium. Landau first established the theory that superconductivity was due to phase transition corresponding to the appearance of a new state, although he was unable to define the exact nature of this new state. From already known physical properties, he and V.L. Ginzburg were able to deduce the structure of this state with remarkable precision: their equations, written in 1950, are still used today.

The next step was taken in 1957 when three American scientists: John Bardeen, John Schrieffer and Leon Cooper published the first comprehensive theoretical explanation of superconductivity. They established that electrons in a superconductor are grouped in pairs (known as "Cooper pairs") in contrast to the usual state in which two negative charges repel each other. This theory won its inventors the Nobel Prize in 1972.

The American "BCS" trio, as they came to be known, had thus discovered that the wave function,

1. The physicist Heike Kamerlingh-Onnes (1853-1926).
2. Piotr L. Kapitza (born 1894).

which according to Ginzburg and Landau represented the superconductive state, was one of the wave functions associated with an electron pair.

To recapitulate: electrons in a superconductor are grouped in pairs and each pair of electrons is associated with a *wave function* which, like any wave function in quantum mechanics, represents the probability of finding these electrons at a given point, and with a phase, which normally has no direct consequence on properties observed macroscopically, but which nevertheless does react, with respect to interferences, in a manner similar to the phase of optical waves.

This leads us to two very basic points: the first is that if all the electron pairs have the same characteristics at a given point, it is therefore possible to have access in the macroscopic field to microscopic quantum parameters. The second point was establishing that these pairs cannot be broken up without a minimum level of applied energy. They can in fact be broken up in two ways: through energy supplied either by means of luminous rays or by phonons (quantities of acoustic energy) leaving quasiparticles or unmarried electrons outside a forbidden band which then blocks the conduction band.

Although most of the electrons in a superconductor are arranged in pairs, unmarried electrons possess a higher degree of energy and react in much the same way as the electrons in a normal metal. However, given that their mass is very different from that of electrons in a vacuum, they are known as quasiparticles.

In 1962 a young English physicist, Brian Josephson, published an article setting out possible new effects in the field of superconductivity. He established that two pieces of semiconductor could be connected by a tunnel junction which would normally present a resistance of several ohms, but that this resistance could be cancelled out in the superconductive state; the barrier could be crossed by a supercurrent due to the tunnel effect of the Cooper pairs in the absence of voltage across the device's pins (the Josephson continuous effect). In the presence of a potential difference across the pins of the Josephson junction, the supercurrent begins to oscillate (the Josephson alternative effect).

Josephson theorized the possibility of a phase sensitive switching system: his first discovery was that in the presence of two superconductors of phase $\varphi 1$ and $\varphi 2$, respectively, the circulating current would be a sinusoidal function of the difference in phase between them. His second discovery was that if the circulating current is a direct function of the phase difference, it becomes possible to visualize the phase difference even if it is not possible to visualize the phases themselves. This possibility gives us a relationship between a macroscopic phenomenon (the electrical current) and a microscopic property (the phase).

The third effect of the Josephson theory is as follows: we know that quasiparticles can pass from one superconductor to the other by means of the tunnel effect. But the electron pairs gain 2 eV of energy which they must lose when emitting a photon to avoid breaking up. Josephson established that *when a Cooper pair passes from one superconductor to another it emits or absorbs a photon in direct relationship to the voltage across the pins of the two superconductors and the frequency of the electromagnetic wave emitted or absorbed.*

In applications of the phenomenon of superconductivity, the Josephson effect is often associated with flux quantification, which may be described as follows: if we take a hollow semiconductor we find that at each point the pairs are represented by a well-determined density and phase. A relationship exists which links the phase and the magnetic field inside the ring and shows that the flux outside the ring is directly proportional to the phase variation built up during movement around the ring. As this phase is a regular phenomenon it stays identical even when it increases by an angle variation of 2π, but φ o, which corresponds to the flux inside the ring when the phase has been increased by 2π, does not have the same value as a phase equal to 0. Flux quantification deals with the difference between these two values. Any phase inside the ring will be equal either to 0 or φ o, a phenomenon which makes it possible to measure very small magnetic fields.

The applications of superconductors are many and varied. In the field of electricity, they are found in electromagnets, magnetic sustentation, high-tension cables, voltage calibration and in other applications. In the field of electronics they are basically used to measure extremely weak magnetic fields; in geology for detection of hot water sources and oil reserves and in the military domain to measure anomalies in the earth field. They are also useful in medical electronics: for example, Dr. Cohen at the MIT National Magnet Laboratory has obtained some success with a superconductor radiometer which registers the gradient (variation of field between two points) of the magnetic field produced by the beating of the heart. SQUIDs (Superconducting Quantum Interference Devices) are often used to obtain such extraordinarily small measurements. These superconducting detectors are placed in a cryostat containing liquid helium. As the heart's magnetic field is some million times weaker than that of the earth, the success of two researchers at the University of Syracuse in the United States in 1963 in detecting a biomagnetic signal from the heart can count as a very considerable achievement.

The magnetic fields appearing at the surface of the skull in response to a sensory stimulus are a thousand times weaker again than the magnetic field of the heart. However three researchers at New York University (S.J. Williamson, L. Kaufmann, and D. Brenner) have initiated research in this field.

One rather overdue and very useful application of this system is in cases of silicosis of the lungs; SQUIDs can be used to draw up a chart of the magnetic bodies in the lungs and to measure the size of any tissue lesions.

However SQUIDs do have one major drawback: it is difficult to interpret the data obtained.

The first computer-switching research was begun by J. Matisoo at IBM, with T.A. Fulton, P.W. Anderson and R.C. Dynes at Bell concentrating on ways of increasing the rapidity and capacity of central memories.

Finally experimental research is continuing in the development of UMF detectors and microwave and infrared mixers for radioastronomical applications. A number of different systems are in competition: Josephson junctions (using the current tunnel between two superconductors), quasiparticle junctions (using the tunnel effect of quasiparticles between two superconductors), Super-Schottky diodes (with a superconductor/semiconductor structure), SIN diodes (superconductor-insulator-normal metal) and ordinary nonsuperconductive diodes.

Another field of application may well be in space. Today's satellites could use high field magnets and SQUIDs, although there is still a problem because of the large bulk of the necessary cooling systems.

1

2

1. The physicist Philip W. Anderson.

2. Bell Telephone Laboratories scientists test a superconducting electromagnet; the small cylindrical object is shown as it is removed from a liquid helium bath (temperature minus 45 degrees F). A later experimental supermagnet made of niobium-tin wire produced a field strength of 78,000 gauss.

Electronics and super- conductors

by P.W. Anderson, Consulting Director, Physics Research, at the Bell Laboratories.

In this article, Philip W. Anderson, who received the Nobel Prize for Physics for his work on superconductivity in 1977, explains the basic principles of this phenomenon. Superconductivity is a property of matter characterized by the absence of electrical resistance at very low temperatures which can have astonishing consequences in various fields of research such as particle physics or nuclear fusion.

*The work at Princeton University was supported in part by the National Science Foundation Grant No. DMR 78-03015, and in part by the U.S. Office of Naval Research Grant No. N00014-77-C-0711.

History

Most metals, when cooled to within a few degrees of absolute zero, undergo a sudden transition into a state with no measurable electrical resistance, the superconducting state. This fact, discovered in 1911 by Kamerlingh-Onnes shortly after he first liquified helium, remained entirely mysterious in the face of concerted attacks by the greatest theoretical physicists of the time (for instance, Heisenberg and Feynman) for forty-six years until it was explained in 1957 by J. Bardeen, L. N. Cooper and although J. R. Schrieffer (the ubiquitous "BCS"). Crucial advances in its phenomenological understanding had been made by among others F. London (1935), V. L. Ginzburg and L. D. Landau (1950), A. A. Abrikosov (1956), and A. B. Pippard.*

The direct use of the superconducting property is already part of the technology of energy but not of electronics, in the production of high magnetic fields with little expenditure of energy. This property is very widely used for research purposes, as, for example, in the physics of high energy particles. Pilot projects on power transmission, suspension and propulsion of electric trains, and electrical machinery have also been built; and possibly most important of all is the prospective use for plasma confinement in fusion reactors.

Most *electronic* interest *per se* in superconductivity, however, stems from further developments generally described as "the Josephson effect." Predicted by B. D. Josephson in 1962, experimental confirmations of various aspects of these effects were made by J. M. Rowell and myself in 1963, S. Shapiro later the same year, and A. Dayem and myself in 1964. An important generalization of the idea, the 2-junction interferometer, was demonstrated by Jaklevic, Mercereau, Silver, and Zimmermann shortly thereafter. The use of this idea for sensitive electrical measurements was pioneered by John Clarke in 1967, and the SLUG and SQUID instruments for delicate measurement of electromagnetic fields based on these developments are now widely available commercially. These are used in areas as diverse as solid state physics, brain science, and archeological exploration.

The A. C. Josephson effect is the present fundamental atomic standard of electrical voltage, following the pioneering work of Langenberg et al.:

the relationship $V = \dfrac{h\,\nu}{2\,e}$ converts

voltage into the more accurately measurable frequency or time standard, or, conversely, may be used as a measurement of the atomic constant h/e.

But by far the most important prospective use of the Josephson effect is as the fundamental component of large computing machinery. This was envisaged quite early by Josephson and myself, but the major thrust has been at I.B.M. under J. Matisoo particularly. J. M. Rowell and T. A. Fulton at Bell Labs have led a less ambitious program. It is proposed by I.B.M. to have a pilot full-scale computer based on this principle in operation in the mid-1980s.

The key advantage is in speed: the low power dissipation and high speed of the Josephson device allow compact design, saving on both signal travel time and operation time, and giving the possibility of a one-to-two order of magnitude improvement over semiconductors. The energy savings with such computers are also not to be ignored.

A Brief Glimpse of the Operating Principles of Josephson Technology

Superconductivity and the Josephson effect are, even more than the maser and laser, pure quantum effects incomprehensible in a classical framework. It is essential to use both the idea that electrons are described by a field Ψ (r, t) that may have the same kind of coherence properties as the electromagnetic light field in a laser; and also to use the idea of quantum tunneling of electrons through an insulating barrier between two metals which was demonstrated by Giaever.

The particles whose field is important in superconductors are pairs of electrons with opposite spin. It was the great contribution of BCS to show how bound pairs of electrons could condense and form a coherent state in the superconducting metal at low enough temperature. All the static properties of the superconductor are adequately described by their theory and its key concept of an "energy gap," 2Δ, equal to the binding energy of a pair of particles.

But only later, when Gorkov related BCS to the Ginzburg-Landau phenomenological theory and when it was discovered that flux is quantized in units of $h/2e$ by Döll and others, did attention focus on the field of electron pairs Ψ pair (r,t). My suggestion that this pair field takes on a definite value at each point in a superconductor, and can be treated like any other classical state of field variable (like electric field $\vec{E}$ or magnetization $\vec{M}$, for instance) intrigued Josephson and he searched for a means to check this experimentally. He showed that there would be a term in the tunnel current in a super-conductor-oxide-superconductor sandwich depending on the relatives phase of Ψ on the two sides:

writing $\Psi_{1,2} = \Psi_{1/2}\, e^{i\phi_{1/2}}$, the current is given by the Josephson equation

$$J = J_1 \sin(\Phi_1 - \Phi_2)$$
$$J_1 \simeq \pi\Delta \text{ (volts)}/R_{junction}\text{(ohms)}$$

It soon was realized that almost any weak junction between superconductors, such as a thin film bridge or point contact, would have the same property. Each of the possibilities has technical advantages for some uses, but the tunnel junction is faster and more reliable.

The great interest and use of this equation is that Ψ is very sensitive to electric and magnetic fields. This sensitivity appears in three ways. The first is by the flow of current: just the relation (1). The second is that if two superconductors are at different voltages V_1 and V_2, the relative phase changes at a rate

$$\frac{d\phi}{dt} = \frac{2e}{\hbar}(V_1 - V_2)$$

and hence the Josephson supercurrent oscillates at a frequency given by V_1-V_2. Hence the use as a quantum voltmeter.

The static value of the phase varies in space when there is a magnetic field, in such a way that a phase change of 2 π occurs for every quantum unit of

magnetic flux enclosed. Since $\dfrac{h}{2e}$ is very small (5×10^{-8} gauss-cm^2) tiny fields are very effective. The sensitive measurement possibilities are obvious.

Equally obviously, the field due to a Josephson current is easily capable of turning another or the same one on or off, so that obvious switching, logic and memory functions are possible.

Two basic approaches can be envisaged. The IBM one is the simple use of the magnetic field to switch the junction on or off: a basic "gate" on the principle of the old concept of the "cryotron" and identical in function with semiconductor circuits. Bell couples junctions not through the magnetic field but by including them in the same circuit loop and adding their currents: a current-coupled concept using Eq. (1). But the two principles are by no means exclusive.

The energy involved in a Josephson current is tiny: $U = \hbar J_1/2e \sim 10$ ev for a 1 ohm resistance level. Cooling of a circuit operating even at ultrafast speeds is not a problem with liquid He; and cryogenics is in no way a barrier to this technology. A real problem encountered today is with reliable junction manufacture on a megajunction scale, and one to be anticipated is the whole input-output question for an intrinsically low-energy, low-impedance system. My own prejudice is that this technology has probably already passed the crucial point at which feasibility is clear, and engineering improvement on a basically sound idea can begin to keep it well ahead of its older competitors.

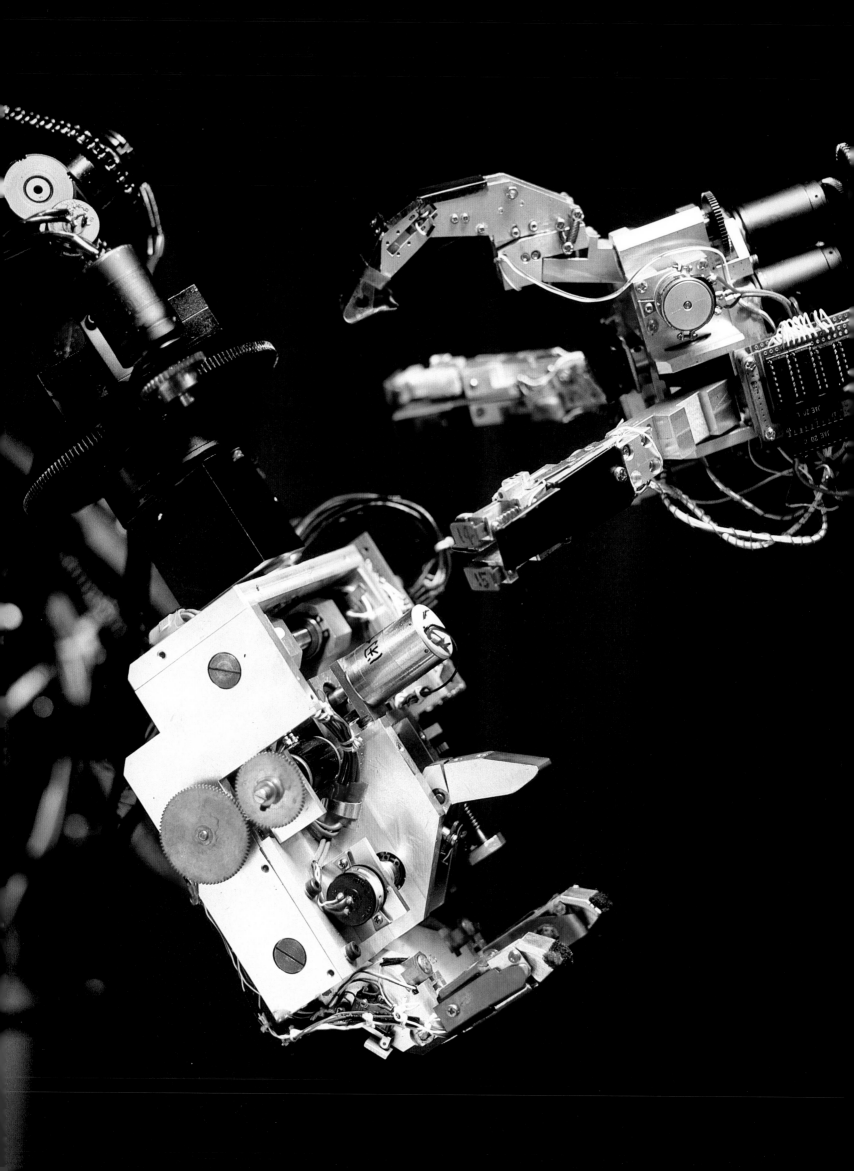

6
Industrial applications and scientific and technical research

Robots and Automation

Whenever the word automation is mentioned, people tend to think of robots, whose potential and limitations have been the subject of much discussion since the 1960s (a period of heated debate on the concept of artificially-created intelligence). However, the idea is not new. Recent science-fiction literature and films from Fritz Lang's *Metropolis* to Toe Animation's *Goldorak* are at least partly inspired by earlier classical flights of fancy as exemplified in the story of the two female "robots" created by Vulcan to help him in his labors and the creation of Galatea to be Pygmalion's ideal wife and companion.

The first complete robot system was the robot vehicle "Shakey" created at the Stanford Research Institute (1968). Later, Brookhaven National Laboratory designed a plastic robot with artificial organs in order to observe the maximum levels of radiation to which man could be safely exposed. The robot was built to exhibit the same radiation absorption capacity as the average human being. Next came the computer-controlled "Sim One," designed to simulate patient response. Sim One was developed at the Southern California Medical School by Doctors Stephen Abrahamson and J.S. Denton with the help of Paul Clark of Aerojet General Corporation. The robot's reaction to various forms of medication and treatment was indentical to that of a human being. However robots soon lost the "human" shape of the first models: the British army in Belfast, for example, uses a bomb-detecting robot that consists of a video television camera mounted on a flexible arm together with a remote-controlled bomb-dismantling system.

It was a short step from arms for robots to prostheses for people. The Stanford Research Institute developed one of the first artificial arms for NASA, in collaboration with the Naval Prosthetics Research Laboratory in Oakland: certain predetermined morse code type signals made by the patient's shoulder movements were fed into a computer which in turn activated a certain arm function. The next step was finger control — contractions of the fingers were governed by microprocessors — and special robotized chairs for quadraplegics. Other electronic organs and devices have also been developed: pacemakers, artificial hearts, artificial kidneys, special organs in the bladder to control incontinence, word

synthesizers, etc. In 1964 Ralph Hotchkiss, an American researcher, invented special glasses for the blind, in which light signals were transformed into sound signals by means of a photoelectric cell, and in 1971 Bionic Instruments invented a special laser cane for the blind. The invasion of robots into fields that up to now have been uniquely populated by man is arousing some concern, particularly as robots are now able to move about, recognize shapes, perform a wide range of different activities and even simulate deductive reasoning. Their "sensors" are in many ways more efficient than human senses, in that their visual organs (camera, lasers for example) give them the possibility of using color spectra invisible to the human eye such as infrared and ultraviolet while their "ears" are able to operate at ultrahigh audio frequencies inaudible to man. In addition, they may imitate the spoken word by means of word synthesizers and simulate the senses of taste and smell by means of mass spectrometers.

This army of robots is being used to perform such repetitive and very delicate tasks as automobile assembly, production line work and jobs that are too dangerous for people to tackle. One example would be monitoring and initial clean-up after the Three Mile Island disaster.

Robots have their own hierarchy, from the "dumbest" (the digitally controlled machine) to the most sophisticated (those equipped with sensors and which improvise alternative strategies and gestures according to the information received from the environment). Between these two extremes is the robot which is capable of leaning.

The Japanese industrial sector is both the biggest producer and consumer of robots, with a total inventory of around 60,000 in 1980 (a growth rate of 30 percent per year) set against a total world inventory of 64,000 robots. Thirty-five percent of Japanese robots are put to work in the automobile industry.

Automatic systems have evolved a great deal since the earliest models. At first, they were controlled by banks of electromagnetic relays linked together by extensive cable networks. Thanks to the incorporation of transistors, solid state switching systems were developed allowing savings of both volume and energy. The final development was programmable automatic devices specialized in sequential-logic operations; these made it possible to

Hitachi conceived intelligent robots with TV camera "eyes", process control computer "brains" and assembly "hands".

69

Industrial Applications and Sciencific and Technical Research

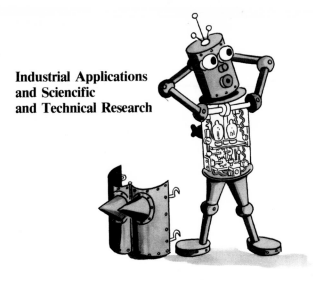

„Ich habe dein Ehrenwort - das Bild bekommt keiner zu sehen!"

1. Humorous sketch by Will Halle.
2. Welding robot.
3. Master Control Panel is the brain center for robot welding of K-car bodies at the Chrysler Assembly Plant in Detroit.
All welding operations are computer-programmed and the robots automatically adjust the welding sequence to accommodate two or four-door body styles.
4. Bodies for the K-cars move through a robot welding line which represents the latest manufacturing technology in the auto industry.
5. Current research by General Motors is concentrating on the development of computer vision systems that will allow robots to locate overlapping or partially obscured parts and those located in close proximity to other types of parts. The inset top left shows the parts as seen by the electronic eye.
6. The PUMA (Programmable Universal Machine for Assembly) robot conceived by General Motors is capable of constructing an assembly made out of small lightweight components.

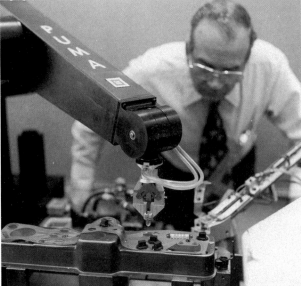

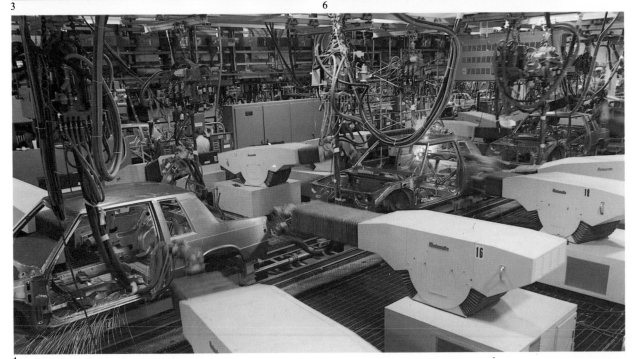

handle certain sequences of an industrial process without having to modify the whole program. A complementary development was the advent of the small computer.

At the present time, the Japanese are investigating the possibilities of the "workerless factory"; this does not, of course, mean that all human contribution is eliminated, since such tasks as production line design, products, programming, and maintenance will still be undertaken by human workers. Researchers are now concentrating on shape-recognition processes: analyzers based on laser-beam technology are already used to detect cancerous cells in smear samples or defects in mass-produced manufactured parts. Nevertheless we are still far from creating a system with the same optical or touch ability as the human being.

Throughout the world a number of different teams are working on increasing the sensitivity of touch sensors. It is mainly due to the advanced technologies developed by industry that medical research (in particular in the field of prosthetics) has made such extraordinary progress. One of the main problems in developing an artificial hand, for example, is the interface (the junction between the biological and the artificial element). French researchers in Toulouse have developed a prototypical artificial skin consisting of a matrix of metallic electrodes covered with conductive elastomer to register the pressure applied to the skin surface. This system makes it possible, for example, to control the pressures applied by an artificial system to the human body.

Electronics is not an indispensable component of automation, as has been proved by whole generations of very "intelligent" mechanical, hydraulic, and electromechanical devices, such as the transfer machines used in the automobile industry. However a further step in the evolution of automation will be the introduction of systems capable of responding to control signals generated at random by monitoring devices. Systems based on step-by-step mechanical logic cannot meet such requirements; neither their control nor remote monitoring capacities allow for further development. They operate within a fairly narrow range of possibilities. Program versatility is the prerogative of computer processing, of electronics and its evolution into microelectronics, that has in turn led to microprocessors — a powerful factor for function decentralization in the "intelligent" automated system.

Power electronics is the intermediate link between the computer's capacity for carrying out intricate processes at very low power levels and the many different machines that bear the brunt of industrial production and the provision of a wide range of services. This branch of electronics is not well known by the general public, which tends to confuse it with "large-scale electronics" (engines, trains, subway systems, etc.). But it is an essential component of automation and also has the advantage of providing systems that are much more reliable than those using electromechanical relays even though they have proved to be most realiable in the past. Thanks to a good price/performance ratio, excellent durability, compactness, and robustness, power electronics has achieved a position from which it will not easily be dislodged, particularly when we consider that it will eventually make considerable contributions to energy savings through its improved power regulation capabilities.

Transistorized circuits have replaced electromechanical relays for control and command purposes. Programmable control systems (basically consisting of microprocessor and calculator equipment) auto-matically control consumption of power. Power levels are preprogrammed to suit circuit requirements, while automatic control of all the circuit perimeters allows substantial productivity increases. When brought together in the form of a complete production unit they may be used to automatically control all operations. Power regulation mainly concerns the control and automatic stabilization of circuit currents using either regulator coil saturation techniques or direct control of a power transformer's regulation stage.

Measurement and Control

Electronics has helped develop a whole series of measurement and control systems: telemetering or remote monitoring (by which the pulse and tempera-

Plasma reactor: here amorphous silicon is grown by reaction in an electrical discharge plasma.

ture of a man in orbit above the earth, traveling at 28,000 km/hr, may be measured and recorded), range-finding for measuring distances, and metrology, a technique for integrating our system of weights and measures to primary measurement units. One of the most important and effective inventions in the field of range-finding was the range-finding laser developed in 1962 by the National Bureau of Standards. The principle behind this technique is as follows: the laser head emits a train of very short duration pulses which in turn pass through an optical lens system. The purpose of the lens system is to diminish beam divergence at transmission and to reconstitute the received signal. When a pulse returns through the lens system its arrival is recorded by a detector circuit. A signal from the detector passes through an amplifier and closes an electronic gate that the same pulse left open at transmission. A counter measures the time taken by the pulse to make the round trip, and the distance can be read directly from a display. There are, however, some problems due to diffusion (e.g., mist or cloud that causes interference with the echo), deviation of the beam by the atmosphere, and loss of coherence. Another drawback is that as the lens contains no

reflecting elements, it can only send back a low energy level. In 1964 the first satellite-based range-finding experiment was carried out in the United States and obtained an accuracy of approximately 100 meters. In 1965, using echos received in France, it was possible, for the first time in history, to determine orbit parameters by laser range-finder.

In the field of metrology the more robust, economic, and less bulky neon-helium lasers are preferred. They are used in road digging to sight-position explosives and to indicate drilling direction. Bitumen-spreading machines can be fitted with a photoelectric receiver system that automatically guides the machine along the axis of a laser beam and permits extremely straight and accurate road-laying operations.

The laser makes it possible to measure the level of steel smelted in a convertor, to undertake surface measuring for such products as corrugated sheet metal, and interferometry.

Gyrometer-lasers are used to measures angles (helium-neon transmitter with a modified reflecting cavity) and in granulometry (control of the cement crushing process, for example).

A teledetection system can be used to evaluate the status of earth resources or to improve the accuracy of meteorological forecasts. There are two different processes in use:

— passive detection: pick-up of the natural energy emitted or reflected by a body. This is based on the principle that each living body has a temperature above absolute zero and therefore radiates electromagnetic energy that may be diffracted or diffused. These electromagnetic waves can be recorded on photographic film or electrically detected by scanners.

— active detection: a signal must be beamed at the target before it can be detected. This signal may take the form of an ultraviolet laser beam or, as in the case of radar, a train of UHF or SHF pulses.

The control of nuclear fusion requires enormous energy sources. Lasers are used to produce dense and hot plasmas for nuclear fusion. They diagnose such parameters as refractive index, electronic density, collision frequency, etc. They may also be used to heat the plasma through energy transfer. One of the basic laws governing nuclear fusion is Lawson's law on the confinement of the plasma. According to this law, before nuclear fusion can take place, it is necessary to confine the plasma (the nuclei stripped of their peripheral electrons) for a period of time whose length is inversely proportional to the density of the matter (the product τ of the confinement duration, multiplied by the density $d = 10^{14}$). This confinement process can be achieved either by the "Tokamak" system (a magnetic cylinder in which particles of deuterium, tritium, etc., are subjected to a magnetic field), or by the "Shiva" system (a laser system with multiple arms, like the Hindu deity of the same name).

In 1967 the French Marcoussis Research Center and the C.E.A. (Atomic Energy Commission) succeeded in creating a laser source that delivered 500 J per nanosecond, a result more powerful than any-

1. Drawing of the moon, by Hevelius 1645.
2. The visionary French filmmaker Georges Méliès made the first science-fiction film in the history of the cinema, "Journey to the Moon", in 1902.
3. The Nançay (France) radiotelescope, third largest in the world after Arecibo (USA) and Bonn (West Germany).

3

thing developed up until then anywhere in the world. In 1969 the Limeil Research Center, also in France, succeeded in producing a stream of neutrons by means of the interaction of laser waves with the matter; a little earlier the Bassov team in the USSR had also produced neutrons but not on a continous basis. In 1974, Dr. Bruckner drew up a mathematical model defining the procedure leading to an increase in density by illumination of a deuterium-tritium sphere which allows this increase in density by means of implosion.

In industry, exoelectrons or Kramer electrons may be used to examine the surface of metals. *Exoelectrons* are electrons emitted by the atoms of a metal surface, particularly when the structure of this surface has undergone a change in state and become worn or cracked. Numerical data can be compiled and analyzed with this technique, which allows early warning and monitoring of defective areas. Exoelectronic pictures are often obtained by scanning the metal surface with a beam of ultraviolet rays. This phenomenon was studied for the first time in the 1940s by German physicist Johannes Kramer following observations he made while examining a faulty Geiger counter. The German scientist later proposed the idea of spontaneous electron emission.

Exploration of the Universe

Radioastronomy. In 1928 Karl Jansky, then aged 22, started work in the Bell Laboratories. He began to investigate mysterious noises that disturbed radio transmissions, and he located a particular signal. He then set out to find its exact origin. When Jansky discussed his problem with a colleague who was familiar with astronomy, the two men concluded that this signal must come from the Milky Way, as it intensified whenever Jansky's antenna was directed toward it. In 1932 the American radio amateur Grote Reber built the first radiotelescope using a VHF receiver coupled to an antenna 8 meters in diameter; he drew a detailed map of the sky and in 1936 (three years after the publication of Jansky's article), pinpointed this source of radiation in the Sagittarius region of the galaxy.

For the first time, space had been studied, not through optical instruments, but through the radio wave spectrum. Strangely enough, scientists did not immediately pounce on the discovery and right up until the time of Jansky's death in 1950, no really consequential application had been found for his work.

It was not until after World War II that radioastronomy received its first real boost, thanks to the experience acquired during war work on radar. In 1947, while he was at work on a radar set, a researcher by the name of J.S. Hey accidentally picked up radio signals coming from Cygnus. Unidentified radioelectric waves had previously been detected by radar receivers in 1942, and their source pinpointed as the sun.

The first workers used the old radar antennas left over from military stores, but large radiotelescopes were also built at Jodrell Bank in England, Parkes in Australia, Green Bank in the United States, Nançay in France, Arecibo in Puerto-Rico, and Bonn in Germany. It is probable that quasars (1963) and pulsars (1968) would never have been discovered without the contribution of radioastronomy, neither would A. Penzias's 1965 discovery of the isotropic radiation of the universe have been possible. Optical observation of the planets can give us information about their surface and their lower atmospheres, but not about their magnetic fields. In 1968, English researchers Tony Hewish and Jocelyn Bell announced the discovery of a new kind of radio source (pulses of a few milliseconds in duration), sent out at constant intervals: these were called *pulsars.* Radioastronomers were the first to take an interest in low-noise receivers, parametric amplifiers, and masers. Giant radiotelescopes and interferometers for good spectral resolution were constructed by a number of different researchers in various countries: by P. Wild in 1968 in Australia, and by Martin Ryle at Cambridge. Radioastronomy research at Cambridge dates from J.A. Ratcliff's decision to continue at Cavendish the research on the ionosphere that had been neglected during the war. He hired Martin Ryle and advised him to concentrate on the recent discoveries of radio waves originating from the sun.

Attention was later turned to the detection and analysis of other radio waves discovered by Hey, J. Bolton, and Grote Reber. The Mullard Observatory team in Manchester, lead by Bernard Lovell, used army surplus equipment, as did the team working at Jodrell Bank.

Research continued in the field of EHF radioastronomy following Charles Townes's discoveries in 1967—68 which showed that thermal transmissions from the sun, planets, and satellites reach maximum strength at these Extra High Frequencies. At the Bell Laboratories, A. Penzias and R. Wilson developed the first high performance extra high frequency receiver (they were awarded the Nobel Prize in 1978 for their work on fossil radiation). The basic component of these EHF receivers was a very advanced miniscule diode mixer which contained a semiconductor diode and was used to bring together the incoming radio wave and the signal from the local oscillator (generally a klystron). These mixers were kept cool in a low-temperature housing. The first specialized EHF system was the MacDonald (Texas) 5-meter diameter radiotelescope built by NASA in 1963. Radio waves have been instrumental in discovering interstellar hydrogen, quasars, pulsars, and isotropic radiation of the universe. Millimetric (extra high frequency) waves have been used to detect interstellar molecules, although research in this field is still in its infancy.

Space

All the electronic measuring and control techniques developed so far have been used in one of the greatest adventures of our era: the conquest of space. On October 4, 1957, the Russians launched "Sputnik," the world's first artificial satellite, from their base in Rywratam: Sputnik was a polished sphere 58 cm in diameter, weighing 85 kgs, with 4 antennas (15 and 7.5 meter wavelengths). It disintegrated on January 4, 1958. In October 1958 the first space probe, Pioneer 1, was launched. In January and then October 1959 the first interplanetary and lunar probes, Luna 1 and Luna 2, left the earth. April of 1966 saw the world's first extra terrestrial satellite (Luna X) and July 14, 1969, the beginning of the Apollo II mission and the landing of the first men on the moon.

It would take too long to give a detailed account of the history of the exploration of space. Let us just point out that satellites have made it possible to develop completely new transmission (see the chapter on Telecommunications) and exploration techniques. One of the most remarkable applications of this research has been the implantation of a network of meteorological satellites.

In 1966, the first geostationery observation stations were put into orbit; they are fitted with measuring equipment designed to work independently.

1. Konstantin E. Tsiolkovski (1857-1935) was deaf. This self-taught genius was the Russian pioneer of space and space rockets and in 1896 wrote an essay on communication with intelligent beings from outer space.

2. Columbia climbs toward space. Aboard the space shuttle, Astronauts Joe Engle and Dick Truly (november 1981). On its second mission, Columbia is carrying a payload of science and application experiments and a Remote Manipulator Arm — a mechanical arm designed to handle cargo in orbit.

3. Werner von Braun present at the launch of Saturn SA6 at Cap Canaveral, May 28, 1964.

4. Cap Kennedy. Launch of Apollo 17, December 7, 1972.

2

3

4

Data collected is first stored and later transmitted to the ground via data relay satellites.

On April 12, 1981, twenty years after Yuri Gagarin's historic space flight aboard Vostok, the Americans launched the space shuttle Columbia with John Young and Robert Crippen aboard. Columbia was the first spacecraft in the history of the world to reenter the earth's atmosphere and, unlike the Apollo, be retrieved intact for future relaunching, thanks to two essential innovations: its refractory armor made from borosilicate tiles capable of withstanding temperatures up to 1,260° (temperature reached during atmosphere reentry), and its automatic computerized guiding system (five computers, one of which is an emergency standby computer). This shuttle is the first of a series of four, with "Challenger" (due for launching in September 1982), "Discovery" (December 1983) and "Atlantis" (1985). A number of European countries are taking part in the "Spacelab" project, a habitable laboratory carried in the main bay of the shuttle, while the Canadians have supplied the 15 meter long remote-manipulated arm that can carry a weight of 14.5 tons. The arm can be used to bring damaged or malfunctioning satellites back to earth and put them back into orbit after repair. It should also be possible to position very heavy reconnaissance satellites such as the Big Birds, which are designed to provide very detailed photographs of earth (the lens system can register objects on the earth's surface measuring as little as 15 centimeters).

Three other countries: the Soviet Union (Raketoplan), France (the manned Hermes and the fully automatic Minos), and Japan are now working on projects for re-usable spacecraft.

It appears that the Soviets are concentrating their research on a permanent orbital station to serve as a military observation platform and a launching pad for spacecraft destined for Mars. Two Soviet cosmonauts, Valery Ryumine and Leonid Popov, broke the world endurance record when they spent six months aboard Saliut 6.

One of the most extraordinary recent technical feats in the field of space exploration is Voyager II's retransmission of photographs of Jupiter and its satellites and Saturn and its rings located some two billion kilometers from Earth. Voyager II is continuing its course towards Uranus (January 1986) Neptune (August 1986) and Pluto (1990).

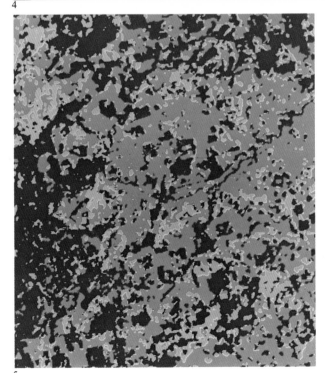

1

2

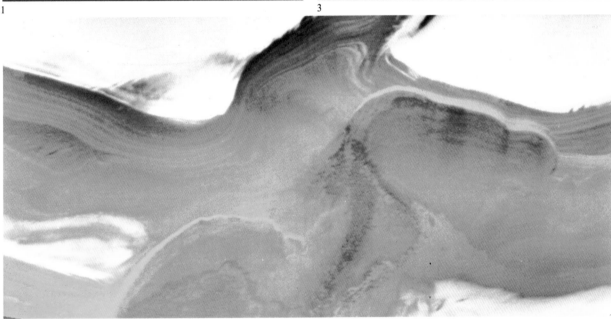

3

1. Sunrise on Venus as seen from the Pioneer spacecraft.
2. Saturn seen from Pioneer, September 1, 1979.
3. The surface of Mars, by Viking 2.
4. Jupiter is the largest planet in the solar system: a gaseous world as large as 1300 earths, marked by alternating bands of colored clouds and a dazzling complexity of storm systems.
5. For several years various states have made extensive use of Landsat data in studies of agriculture, range lands, forestry, water resources, noxious weed control, surface mining and coastal zone management. This demonstration project is centered on a 220,000 acre test site in the eastern part of Douglas County.
6. The Ball Corporation (Muncie, Indiana) Agricultural projects, wherein the system approach was employed, a technique developed of necessity for managing extremely complex aerospace programs involving integration of a great many individual systems: Wheat and green crops dot once-barren areas of the Sahara Desert, made fertile by the largest center pivot-irrigation project ever undertaken in the North African desert.

4

5

6

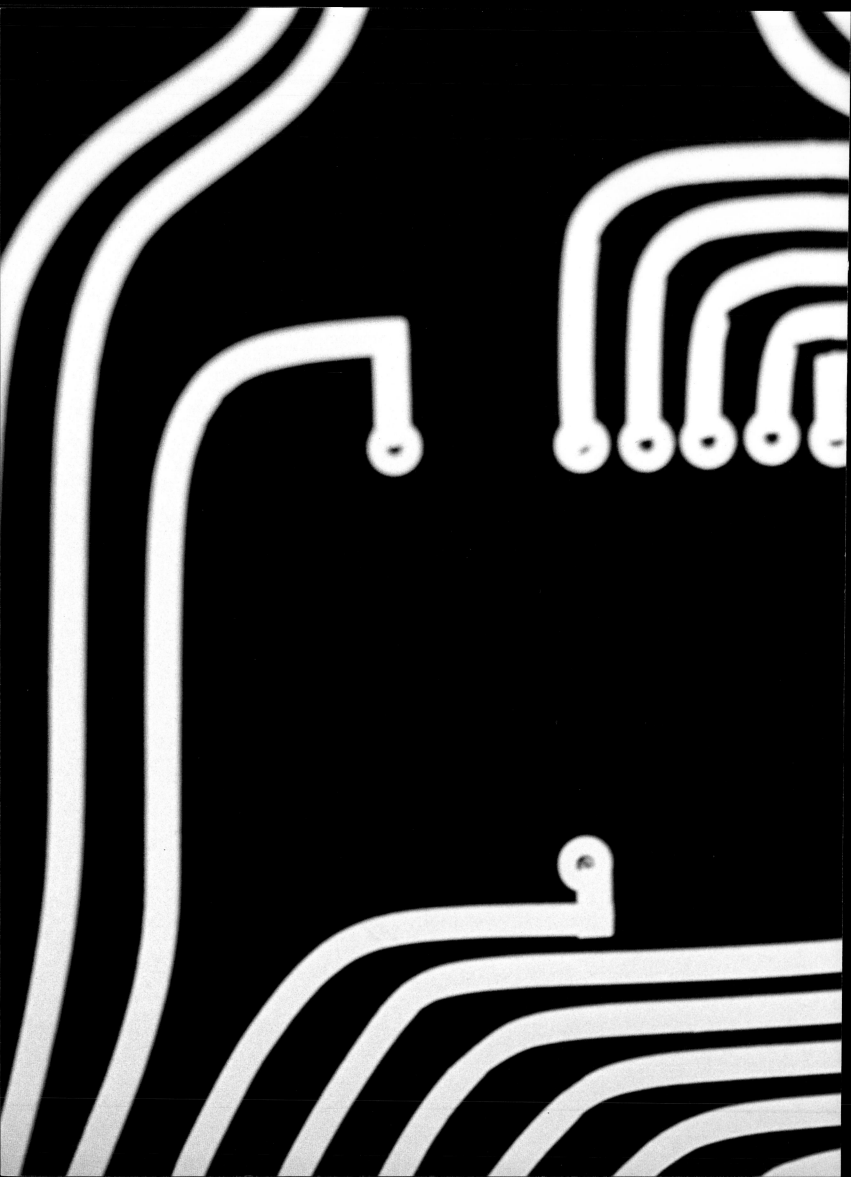

PART TWO

The 1948 revolution: the transistor
The 1968 revolution: large-scale integration

The coherent order of civilization is in itself enigmatic. The mystery of life does not reside in chaos but in organization. As Spinoza showed, that which exists is never accidental. The organic nature of a civilization has genuine poetic power and evokes the idea of internal necessity.
André Malraux

7
Semiconductors

Without semiconductors there would be no pocket calculators, mini-cassettes, miniature radios, office computers, industrial automation systems or scanners. Neither would the "chip" (the basis of all microelectronics systems) and its most famous application in microprocessors and their memories exist. Microprocessors are miniature processing units that can record and answer queries by means of high capacity memories. The semiconductor is therefore the fundamental component of modern electronics. But what is it exactly? It is an "imperfect insulator," a crystal that becomes conductive only under certain conditions: thermal agitation, impurities, crystalline defects or, as scientists put it, stoichiometric faults (deviations from nominal chemical composition). Semiconductor technology has evolved from a number of different branches of science; chemistry, crystallography, metallurgy, and solid state physics. The famous lead sulphide crystal set used since the early days of radio could be seen as the first joint application of these sciences, the first electronic development (or "radioelectric" as it was then described.) In 1874, researchers noticed that by establishing a contact between a metal wire and lead sulphide, electrical current would flow in only one direction. This represented a big step indeed in the field of radio signal "rectification" although the galena crystal circuit was not particularly reliable. There were still many more complex problems to be answered particularly in amplification of the rectified signal and it was only after World War II that these problems began to be looked at in depth.

There are two kinds of semiconductor: those composed of simple elements (selenium, silicon, germanium, tellurium) and those made from compound elements (gallium arsenide, indium antimonide, gallium phosphide, etc.). Semiconductors represent the main active element in transistors, optical electronic devices (such as light-emitting diodes and junction lasers), some types of particle detector, extra high frequency generators, and many other systems. Historically speaking, the first three semiconductors used were selenium, germanium, and silicon, which was to the electronics revolution what steel was to the industrial revolution. Selenium was confused with tellurium until 1817, when a Swedish chemist, Jons

Jacob Berzelius, identified it as a separate substance which he called selenium (*tellus* is the Greek word for "earth," *selenium* is derived from the Greek word for "moon"). In 1851, the German physicist and chemist Johann Wilhelm Hittorf discovered the electrical properties of selenium, and finally in 1873, the Englishman Willoughby Smith discovered that selenium's electrical conduction characteristics change under the effect of light, thus foreseeing one of the basic effects of electronics: photoelectricity.

Selenium is rarely found in its pure state. The main producer countries are the United States, Canada, Sweden, Japan, and Zambia. However it is relatively expensive, and there is a tendency in the electronics industry today to replace it with germanium or silicon (for signal convertors), and with cadmium sulphide (in photoelectric cells). Its most common applications are to be found in photographic exposure meters, for intruder alarms, and in copying machines. The theoretical existence of germanium was established by the brilliant Russian chemist Dimitri Ivanovitch Mendeleieff, when he drew up a periodic table of the elements according to their atomic weights; he generalized the laws of analogy existing between the weights of certain elements and their properties, and thus predicted the existence of an element situated between silicon and tin. Fifteen years later his hypothesis was confirmed by the German scientist C. Winckler, who isolated the element germanium and gave it the name of his native country.

Germanium is a rare element never found in its pure state. It exists as a component of certain minerals found particularly in South-West Africa, in the Congo, and in tiny quantities in Kansas and Oklahoma. It may also be extracted from bituminous schists; this is currently being done in both Japan and England, no doubt also in Russia. Germanium was the first semiconductor to be used in research work and in the development of the transistor, probably because the laboratory preparation of pure germanium and monocrystals is relatively easy. In fact, from the beginning of this work, scientists realized that the degree of purity and the subsequent preparation of the crystals involved important fundamental parameters. Another fundamental electronic

The atmosphere in a silicon-manufacturing laboratory is reminiscent of hospital operating rooms: a system of filters absorbs all impurities and the smocks, headgear and footwear used must all be sterilized after use.

Semiconductors

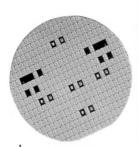

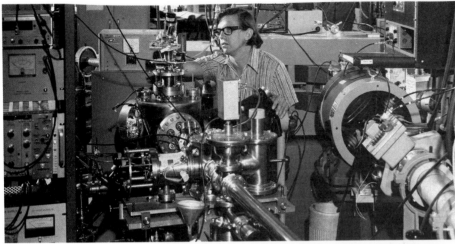

1. Silicon wafer after formation of electrodes through evaporation.

2. The chemist and Swedish Baron Jöns Jacob Berzelius (1779-1848), one of the creators of modern chemistry. He isolated a large number of pure bodies such as selenium, calcium, barium, strontium and thorium.

3, 5 and 6. The silicon bar is baked in an oven with an oxide layer at a temperature of 1,800 °C. The oxide coating protects the silicon against foreign atoms. Tens and even hundreds of thousands of transistors will be integrated into this 10 cm diameter silicon wafer which is divided up into 44 mm-sided squares.

4. Semiconductor devices depend on "doping", which means the introduction of minute amounts of another material into a semiconductor in order to obtain the desired electrical properties. This is an "ion-implantation" test: the semiconductor is bombarded by high-speed charged particles or ions of the desired material.

property is mobility; the electrons in germanium are much more mobile than those in silicon. Nevertheless, silicon eventually took over first place and is now the base for almost all electronic components; this is because germanium does have one important defect: extreme sensitivity to temperature. However, one of its most interesting properties is its photoelectric characteristics. It is opaque in visible light but transparent in the infrared spectrum; this is why it is used in photodiodes and infrared detectors.

Silicon is found all over the earth; it ranks only after oxygen as the most common element in the

earth's crust. It is never found in its pure state; but it exists in stones, water, the atmosphere, in many plants, and in the skeletons and tissues of a number of animals. Once again, it was the Swedish chemist Berzelius who in 1817 discovered silicon, although its existence had already been hypothesized thirty years earlier by French chemist A. Lavoisier. In 1823 Berzelius finally classified it as an element. A process for reproducing silicon in crystalline form was devised by H. Sainte-Claire Deville in 1854. Silicon possesses good mechanical and thermal qualities, and silicon oxide is very easy to obtain. The oxide, which acts to protect the material and to facilitate the introduction of localized impurities, has excellent insulating qualities which are applied in MOS (metal-oxide semiconductor*) technology, one of the most important microelectronic techniques. The most delicate problem concerns the production capacity of industrial plants that specialize in the transformation of silicon into its very pure monocrystalline form.

The existence of semiconductors was therefore known more than a century before they were actually used in electronics. Their chemical properties, atomic structure, and other parameters had been explored for many years; but guidelines concerning the direction of future semiconductor technology and their use in industrial production had to be laid down before they could be used effectively. Their physical properties were also explored before World War II; Pierre and Jacques Curie were the first to systematically study the properties of dielectric materials (as they were then called) in an electrical field, and to establish the basic laws governing these properties. Large laboratories in the USSR (Joffé's school at the Leningrad Physics and Technical Institute) and other countries began to take an interest in semiconductors. However, it was not until 1928 that a solid technical base was laid down by Felix Bloch, Leon Brillouin, and a number of other scientists when they used quantum theory to examine the behavior of electrons in electronic fields. After 1933 (the 86th and 87th International Chemistry and Physics Congresses) Bloch, Brillouin, and Joffé laid

down the basic principles, and established the hypothesis of free electrons.

The idea of a positive charge possessing approximately the same characteristics as the electron and opposing the negative charge, was put forward for the first time around this same period. The charge in fact corresponded to a defect (missing electron or vacancy) in the electronic structure, called a "hole." In the presence of an electronic field a vacant "hole" will be filled by an electron. This electron has to vacate its own "hole" which in turn is filled by another electron which leaves its hole and so the process continues all the way down the line. The holes, which represent a positive charge, move in the opposite direction to the electrons. The transistor effect, as we shall shortly see, would consist of controlling these positive and negative charges.

The first important experiments in the application of semiconductor technology in electronic systems were conducted in Berlin just after World War I. However, the only real progress made in the years between the two world wars was in the field of detectors. At the same time, R. W. Pohl was studying the problems posed by photoelectricity and the properties of solids; he established a relationship between the changes in light emission and the conductivity of zinc sulphide powder, as Willoughby Smith had done earlier with selenium. Then, using a diamond crystal, he measured the speed of electrons liberated by light using the Hall effect. This Hall effect dates from 1880 and defines the action of a magnetic field or an electrical current in a solid. If this magnetic field is not parallel to the electrical field, it causes a displacement of charges and the appearance of a characteristic lateral voltage. This voltage is proportional to the electrical current (to the number of charges and their speed), and its direction depends on whether the electrical charges are positive or negative. By combining the measurements obtained through the Hall effect and the conductivity of the solid, it thus became possible to determine the nature of charges, their number, and their speed. However, this measurement was a macroscopic one and was thus inadequate for giving exact microscopic indications on the diffusion and the exact path taken by electrons or holes. In the case of insulators and semiconductors, it had been known since the 1930s that the Hall effect gave useful, although incomplete information on the nature and displacement of charges, but that this displacement was complicated by complex phenomena impeding their propagation and whose precise nature was not yet known. These phenomena have since been the subject of much research.

In 1933, Pohl showed that under the influence of an electrical field and in certain conditions it was possible to modify the color of certain crystals, that is, to modify their properties. Pohl worked on the amplification effect which in those days was done exclusively with tubes: triodes, pentodes, etc. He believed that one day it would be possible to control the current in a solid just as in a vacuum and thereby replace the bulky, fragile and expensive tubes with semiconductor devices. In 1938, Hilsch and Pohl developed a system that had many of the same elements as the transistor: they fitted two electrodes to a potassium bromide crystal and then, between the two, inserted a metal wire near the cathode to control the current flow. On the eve of World War II it was found that silicon crystal diodes exhibited excellent signal detection characteristics when used for high frequency radar applications. During the war, research teams at RCA, MIT, Westinghouse and Bell were all working on improving selenium detec-

Instrument for measuring secondary electron deflection (red dots on the diagram).

* MOS technology is discussed further.

tors; Stephen Angello at Westinghouse succeeded in obtaining the purest sample (it contained only 1.2 percent impurities). However, although the rectifying properties of semiconductors were known before 1948, no one had yet succeeded in using them as amplifiers.

The systematic research carried out on germanium and silicon during World War II work on radars led to a decisive change. Even before the war a team of chemists and metallurgists (R.S. Ohl, J.H. Scaff and H.C. Theuerer) had begun work at Bell on techniques for purifying silicon: the work was inspired by Bell Research Director Mervin Kelly's intuition of all the possible technical consequences of silicon in radar detection. From 1942 onwards, the Bell team worked on all the problems associated with silicon purification; while Purdue University, under the direction of Karl Lark-Horovitz, began similar research on germanium.

Conduction can occur in two different ways. If it is produced by excess electrons that are not in valence bands, it is known as an n (negative) type. This is what happens, for example, when impurities such as arsenic atoms (valency of 5) are introduced; the arsenic atom is a donor because it donates its fifth electron to produce conduction. In contrast, if conduction is produced through a hole corresponding to an electron missing from the valence bands, it is known as p (positive) type. In this case, the impurity could be a gallium atom (valency of 3) which takes an electron from the germanium and thus acts as an acceptor.

When conduction is produced by means of a hole, it is known as p (positive) type. Here the impurity could be a gallium atom (valency of 3) which takes an electron from the germanium and thus acts as an acceptor.

Basic semiconductor theory was laid down in the 1930s by English scientist Nevill F. Mott and A. H. Wilson, German scientist Walter Schottky and Soviet researchers Jacob Illitch Frenkel and Alexander Sergueievitch Davidov.

A. H. Wilson was the first to suggest applying the band theory, which Felix Bloch had applied to metals, to semiconductors and to show that the concentration of carriers depended on the nature of the impurities. When atoms combine to form a crystal, permitted energy levels arrange themselves in continuous bands. In insulators and semiconductors there is a gap in energy between those in valence bands and the higher conduction band and nearby bands which are normally not occupied.

In an intrinsic semiconductor the electrons in the conduction band and the holes in the valency band are created in equal numbers by means of thermal agitation. Electrons also may be excited from the valency band by absorption of light of a suitable frequency to give photoconductivity. In extrinsic semiconductors, the current carriers (electrons or holes) are introduced by impurities.

Mott and Schottky's contribution was to establish the theory of the rectifier effect obtained when a contact is established between a normal metal and a semiconductor. Schottky, who worked with Siemens in Germany, has made a number of contributions to the history of electronics. His career covers early research into thermionic tubes (invention of the tetrode) right up until the development of semiconductors. As a theorist concerned with codifying physical laws on the basis of observed electronic phenomena, he has been responsible for making basic contributions to both of these fields.

J. Frenkel worked out fundamental equations (1933) which would later be used for p-n junctions and the junction transistor. These equations were derived to account for experiments on the changes in contact potential with light. Davidov used these equations for p-n junctions (1938) but because he used incorrect assumptions about the boundary he missed carrier injection.

Another essential step in semiconductor technique was taken in 1948 when Gordon Teal at Bell obtained the first pure single germanium crystals. Up until then work on transistors had been done with polycrystalline ingots. Teal's determination to obtain a pure single crystal, despite the scepticism of his colleagues (he was forced to carry out his research outside regular working hours with borrowed equipment) later made Shockley's theoretical junction transistor a reality. Grown junction pnp and npn transistors were made by adding appropriate impurity elements to the melt as the crystal was being drawn. Teal later joined Texas Instruments as Director of Research and played a leading role in the development of the first grown-junction silicon transistors.

Monocrystals are basically produced by "drawing off" from the melted material, which is continuously heated to near melting point. A machine dips a fine monocrystalline nucleus into the liquid. In order to ensure uniform growth of the sample, the machine also rotates the monocrystal. Carefully calculated quantities of impurities may then be introduced into the melting bath or by a diffusion method that involves heating the solid in an atmosphere containing the impurity. This diffusion technique has become one of the most important parts of present-day silicon technology.

In 1952 a young metallurgist at Bell Labs, W. Pfann, discovered the zone refinement technique. Purification of the semiconductor and controlled introduction of impurities are the two fundamental operations used in making a semiconductor, and call on very sophisticated chemical methods as well as a physical method known as zone fusion. Pfann observed that when a melted body is solidified the impurities tend to remain in the liquid and he realized that by melting one zone and moving it he could transfer the impurities in the liquid.

In France, research progress owed much to Yves Rocard, already well known for his intuition, the extreme ease with which he assimilated scientific problems and his almost miraculous quickness of mind. During the 1930s he began to explore the problems involved in low-visibility aircraft landing systems, and his research on radar was well enough known for R. V. Jones to mention him in his book on the secret war. In 1946 he handed over an area of research on infrared detection using lead sulphide photocells to one of his students, a young man named Claude Dugas, and asked him to familiarize himself with solid state techniques. Dugas was sent to Seitz in the United States where he met Pierre Aigrain, a brilliant young ex-naval officer, who had left the Navy to pursue his interest in science. In 1948, Dugas and Aigrain set up the semiconductor laboratory under the direction of Yves Rocard at the Ecole Normale Supérieure, one of the most prestigious French professional universities. Dugas eventually went back into industry at the request of Maurice Ponte in 1952, but Aigrain remained to train most of the young researchers of the time. After 1948, fundamental and applied research into simple monocrystalline semiconductors multiplied in universities throughout the world, followed by research into more complex semiconductors such as gallium arsenide, previously discovered by H. Welker. The pace has kept up ever since.

Recently new norms have been established for semiconductor techniques that bring very precise physical (for example, optical) properties into play. But silicon used in photodetection, i.e. the transformation of incident photons into an electrical signal, is not an effective light generator, and so modern research is turning to new materials such as binary alloys or pseudo-binary alloys.

Research has recently been oriented towards new materials – binary or pseudobinary alloys. Hermann Welker at Siemens has emphasized the importance of group III and V compounds (gallium arsenide, indium phosphorous, etc.) for semiconductor electronics and was the first to set up a research program to study their properties. In his first article he showed that the crystalline structure and chemical composition of these alloys are very similar to Group IV components such as silicon and germanium. They have the advantage that the band gap between the valence and conduction band can be varied over wide ranges by varying the composition and also because in some compounds the electrons have exceptionally high mobility.

The most well-known of these binary alloys are gallium arsenide and indium phosphide which combine excellent electronic mobility with good optical properties. This research should lead to new developments in microelectron optics, light-emitting diodes, lasers, optical switching systems, photodetectors, transistors, etc., with the possibility of integrating all generating, detection, amplification and optical switching functions on one circuit.

In 1960, two English physicists, J. W. Allen and P. E. Gibbons, succeeded in developing the first gallium phosphorous contact-point light-emitting diodes (LEDs).

In 1979 D. A. B. Miller and S. D. Smith at Heriot-Watt University in Edinburgh and H. M. Gibbs at Bell simultaneously published papers showing the feasibility of a new design for optical transistors, or *transphasers*. The Edinburgh team used a semiconductor crystal made of indium antimonide, while Gibbs used gallium arsenide. This was the first step toward the development of optical computers that can work at much higher speeds, since photon beams are faster than electron beams. Another advantage of these transphasers will be their easy compatibility with optical fiber transmission; gallium arsenide and indium phosphide, both of which are superior to silicon in terms of electron mobility, can work at frequencies particularly suitable for optical fiber transmission systems. Gallium arsenide's remarkable electrical properties give it great advantages at extremely high frequencies (field effect transistors). It forms the basis of semiconductor laser technology and responds to the high transmission capacities demanded by such newly developed services as telematics. Gallium phosphide is used in the industrial production of green light-emitting diodes, while indium phosphide is just beginning to be used for light-emitting diodes or detectors. In article published in May 1980 by a CNET review in France, Jean-Pierre Noblanc gives a table with some comparative physical characteristics of germanium, silicon, indium phosphide and gallium arsenide:

The purest substances in the world – 99.999999 per cent pure – are being made at Bell Labs by means of a new and extremely simple refining method developed there. W.G. Pfann (left) inventor of the so-called "zone-melting" process and J.H. Scaff, who is holding a large single crystal of germanium purified by means of this technique.

	Ge	Si	InP	GaAs
Forbidden energy gap (eV)	0.67	1.1	1.3	1.43
Associated wave length (μm)	1.85	1.13	0.95	0.87
Electron mobility (cm²VS)	4,000	1,500	5,000	8,500
Thermal conductivity (W/cm°K to 300 K)	0.7	1.45	0.7	0.5

(J.P. Noblanc, May 1980: 44).

1

Pseudobinary alloys such as gallium – indium arsenide and phosphide, or gallium and aluminum arsenide are still at the research stage. Introduction of aluminum, for example, changes the spectral field, and in this way the mobility, forbidden energy gap*, optical index, and crystalline parameters can all be modified to suit the purpose for which the crystal is designed by means of changing the proportions of the compound. By changing the composition of the active zone, it is possible, for example, to vary the wavelength of a laser and adjust it accurately for minimum fiber attenuation. However, it must be kept in mind that improving one property can deteriorate another.

We should also mention another special kind of semiconductor, the amorphous semiconductor. Following the announcement by American physicist Stanford R. Ovshinsky that semiconductor glass exhibited rapid switching and certain memory characteristics, scientists began to work on a series of amorphous semiconductors (noncrystalline composition).

Although these materials have been known for over a century, ever since the first synthesis of sulphur arsenide semiconductor glass, their disorderly atomic structure hindered research efforts. The atoms are distributed in a disorderly manner so that the laws articulated by Wilson, Bloch, and Brillouin cannot be applied. Ovshinsky's team has produced devices in which variations are caused by the addition of energy in the form of light pulses; these devices have been used to produce optical memories and new printing processes. They have also led to the development of other devices in which the change in structure is produced by a voltage pulse, as in the case of memory switching systems. Research on amorphous semiconductors is still in the early stages, but we can already foresee a number of interesting technological applications that may eventually open up some completely new fields.

The case of hydrogenated amorphous silicon is particularly interesting and physical research is already fairly well advanced. It could lead to the development of a number of different devices in the coming years particularly in the field of solar energy.

2

* Electrons move in a conduction band, and the holes in a valency band; the so-called *forbidden energy gap* corresponds to the minimum energy required to raise a valence electron into the conduction band and also the amount of energy an electron must gain or lose while making the transition.

1. Semiconductor laser with optical fibers.
2. Light-emitting diode made from gallium arsenide.
3. A low noise field-effect transistor made of gallium arsenide with a submicronic grid.
4. Gallium arsenide integrated circuit.

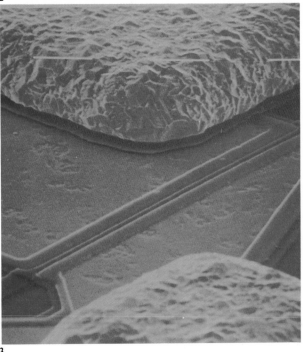

3

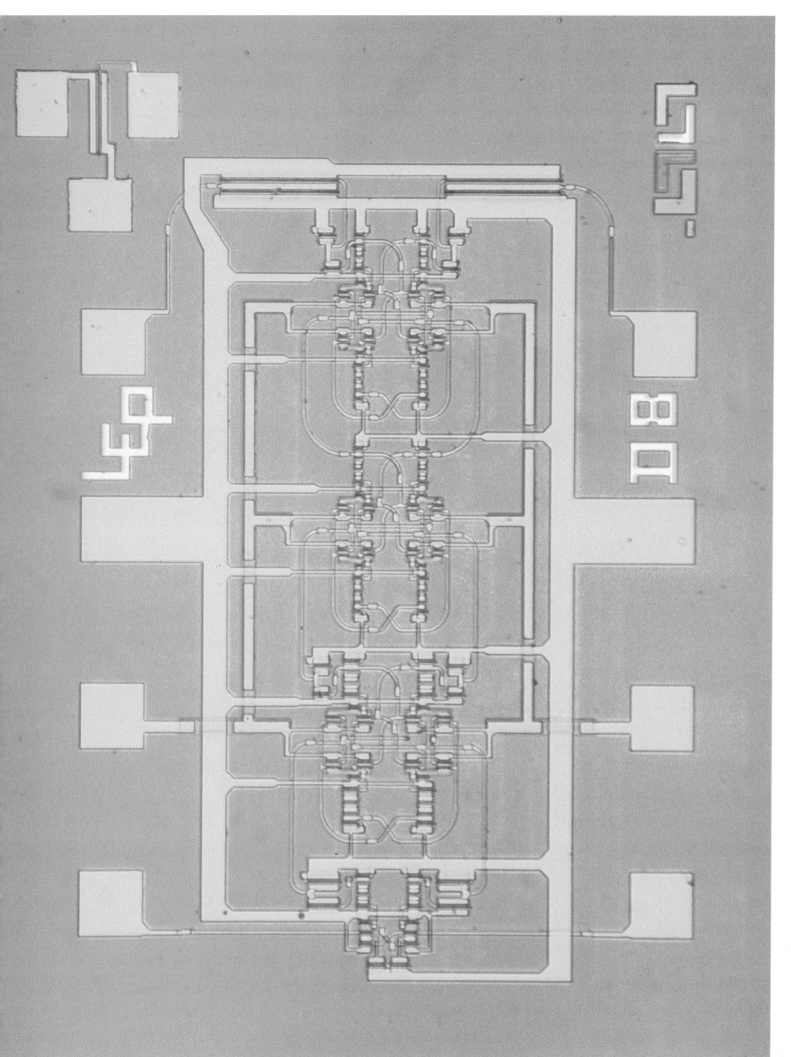

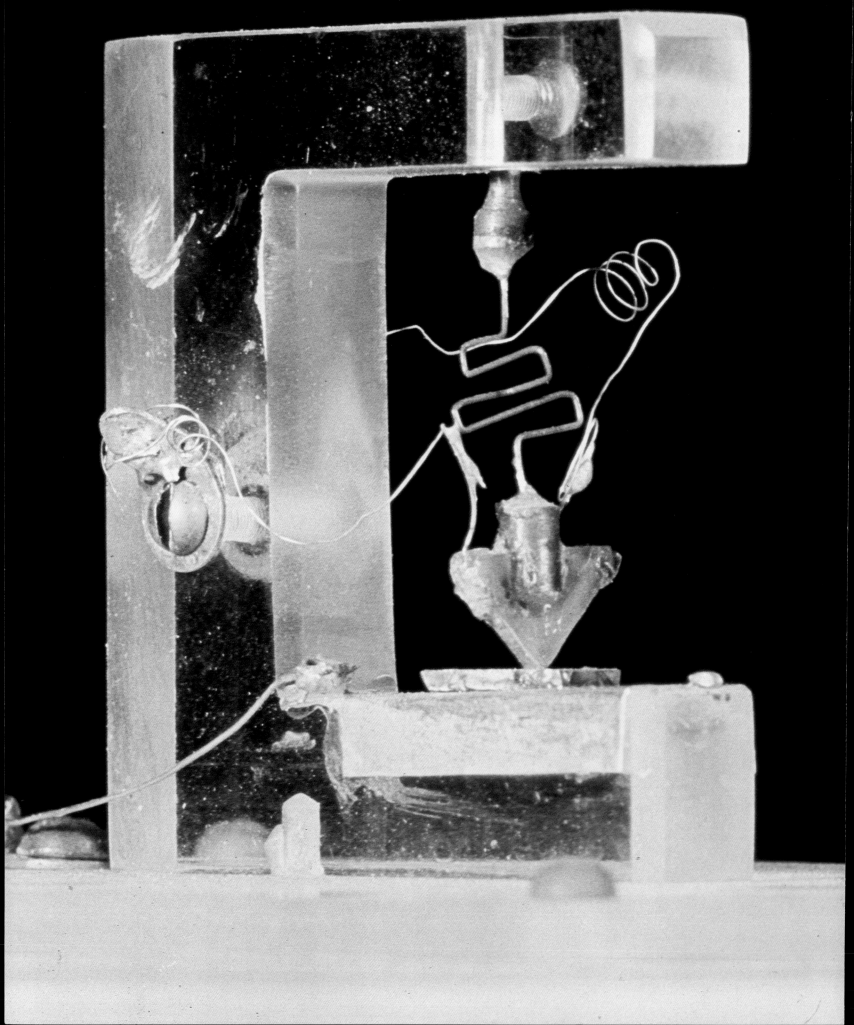

8

Shockley, Bardeen and Brattain's discovery at Bell Laboratories

Transistor is a term that has been a part of our language ever since it was first applied to the small portable radio that inherited the name of its vital component. This indicates the importance of the invention made possible by earlier mastery of semiconductor techniques whose operation was first described by means of quantum physics and whose theoretical bases, articulated by Félix Bloch and Léon Brillouin, were subsequently developed by many different scientists. Without the invention of the transistor, system miniaturization would never have taken place, and electronics as we know it today would not exist.

Here again we cannot talk about "revolution" unless we are prepared to ignore the background of research, trial and error, and preliminary experimentation that lies behind it. The idea behind the transistor (a solid three-electrode system that amplifies signals received) had been in the air since the 1920s. In 1923, a young Russian physicist, O. Lossev, working at the Nijni-Novgorod radioelectronics laboratory, discovered that a zincite (red zinc ore) crystal could function as an autodyne detector or an oscillator when connected to a voltage source of several volts and connected into an oscillating circuit. In 1928, Julius Lilienfeld (Leipzig), described the operating principle behind the field effect transistor. In 1930, H.C. Weber (Industrial Development Corporation, Salem) patented a system of controlling the transmission of electrons in solids. In 1934, a German scientist, O. Heil, took out an English patent describing the field effect transistor; and in 1936 G. Holst and W. van Geel (at Philips) described the current handling characteristics of a semiconductor rectifier. In 1938, as we have seen, R. Hilsch and R.W. Pohl wrote an article describing a three-electrode device using potassium bromide. In 1939 A. Glaser, W. Koch, and H. Voigt took out an American patent for a system very similar to van Geel's except that they modified the insulator to improve conductivity.

Like the transistor, this device consisted of three superimposed layers, two of which were semiconductive. In 1943 and 1945, W. van Geel took out further patents that incorporated a number of refinements.

Why did none of this research achieve any definitive conclusion? This is easily established with hindsight: to obtain a transistor effect, it is necessary to use very pure monocrystalline materials with a controlled impurity content. Nothing like this existed in the technology of the pre-World War II period. The materials used by scientists and researchers before 1940 could not be used to produce transistors for reasons connected with the laws of physics, reasons perfectly clear to us today. Another stumbling block was that there was no really effective liaison between the latest therories of Bloch, Brillouin, and other theoretical scientists and the laboratory work and experimental demonstrations being done by practical researchers. It is also true that on the eve of the war it was far more important to concentrate on such imminent technical priorities as radar to the detriment of other fields.

Bell was finally responsible for uniting the men and conditions necessary to produce the transistor. Likewise, the three inventors of the transistor were all Bell men: the experimental researcher Walter Brattain joined Bell in 1929, William Shockley in 1936, and John Bardeen in 1945. Bardeen has twice been awarded the Nobel Prize: once for his work on the transistor and the second time for his work on superconductivity. The replacement of tubes by solid state amplifiers was a favorite dream of theorists and experimentalists for many years. One of the first suggestions was to control electron flows by means of a Hilsch-Pohl type device with a grid (as for the triode) but the extreme thinness of the space charge zone — in the order of 10^{-4} cm — proved to be an obstacle. Another suggestion was to control the conduction of a semiconducting plate by applying a transversal electrical field (Lilienfeld's field effect). This was the idea which Welker and Mataré at CNET in Paris and William Shockley at Bell decided to develop immediately after the war.

However the first transistor discovered operated according to a different and hitherto unsuspected principle. This was the point-contact transistor, discovered by Bardeen and Brattain during basic research into the surface properties of semiconductors; in this transistor electrons and holes functioned simultaneously. This was quite different to the

The first point-contact transistor.

87

Shockley, Bardeen and Brattain's Discovery at Bell Laboratories

1

OFFICIAL FIRST DAY COVER

25 YEARS
OF TRANSISTORS

WALTER BRATTAIN, WILLIAM SHOCKLEY AND JOHN BARDEEN, AWARDED
NOBEL PRIZE FOR TRANSISTOR INVENTION AT BELL LABORATORIES

Progress in Electronics

SERIES OF 1973

2

1 and 3. The point-contact transistor and the junction transistor. Transistor is an abbreviation of the term "transfer resistor".
2. The three inventors of the transistor: Shockley, Bardeen and Brattain.
4. The transistor eliminates one of the main disadvantages of the tube i.e. its large size, and opened the way for microelectronics.

3

4

system envisaged by Shockley, who was working on the possibility of a unipolar field effect transistor, rather than a bipolar one. Shockley's transistor was to use only one sort of conductor: *either* the electrons *or* the holes and not both, as in the point-contact transistor.

A number of different researchers were jointly responsible for this very basic discovery which marked a new turning point in the history of electronics.

Walter H. Brattain, the experimentalist, was the first to join Bell, in 1929. While studying at the University of Minnesota he attended lectures given by leading quantum mechanics theorists such as Erwin Schrödinger and Arnold Sommerfeld. He was first involved in electronic tube research but his experience in solid state physics soon led him to take an interest in semiconductors, and from 1931 onwards he began to study the operation of copper oxide rectifiers. Brattain conducted the key experiments which later led to the discovery of the transistor. After that he worked specifically on germanium research.

William Shockley was the leader of the team and also the youngest man in it. He joined Bell in 1936 and began working on tube research with C.J. Davisson (1937 Nobel Prizewinner). However, John Slater, his professor at MIT introduced him to semiconductors. After 1939 he worked with Brattain and began to consider the possibility of a copper oxide amplifier. Shockley had made an extensive study of space-charge layer theory, which concerns the surface layer of a semiconductor near its junction with the metal, and he observed that the zone lost its carriers in the presence of an inverse field potential. He began to wonder whether this phenomenon could be used as a regulation valve to control electron flow. Brattain was at first sceptical but realized the possibilities at stake and initiated a series of experiments based on Shockley's ideas.

John Bardeen joined Bell at the end of 1945. He had known Shockley in Boston when the former was working at Harvard, and Shockley at MIT. Like Shockley, Bardeen is a theorist, although he collaborated very closely with Brattain and Gerald Pearson, the two main experimentalists on the team, suggesting experiments and helping to interpret the results. Later he emphasized the extent to which research at Bell Labs differed from previous attempts to make a semiconductor amplifier in that it was scientific rather than empirical. *At each stage we tried to understand what was going on in terms of basic theory. If an experiment didn't turn out as expected we wanted to know the reason why. A lot more was involved than putting two contacts in place of one on a germanium block.* (Letter to the author dated January 1982.)

During and after World War II, a number of other researchers were working on concepts that were important to the work being done by Brattain, Shockley, and Bardeen: the chemist Russel S. Ohl explored the rectifying properties of silicon; the two metallurgists, J. H. Scaff and H. C. Theuerer purified silicon and helped identify the impurities that account for the difference between n and p zones. Scaff and Theuerer discovered that Group V elements such as phosphorus or arsenic provide excess electrons (n zones) and that Group III elements such as boron or indium give an excess of hole (p zones.) Some ingots were half-p, half-n ingot, giving a p-n junction between them.

After the war, a group was formed under the direction of Shockley and Stanley O. Morgan to exploit understanding of solid state physics developed during the thirties. It was during this period that Bardeen joined the Bell team.

Shockley suggested working on the development of a field effect device in which a charge would be introduced near the semiconductor surface by a voltage applied to a capacitor close to the surface. For an n-type semiconductor a negative voltage applied to the plate would be expected to form a positively-charged space-charge and thus reduce the conductance. However all the experiments done to try and detect this effect failed. Shockley had built his field effect theory on Mott and Schottky's theories on rectifiers, which were based on a space charge layer formed near the metal-semiconductor interface.

Bardeen was intrigued by the failure of these experiments and he then advanced the theory of surface states: for the semiconductor surface to be in equilibrium before any electrical contact, there may be electrons in surface states with the charge neutralized by a space-charge layer with a total charge equal and opposite to the charge of the surface electrons. This theory made it possible to envisage the existence of a space-charge layer on the free surface, an idea that could be tested experimentally. Brattain immediately initiated a series of experiments on the surface properties of germanium and silicon.

One of the most significant of these experiments involved studying the change of contact potential of a semiconductor surface when exposed to light. This provided information on the surface space-charge layer. Brattain obtained two very significant results:
1. there could be an inversion from type-n conduction in the volume into type-p on the surface.
2. A field could penetrate the surface and alter the space charge zone if it was applied by means of an electrolyte in contact with the surface.

Bardeen and Brattain then tried using a rectifying point-contact which would make a low resistance to the inversion layer and at the same time a high resistance to the electrons in the interior. A field was applied across an electrolyte. A positive voltage applied to the electrolyte decreased the hole concen-

tration near the surface and the current to the point contact biases in the reverse direction (negative). Further experiments led to the discovery of the first transistor, a bipolar type, which was not based on the field effect but on the injection of holes, minority carriers, into the germanium block. When it was attempted to apply the field across a gold contact evaporated on an oxide layer, the gold made contact with the surface. A positive voltage increased rather than decreased the current to the positive contact.

They then had the idea of placing two electrodes next to each other on a germanium ingot: a current was injected into the base by a first emitting point and collected by a neighboring oppositely-polarized point; this current, could be controlled by means of a small voltage applied between the emitter and the base. An input signal could be amplified by a factor of about 100 to a load connected between the collector point and the base.

The study of point transistors gave a better understanding of the functioning and behavior of carriers in semiconductors. It made it possible to directly measure the flow of holes and electrons, which up until then had only been observed indirectly. The point-contact transistor had its first public demonstration in New York on June 30, 1948. It was an exceptional discovery in that it was the first solid state amplifier ever invented. However the technique was still both difficult and expensive and it was only adopted for limited telephone applications.

Two Purdue students, Seymour Benzer and Ralph Bray, afterwards claimed to have come very close to the same discovery. Bray commented: *If I had just put my electrode close to Benzer's electrode we would have gotten transistor action* (quoted by E. Braun and S. MacDonald in *Revolution in Miniature).* However, it is unlikely that this would have happened. First, because of the fineness of the wires required and secondly because without appropriate treatment of the surface the inverse collector current would have been too tiny to produce a strong enough field to attract the holes of the emitter. Ben-

1

2

1. First all-transistor radio in Japan, August 1955. The term "transistor" passed into the language to describe the portable radio which contained these components.
2. Pilot shop in the new Allendown Works in Western Electric where the first production transistors in the world were made in 1951. Here girls are performing assembly operation with hand-operated fixtures. Today's modern production lines use sophisticated high-speed equipment.

zer and Bray had observed an excess conduction when a point-contact is biased positively but they did not attribute it to an increase in the concentration of electrons and holes near the contact (they would in that case have discovered the principle of injecting a minimum number of carriers, the basic effect of the bipolar transistor) but to an increase in the mobility of electrons in the electrical field near the contact.

Bardeen further states: *Since discoveries and inventions are generally made when the time is ripe, we were surprised but gratified after we announced the announcement of the transistor, to note that no-one else in the world was close to a bipolar transistor.*

A month after the December 17, 1948, discovery, Shockley, in trying to get a better theoretical understanding of the point transistor, conceived the idea of a second device, the junction transistor, in which the process would take place not at the surface but in the body of the semiconductor. Shockley suggested replacing the point contacts by p-n junctions.

In an npn junction transistor a thin layer of p-type conduction separates two regions of n type conduction, with each of the three zones connected to an electrode. When appropriate voltages are applied, electrons from the emitter are injected into the base and gathered in the collector, biased positively. Voltage applied between the base and the emitter controls the current flow between emitter and collector. As we have already seen, it was because of Gordon Teal's determination to obtain very pure single crystals that he succeeded in producing this device by the grown junction technique in 1949 and the first paper was published by Shockley, Sparks, and Teal in 1950.

For a number of reasons junction transistors allow greater energy savings when compared to tubes: first the transversal section of the transistor is physically small so that no matter what density of current might be present the overall current does not exceed a certain minimum threshold. Secondly the currents (which are very tiny: in the order of one microamp) are controlled by incoming electrical signals whenever the voltage exceeds 25 mV, enabling the junction transistor to function with voltages in the order of 0.1 v. In telephone communications, for example, tubes require enormously more power to transmit a signal. Using tubes would be, to borrow one of Shockley's expressions, "like using a freight train to deliver a pound of butter."

Other teams were ready and able to take up the torch. The first semiconductor alloy junction transistor and the first diffusion transistor were produced at General Electric by R.N. Hall and his team. Hall had been Director of semiconductor research at G.E. for some time and we owe him a number of theoretical and practical contributions in the field. The group which he headed was the first to recognize the importance of diffusion methods for doping semiconductors. Later, Hall's team was one of three which simultaneously observed laser emission from gallium arsenide (the other two teams were at MIT and IBM).

The first silicon single crystal alloy junctions were produced by Pearson at Bell, but the first commercial silicon junction transistors were developed by Teal, who had by then joined Texas Instruments.

A significant advance occurred in 1958 with the appearance of the first silicon Mesa junction transistor by Bell Laboratories. It was given the name Mesa because its shape resembled a kind of hill or "mesa." The novelty of this system was that it was built by diffusion. This new technique made it possible to have ten times more accurate control over the positioning of the base depth to which impurities penetrated, which could be forecasted from such data as temperature, vapor pressure and time. Lithography and oxidation by masks were introduced and developed to produce this transistor. The Mesa could operate at much higher frequencies than other junction transistors, very near the Gigahertz region.

The field-effect transistor, which was the first envisaged by Shockley, was in fact the last to appear. In 1957, John T. Wallmark (RCA) took out a patent on the field effect transistor, but he did not take it any further; two years later Paul Weimer, also at RCA, built a thin-film field-effect transistor with cadmium sulphide. In the meantime, in 1958, Stanislas Tetzner (Compagnie Générale d'Electricité) created the first junction field-effect transistor (JFET), the technotron. In 1962 Steven Hofstein and Frédéric Heiman (RCA) completed the silicon field-effect transistor which opens up a whole new chapter in the history of solid state electronics.

Smaller, lighter, lower energy consumption: these were three very significant factors in the success of the new techniques.

1. Very High Frequency transistor for wide band amplifiers (60 kHz-60 MHz).
2. UHF field effect transistor made from gallium arsenide with beam-lead structure for ultra high frequency applications (up to 20 Gigaherz).
3. Gallium arsenide field effect transistor.
4. Ultra High Frequency transistor.
5. Field effect transistor.

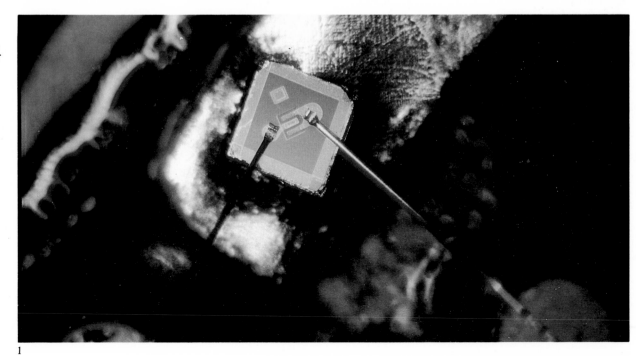

1

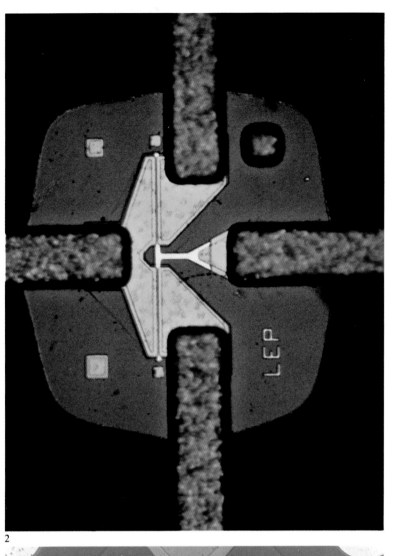

2

3

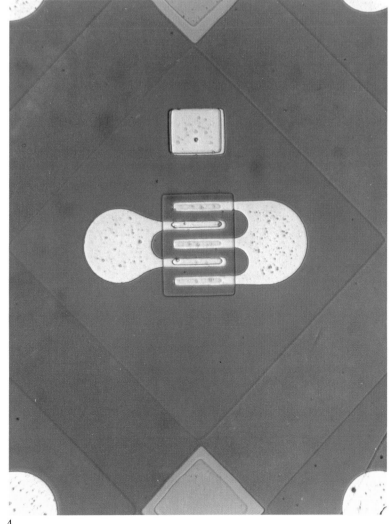

4

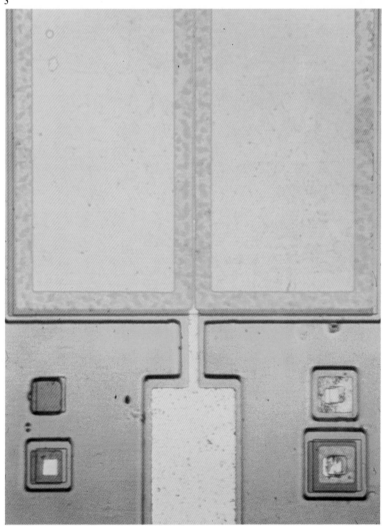

5

Transistors

by Robert Bernhard I.E.E.E. Spectrum

With the discovery of the transistor, a whole new field opened up in electronics. What is the operating principle of bipolar transistors and field-effect transistors? How can they be used?

Transistors have many different forms with different electronic characteristics. However, all transistors are basically three-terminal devices. One terminal controls the flow of charge between the other two, so that transistors may function either as "on-off" switches in digital circuits, or amplifiers that increase current, voltage, or power. In many cases, the electronic characteristics of transistors in integrated circuits differ from their discrete counterparts. Silicon is the basic semiconductor in transistors that are used in most applications. However, gallium arsenide transistors have recently been developed for special applications, such as microwave amplifiers in communication systems, and digital circuits for ultrahigh speed processing of radar data in military systems.

Most transistors in use today are either the *bipolar* type or those termed field-effect transistors (FETs). The three terminals of the bipolar transistor are the *emitter, base* and *collector*. The terminals are associated with three layers consisting of either two n-type and one p-type material (npn transistor), or two p-type and one n-type materials (pnp). The term *bipolar* refers to the fact that current is carried by both holes and electrons. Current conduction between the emitter and collector is controlled or modulated by a voltage applied to the base terminal.

The three terminals of a FET are the *gate, source,* and *drain.* The source and drain terminals are connected by a channel of n-type material (conduction by electrons) or p-type material (conduction by holes). The channel lies directly under the gate. A voltage applied to the gate terminal controls the conduction current.

Field effect transistors are classified either as insulated gate devices (IGFETs) or junction devices (JFETs), depending on whether or not the gate is separated from the channel by a thin insulating barrier. The insulating barrier is usually silicon dioxide. Alternative insulators are silicon nitride or aluminum dioxide. Insulated gate devices are the transistors used most commonly in such mass produced items as hand calculators, and memory or microprocessor chips that contain thousands of transistors on an area about 25 to 50 square mm. The IGFET is commonly called a MOSFET, or metal-oxide semiconductor FET. Most often it is referred to as an NMOS or PMOS device, depending on whether the channel is n-type or p-type. The J FET may be fabricated from silicon or gallium arsenide. The MOSFET can be made of silicon, but not gallium arsenide, since no suitable insulating layer has yet been developed for the newer material.

Bipolar Transistors

The npn transistor is the most widely used transistor in integrated or discrete bipolar circuits. The operation of this device is the same if the roles of electrons and holes are interchanged. Both npn and pnp devices are the solid-state analogs of the vacuum tube triode. The equivalent cathode in the npn device is the emitter; the equivalent grid is the base; and the plate is the collector. The electron current enters the emitter and flows to the base and collector. The base and collector of npn devices are biased positively relative to the emitter. However, electron current in pnp devices flows opposite to that in the npn devices; thus, the base and collector in pnp devices must be biased negatively relative to the emitter. The emitter-base junction of the npn transistor is termed *forward biased* (positive voltage applied to the p layer, and negative voltage to the n layer), while the collector-base junction is *reverse biased* (positive voltage applied to the n layer, negative voltage to the p layer).

A positive voltage applied to the emitter-base junction of the npn transistor results in a flow of electrons from the n-type emitter to the p-type base. The total current that flows is described by the equation,

$$I = I_0(e^{eV/kT} - 1),$$

where I_0 is the

current with zero bias; eV is the difference in potential energy between the Fermi level and the conduction band of the material; k is Boltzmann's constant; and T is absolute temperature. A small increase in voltage, therefore, results in a large increase of current into the base. Amplification is possible, since relatively little power is needed from the input signal to control the current flowing from the emitter through the base to the collector.

The majority of electrons flowing into the base diffuse easily through the narrow base region, and then into the base-collector junction. The diffused electrons are drawn to the collector by the electric field existing between the positive donor ions on the collector side, and the negative acceptor ions on the collector side. The region of uncovered ions on each side is called the *depletion layer*, since the region is depleted of carriers. The collector bias voltage causes the electrons to flow through the collector, and thus deliver power to the external load. In summary, the amplifying action is the result of transferring a current from a low resistance circuit to a high resistance circuit.

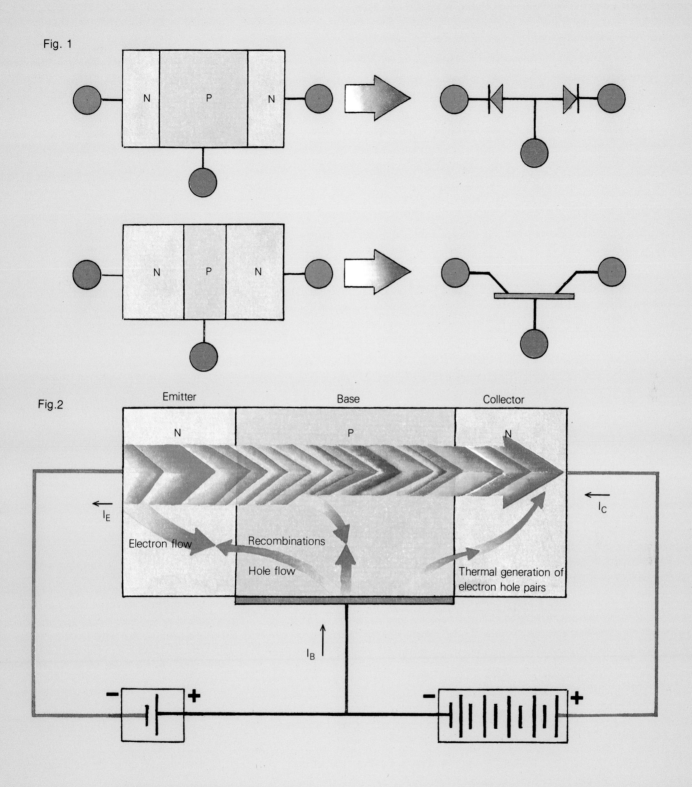

Fig. 1

Fig.2

Emitter Base Collector

N P N

I_E

I_C

Electron flow

Recombinations

Hole flow

Thermal generation of
electron hole pairs

I_B

− +

− +

Fig. 1
Associating two junctions in one semiconductor
crystal may be the equivalent of two diodes
(above) or one transistor (below) depending on
the width of the central zone.

Fig. 2
Although the current in the collector lead
consists basically of electron current, a small
parasite current is present and leaks across the
reversed biased base-collector junction. This is
called the leakage current and is caused by
changes in the temperature of the crystal which
cause electron-hole pairs to be gradually
liberated in the base and collector zones.

Different device characteristics are needed for various applications. For linear circuits, such as amplifiers, the transistor is required to operate in a nonsaturating mode with high current gain and high breakdown voltage. The term *saturation* refers to the maximum value of the current flow in the device for a given set of circuit parameters. As input voltage is increased, a point is reached after which the device goes into breakdown, where the current increase is no longer limited by the device. Breakdown can result in permanent damage to the device, unless the external circuit provides a limit on the current.

Digital logic circuits require higher switching speed and lower saturation voltage. Lower breakdown voltages are more acceptable in digital circuits.

Field Effect Transistors

N-channel MOS and PMOS devices are used widely in integrated circuits because of their relatively simple construction. Their relative simplicity allows them to be made more cheaply and packed more densely than the larger bipolar transistors. MOSFETs are fabricated either as depletion devices or enhancement devices. In making depletion devices, a channel is physically constructed between the source and drain. When a voltage is applied across the source and drain, a portion of the channel carriers is depleted and results in current conduction in the channel. Enhancement devices, by contrast, have no channel formed during their construction. Only when a voltage is applied to the gate does a channel of charge carriers develop, and a current results, when voltage is applied across the source and drain.

Enhancement mode MOSFETs are used more widely than depletion mode devices in mass produced integrated circuits because they are self-isolating. They can be fabricated by a single diffusion step that forms the source and drain pockets. All of the active regions are reverse biased relative to the substrate, and so adjacent devices on the same substrate are electrically isolated from each other. *(Reverse bias* is defined as a bias favoring the flow of minority carriers.)* Adjacent depletion devices, by contrast, must be isolated from each other by a bulky isolation diffusion. Enhancement devices can, therefore, be packed more densely than depletion types.

The control voltage applied to the gate of a MOSFET induces an electric field across the thin dielectric barrier, and thus modulates the free carrier concentration in the channel. In enhancement devices, an n-type inversion layer is produced when a gate bias of proper polarity is applied and increased beyond a threshold value.

Further increases in the gate bias lower the resistivity of the induced channel, and result in increased (enhanced) current flow from source to drain. Positive voltages above the threshold value result in increased drain current. The current is described by the equation, $I_d = K(V_{gs}-V_t)^2$, where K is typically 0.3 mA/V^2; V_{gs} is the gate-to-source voltage; and V_t is the threshold voltage.

An increase of the drain-to-source voltage, V_d, causes the drain current to "pinch off," or deplete completely, near the drain. Saturation of the drain current occurs when V_d is equal to or greater than $V_{gs} - V_t$. At such values of V_d, the net gate bias is no longer able to maintain an induced channel.

As MOSFETs are scaled down to ever smaller sizes, their switching speed increases and their power dissipation decreases. The most dense NMOS digital circuits in commercial use have transistors with channel lengths of 1.5 to 3 micrometers. This corresponds to the placing of 100,000 to 450,000 transistors on a chip with an area of about 25mm^2. In theory, physical laws limit the size of the smallest MOSFETs to channel lengths ranging from .2 to .25 micrometers. The limit on size is due to "punchthrough." Punchthrough refers to the breakdown of transistor action that occurs when the channel length shrinks to where the source and drain depletion regions overlap. At that point, the gate bias can no longer control conduction in the channel.

The smallest practice MOSFETs ever made (as of March, 1981) have channel lengths of 9.3 micrometers. Made by engineers at Bell Laboratories in Murray Hill, New Jersey, U.S.A., they have exhibited the fastest switching speeds ever measured for digital devices: from 30 to 50 picoseconds. The MOSFETs were fabricated by a revolutionary new x-ray lithographic process that could make it commercially feasible for the first time to place one million transistors on a single chip.

Silicon Versus Gallium Arsenide

Gallium arsenide technology has been developed primarily by U.S. defense electronics firms because it has an inherent speed advantage over silicon for digital circuits. Gallium arsenide has about five times higher electron mobility than silicon. However, significant improvements must be made in the new technology before it is introduced for wider use.

Gallium arsenide amplifiers have been used for several years in microwave circuits, and gallium arsenide integrated digital circuits are under development to handle data rates of one to five gigabits

per second, far beyond the capabilities of current silicon technology.

Gallium arsenide transistors are basically JFETs. State-of-the art gallium arsenide integrated circuits are currently at the 100 to 1000 gate level of complexity, whereas silicon technology is capable of 5000-gate bipolar integrated circuits, and at least 10,000-gate NMOS integrated circuits.

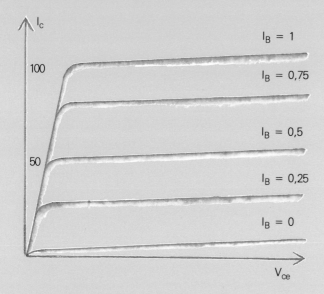

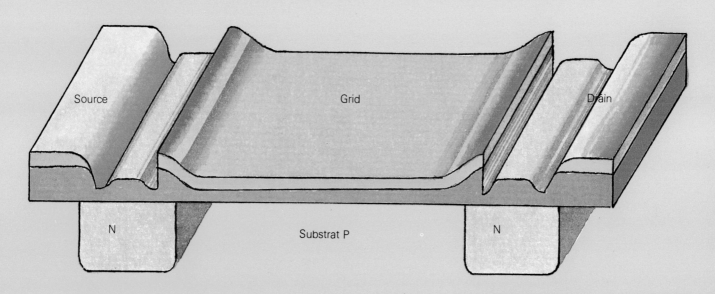

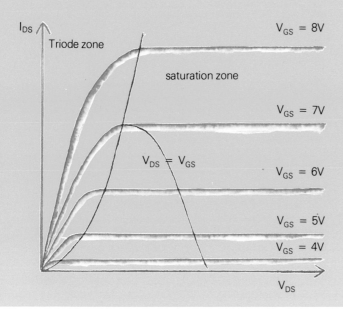

Fig. 3
The characteristic $I_C = f(V_{ce})$ typical of a bipolar transistor.
I_C = collector current, I_b = base current, f = "function of",
V_{ce} = collector-emitter voltage.

Fig. 4
Section of the structure of a normally off n-channel MOS transistor, the most common type of structure.

Fig. 5
Typical MOS characteristics. I_{ds} = drain-source current,
V_{ds} = drain-source voltage, V_{gs} = grid saturation voltage.

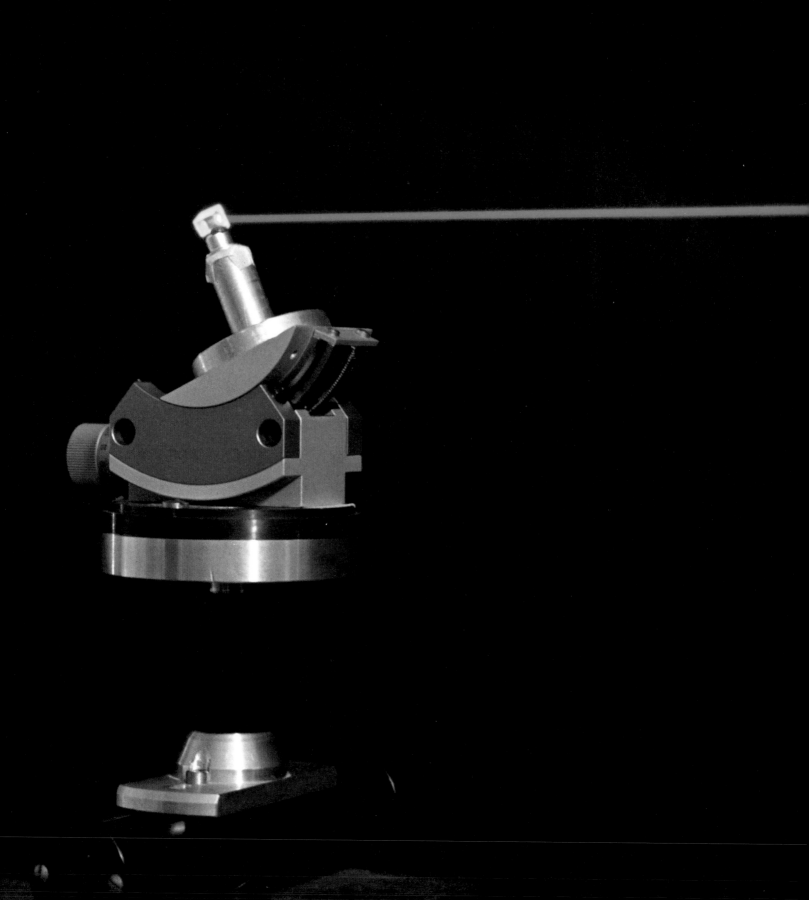

9

The laser and quantum electronics

Over the last few years the laser has become a multipurpose instrument used in hospitals, industrial plants, design agencies, discothèques, telephone installations and the army. Most of its applications are based on semiconductors. World turnover (not including the Eastern bloc countries) is probably in the region of $1 billion, half of which is earmarked for military purposes. A laser physics specialist at the CNRS*, defines it as follows: *An instrument that operates on wavelengths from infrared to microwave and that transports energies varying from the microwatt to the terawatt (10^{12} watts), that can cost either a few dollars or anything up to $10 million, and that possesses limitless applications.* (Séroussi, November 2, 1980: XIV).

The laser can be defined above all as a source of coherent light, coherent in terms of space — instead of dispersing its light in every direction like the sun or an electric light bulb, the laser sends it in a one-directional beam — and coherent in terms of time: the beam is monochromatic which means that only one color of the light spectrum is generated and the period of the wave is therefore constant. This coherent source of light is amplified by *stimulated emission,* by using atoms and molecules to produce energy. The principles of stimulated emission had been known for some time before 1950 when the Russians and Americans first began work on the laser, but its applications had not yet been explored. Maser and laser technology both rely on quantum electronics whose spectral field begins with radar frequencies (hydrogen maser: 21.2 cm wavelength) and stretches to the visible spectrum.

In 1917 the ubiquitous Albert Einstein had shown that in order to obtain an exact description of the interaction of matter and electromagnetic radiations, it was necessary to design a system whereby atoms would be stimulated by an electromagnetic field so that they would supply or absorb energy.**

It took another forty years for the principal conclusions to be drawn from this theory. In his memoirs, Charles Townes wrote:

Picture yourself in the year 1950 as a research administrator or government official who wants to develop an amplifier for communication which would be 100 times more sensitive than anything previously available, or a clock which would be 1,000 times more precise than previous ones, or a searchlight which would be one million times more directional than any previous searchlight, or a light which would be one billion times stronger than any previous light intensity including the sun. Where would you have gone to have such things developed or invented? For amplifiers, of course, to the electronics experts, who, with a lot of work and money would likely improve things by a factor of two, but not 100. For clocks there would be a different set of experts, but none could promise more than a modest improvement in accuracy, because efforts in this direction have already been going on for years without significant breakthroughs. For enormous increases in the intensity of light, perhaps one would try designers of electric arcs or perhaps in desperation the nuclear bombmakers. But in fact, you would probably have either discreetly given up any open plan to obtain improvements of these magnitudes or been removed from your administrative job as a crackpot. Yet all these things, and a good deal more, were at that time just on the verge of appearing from a completely unexpected research direction, the study of responses of molecules to microwaves, known as microwave spectroscopy.

Microwave spectroscopy was an offshoot of the work done during World War II on radar and microwave oscillators. Townes was then one of a number of scientists exploring the possibilities of harnessing smaller and smaller wavelengths. Another researcher, Joseph Weber, was also working on this problem at the University of Maryland. Townes describes how one morning the idea of the maser occurred to him. He was in Washington to promote a microwave oscillator project.

It was clear that what was needed was a way of making a very small, precise resonator and having in it some form of energy which could be coupled to an electromagnetic field. But that was a description of a molecule, and the technical difficulty for man to make such small resonators and provide energy meant that any real hope had to be based on finding

A pulse laser.

* French Scientific Research Center

** When light strikes an object it produces a weak collision force known as the *radiation pressure*. The idea was first advanced by Kepler in 1619 when he observed that the tails of comets pointed in the opposite direction from the sun because of the pressure of solar radiation. The first scientific confirmation of this hypothesis was provided around 1900 by Ernest F. Nichols and G.F. Hull in the United States and P.N. Lebedev in Russia: radiation pressure can be shown by concentrating a light source on a plate suspended by a thin wire in a deep vacuum and observing the torque produced in the wire.

a way of using molecules! Perhaps it was the fresh morning air that made me suddenly see that this was possible; in a few minutes I sketched out and calculated requirements for a molecular-beam system to separate high-energy molecules from lower ones and send them through a cavity which would contain the electromagnetic radiation to stimulate further emission from the molecules, thus providing feedback and continuous oscillation.

One of Townes's students, James Gordon, agreed to work on this project for his thesis along with Herbert Zeiger, one of Townes's assistants. Three years later, in 1954, they succeeded in constructing the oscillator Townes had envisaged, and baptized it Microwave Amplification by Stimulated Emission of Radiation or more simply, the maser. The first maser was an ammonia type. In 1955 Townes went to France to work at the Ecole Normale Supérieure Laboratories (a French university-level college specializing in professional training), with Alfred Kastler, who won the Nobel Prize in 1966 for his work on optical "pumping." As Townes said : *Kastler was not the father of the laser, as a certain section of the French press wanted to prove. His optical pumping procedure, as he saw it, had nothing to do with our research; he even admitted that himself.*

Townes, working with Combrisson and Honig, developed the first solid state maser. He then made a trip to Japan : *At the time, Japan had done very little work on the maser. But at Colombia I had worked with a Japanese scientist, Shimoda, who had taken the basic principles of the work which we did back to Japan with him.* In 1954, Townes first met the Russian team composed of N. G. Basov and A.M. Prokhorov, who had been doing parallel research on the maser; he shared the Nobel Prize with them ten years later.

We later found that the idea that stimulated emission could increase the number of photons in radiation had already been mentioned several times in scientific literature. So what had we really added besides finding a practical system? First there was coherence, a concept which deemphasizes photons, a different habit of mind from that which had been carefully nurtured in physicists for a couple of generations. That stimulated emission was coherent − that is, of exactly the same frequency as the stimulating radiation − was a known result of quantum mechanics, but obscure enough to many physicists that a number of my friends doubted it. Secondly there was the addition of an external resonator for feedback − a resonator which contained the radiation, coupled it more strongly to the molecules, and provided a threshold for the sudden buildup of continuous oscillation.

1

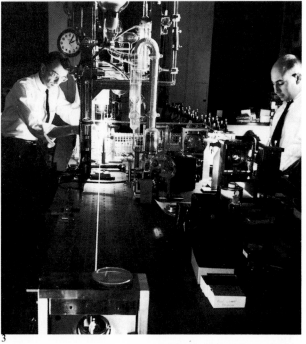

2

3

1. In 1957, three Bell scientists, H. Seidel, G. Feher, and D. Scovil (left to right) demonstrated for the first time that solids could be made to amplify throuth the maser principle. They went on to develop the ruby maser amplifier.
2. The three inventors of the laser − Basov and Prokorov (first and third from the left) and Townes (right).
3. The first gas laser to generate a continuous beam of visible light (1962).

* In the last ten years, researchers have discovered natural stellar and interstellar masers, composed of clouds of gas in the vicinity of nebulas: these naturally-occurring phenomena act as giant cosmic masers. Subsequently the dusty atmospheres surrounding ageing red stars were discovered. In 1965, Berkeley researchers discovered the first of these natural masers when they detected a signal coming from a cloud of hydroxyl molecules in the huge Orion nebula. William J. Wilson and Alan H. Barret (MIT) discovered the first stellar maser in 1968.

The hydrogen maser was developed by Norman F. Ramsay at Harvard University; the solid state maser proposed by another Harvard researcher, Nicolas Bloembergen, was used in the ground-based station for the Echo project and for communications with the Telstar satellite. In the field of spectroscopy, researchers were most enthusiastic about the technique, as Bell Laboratories was interested in the sales aspect.*

However, Townes was already pointing out that in order to obtain very high frequencies it was better to use an optical agent and to concentrate on infrared based tranmission methods. The problem was to obtain ultramillimetric wavelengths (less than 1 millimeter). In September 1957, Townes conceived the first maser function on optical wavelengths instead of amplifying electromagnetic waves in the centimetric spectrum; this was the *optical maser* or laser (Light Activation by Stimulated Emission of Radiation). In 1960 Townes and Arthur Schawlow from Bell Laboratories took out the first laser patent. In that same year Theodore Maiman of the Hugues Research Laboratory developed the ruby laser, followed by the crystal laser (M. J. Stevenson and Peter Sorokin at IBM) and the gas laser (Ali Javan, William T. Bennet Jr, and Donald Herriott at Bell Laboratories). Liquid lasers (using organic coloring agents such as cyanide) and semiconductor lasers (using gallium arsenide) were developed subsequently. Semiconductor lasers were developed as a result of work undertaken in the field of telecommunications at General Electric, IBM, Bell and Lincoln to create a single beam that could carry several million telephone conversations. The active material generally used in semiconductor lasers is the function of a rectifying p-n-type diode. There are three alternative excitation methods: injection, electronic bombardment with high-energy electrons, pumping, where the semiconductor is illuminated by a very intense light whose frequency is superior to that of the laser emission desired.

The semiconductor diode laser undoubtedly has the best sales potential: it is very tiny (less than 1 mm across) and may be used in combination with integrated microelectronic circuits providing direct control. The first semiconductor diode lasers contained only gallium arsenide, and the diode had to be kept at a very low temperature for the laser to function continuously. In 1969, Jaures Alferov and Rudolph Kazarinow combined the injection of electrons with a layered structure acting as a wave guide. A layer of gallium arsenide is placed between two layers of aluminum gallium arsenide. This arrangement allows semiconductor diode lasers to operate at normal temperatures.

After 1960, work on the laser popularized development of the hologram technique; the technique had been discovered by Dennis Gabor in 1948, and ignored up until the 1960s because of the unsuitability of conventional light sources lacking temporal uniformity. Holography was first conceived as an adjunct to the electron microscope: Gabor suggested producing a hologram with the wave associated with the electrons, then reconstituting the front of the wave originating from the irradiated object by means of optical radiation, thereby eliminating aberrations in the electronic lens. In addition, the magnification would exhibit a relationship to the frequency used. The technique developed by G.W. Stroke eliminated blurring in pictures taken by electron microscope. In 1962, Americans Elmeth Laith and Juris Upatnicka produced the first holograph, although it was not an electronic one; it was not until 1968 that a Japanese team (A. Tononura, A. Fukuhara, H. Watanabe, and T. Komoda) produced an electronic holograph that was illuminated by a helium-neon laser, and that could be used to observe particles of gold 10 mm in diameter dispersed throughout a substrate. In March 1974 two scientists working at Ann Arbor University in Michigan (C.L. Ritz and L.S. Bartell) succeeded in taking a holographic shot inside an atom. Holography is used to reconstitue relief images; as we have seen, it appeared fifteen years before the laser but could not be put into use until a source of powerful and uniform light was discovered.

Each element of the hologram contains all the information; that is, a fragment of the image is the same as the whole image, although with less definition.

A procedure discovered by Andre Marechal makes it possible to do analog calculations using the optical method, thereby making it possible to localize a word in a text or a particular place on a card. Acoustical holography is based on the same principle and is used to reconstitue a three-dimensional image of a submerged object. For example, microphones pick up the pattern of sound interference between the reference transmitters and the signals reflected by the object. This network can then be transformed into an optical image using a visual display unit and the results interpreted by gas laser.

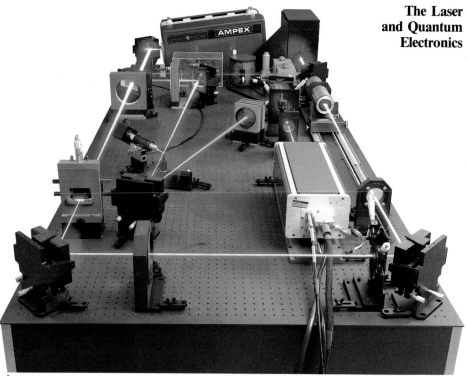

1

2

3

1. Laser recording system.
2. Holographic portrait of the Hungarian-born English scientist Dennis Gabor (1900-1979) who won the Nobel Prize for Physics in 1971.
3.. Jean Mortès, director of the French Museum of Holography, behind a holographic stereogram representing the Venus de Milo.

Coherent light

By Jean Robieux, Scientific Director of the Marcoussis Laboratories (Taken from his paper "Exposé on the Perspectives Opened Up by the Evolution of Laser Research" delivered to the Academie des Sciences on February 10, 1975).

Light is a form of radiant energy emitted by bodies heated to high temperatures or excited bodies. The photons of which it is made up are generated "incoherently." The laser, on the contrary, produces a "coherent" beam. What is the principle behind this?

Laser amplification is stimulated emission amplification. The difference between amplification through emission and stimulated emission was clearly described in a paper presented to the Institute on February 10, 1975, by Jean Robieux.

Inside an atom electrons take up their positions in successive layers, each of which corresponds to a particular energy level. If an atom is not excited by an exterior source of energy, an electron will occupy energy level 1. Under the influence of an exterior force, an electromagnetic wave for example, this electron will be able to pass from level 1 (its fundamental state) to level 2 (its excited state). After a certain time T has elapsed (which corresponds to the length of life of level 2), the electron will fall back from level 2 to level 1 and at the same time emit a photon whose energy $h v$ can be found from the following equation: $h v = E2 - E1$ (h is Planck's constant, v is the radiation frequency, $E2 - E1$ is the difference in energy between levels 2 and 1). An excited atom spontaneously emits an electromagnetic beam. Taking into account the values E1 and E2 for the atom's outside layers, this radiation will often fall within the optical region. When an atom has been excited but is not subsequently subjected to an outside force it spontaneously emits photons. This process is called *spontaneous emission* and has been used for a long time in the generation of light sources. Quantum mechanics teaches us that when an atom containing an electron on level 1 is exposed to an electromagnetic wave, interaction will be negligible if the frequency v of the electromagnetic wave is different from that given in the above equation. If, on the other hand, the frequency of the wave agrees with the above equation the electromagnetic wave will give up energy to the atom. This energy is received in the form of potential energy allowing the electron to move from level 1 to level 2. The electromagnetic wave has therefore lost a part of its energy and will be totally absorbed should it meet up with a large number of atoms with level 1 electrons. This phenomenon occurs when a light wave moves through a medium which has not been excited by a power source. A body in equilibrium at a temperature t, will contain atoms with electrons occupying levels 1 and 2 with a probability ratio as defined by Boltzmann's Law. There will always be more electrons in state 1 than in state 2. Such a body will absorb the electromagnetic wave.

In fact as the electrons pass from state 1 to state 2, they absorb the energy of the wave. These transitions are more numerous than those from state 2 to state 1 which in turn result in the transfer of energy from the medium to the electromagnetic wave. The wave frequency v stimulates the passage of electrons from level 2 to level 1. This phenomenon is called stimulated amplification and is the basis of all laser generation. Stimulated emission refers to the radiation from an atom when it is subjected to an electromagnetic field $h v$. All laser technology is concerned with the concentration of electromagnetic energy present on a very small number of natural modes of cavity oscillation, perhaps only on one.

"Coherence" signifies phase agreement and can be either spatial or temporal. Spatial coherence is relative to the different points of the emitted surface: emission is thus directive, requiring no lens system; the light is contained within a small angled cone whose angle or "divergence" normally equals a milliradian, i.e. the beam of light could light up a circle measuring only 1 meter in diameter at a distance of 1 kilometer. Temporal coherence refers to the fact that the beam is very monochromatic and has a coherent wavelength.

The Fabry-Perot interferometer is almost always used to stimulate the return of an inverted population to its normal state. It behaves like an extremely high-Q electromagnetic resonator. The return of the inverted population to its natural state inside the resonator interior starts off an electromagnetic oscillation which is amplified by successive reflectors on the resonator's end reflectors. Once this amplification is sufficient to compensate for losses, the system will begin to oscillate and an intense pulse of coherent electromagnetic rays will emerge from the end phases. The energy of the beam is controlled by adjustments to the pumping current, the percentage of reflection from the reflectors and the temperature.

P 101 This is a 1/2 scale model of the Laser Geodynamic Satellite (LAGEOS), to demonstrate and employ the capability of laser satellite tracking techniques to make accurate determinations of the Earth's crustal and rotational motion. An important benefit arising from this project could be the development of techniques for predicting earthquakes.

Common types of lasers and industrial applications

Material	Wavelength	Working mode	Pulse length	Energy per pulse or power level	Rate	Application
He-Ne gas (Helium - Neon)	633 nm	Continuous		1 to 250 mW		Various alignment procedures: • Mechanical constructions • Civil engineering - building • Ship building • Underground shafts Metrology: • Shape recognition • Granulometry • Tolerance measurements Holography: • Shape recognition • Data storage.
YAG	1.06 m	Pulsed	50 ns (50.10⁻⁹s)	1 mJ	800 Hz to 30 kHz.	Various evaporation procedures: Resistance adjustments Capacity adjustments
Yttrium aluminum garnet doped in neodymium			$500\mu s$	1 Joule	10 Hz	Various micro-drilling procedures: • Watchmaking • Electronics • Dies.
			A few ms	A few Joules	1 Hz	Micro-soldering: • Watchmaking • Electronics
		Continuous		100 W		Medical (Internal surgery)
Ruby	694 nm	Pulsed	30 ns	100 mJ	1 Hz	Hologram recordings of moving objects
			$500\mu s$	500 mJ	6 pulses per minute.	Micro-drilling Lighting
Glass doped in neodymium	$1.06\mu m$	Pulsed	from 0.5 to 5 ms.	from 1 to 30 J	10 pulses per minute	Spot welding Calibration of gyroscopes Drilling Laser spectrography
Carbon Dioxide	$10.06\mu m$	Continuous		20 to 100 W		Medical (E.N.T. for example)
				from 10 to 10,000 W		Cutting various materials, Cutting compound materials, Welding metals, Welding plastics, Surface treatments.
Argon	514 nm	Continuous		5 to 25 W		Medical (photocoagulation) Raman spectrography Lighting effects for shows

Table prepared by Claude Barthélémy, engineer at the Marcoussis Laboratories (July 1980).

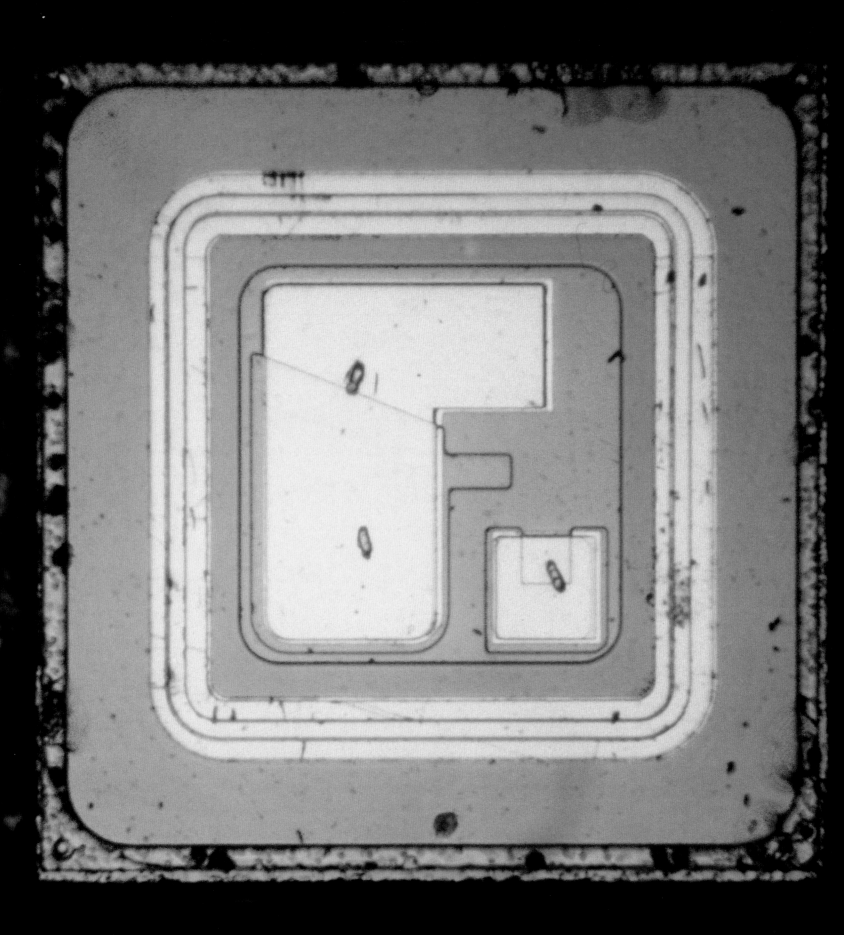

10
Microelectronics: toward maximum integration

Many technicians and companies did not at first realize (or else were unwilling to admit) the true significance of the invention of the transistor and the mastery of semiconductor techniques. Today these techniques are making it possible to integrate an increasingly large number of circuits and functions on a single chip. This extraordinary market turnabout sounded the deathknell for a number of firms and created a great deal of confusion in others. However, it was also responsible for the appearance of a number of new companies. In 1950, just after the invention of the transistor and while tubes were still a major electronics component, 80 percent of American electronic tubes were produced by General Electric, RCA, and Sylvania. Fifteen years later, these three firms accounted for a mere 18 percent of the market, as opposed to companies who took full and immediate advantage of the new technological development (Texas Instruments covered 17 percent of the market) or companies whose whole activity was geared around it (Fairchild Semiconductor: 13 percent).

Although the semiconductor company that William Shockley founded in Palo Alto was the first such specialized company in Silicon Valley,* the real birthdate of the new industrial "Far West" was 1957, with the foundation of Fairchild Semiconductor. This event marked the final ascendancy of commercial imperatives over scientific imperatives. Since then, research orientations, which had of course been affected greatly by the behavior of the market, have become completely dependent on consumer requirements, in particular those of the giant industrial companies. These requirements may be narrowed down into three complementary imperatives: increasing transmission speed, decreasing size, and improving electrical yield (lower currents, less loss of current, less dissipation, increased sensitivity, lower noise levels, etc.).

From the first appearance of radio tubes, research had been going on into reducing tube size; tubes were fairly large in the early days and consumed far too much energy. In 1930, the Loewe Company had succeeded in housing three triode amplifier sections in a single tube only slightly larger than the ordinary tubes used at the time. The transistor had

already eliminated filament heating requirements, and tubes became increasingly relegated to the domains of power transmission and high-frequency circuits. Everything else was transistorized. The first imperative was the assembly of increasingly small components; this, in fact, was the beginning of microelectronics.

Resistances were achieved by etching or by evaporation of metals or alloys, and units were assembled in small modules. This was the beginning of printed circuits. Hybrid circuits made their appearance on the scene between printed circuits and integrated circuits. They first took the form of transistors and subminiature components and then developed into the common integrated circuit. Today micro-assembly of components is carried out using printed circuit methods. The goal is always the same: reliability and the smallest possible dimensions compatible with circuit heating characteristics. Another advantage is that smaller dimensions allow increased transmission speed and operation at extremely high frequencies.

Hybrid circuits are constructed by implanting passive components on an nonconductive substrate by means of vacuum evaporation techniques, and then adding the active components in the form of beads. The beads are linked together by very fine wires and passive components. They are inserted into the insulating substrate by use of various different techniques: chemical, pyrolysis, screen-printing, or diffusion. System miniaturization is a development of particular interest to the military: before the invention of the integrated circuit, every B-29 bomber, whose electronic equipment was among the most complex and sophisticated of its time, was filled with very bulky, space-consuming equipment consisting of thousands of vacuum tubes and tens of thousands of passive components.

The National Bureau of Standards was responsible for one of the first attempts to simplify production techniques. Globe Union Incorporation's central laboratories suggested a project whereby ceramic substrates would carry metallic interconnections and miniaturized vacuum tubes. After the war, NBS and Globe Central Laboratories continued their research. At the beginning of the 1950s, Robert Henry of

Integrated circuit: a silicon bead in which transistors, diodes and resistors have been diffused.

* The first company to specialize in semiconductor devices was probably the Transitron Co. once the second largest producer of alloy junction Transistors.

Microelectronics: Toward Maximum Integration

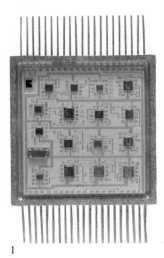

1. This printed circuit contains 16 integrated circuits implanted on a ceramic multiboard circuit.
2. Molecular-beam epitaxy (MBE), a new technique for growing crystals in an ultra-high vacuum, is used to obtain thin layers and multi-layers whose dimensions and composition can be very accurately controlled.

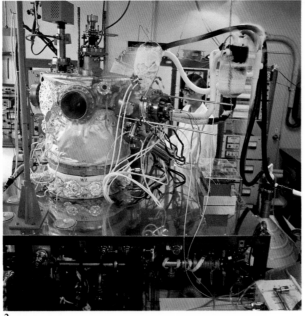

1

2

NBS designed the "Tinkertoy" project; in 1949 Danko and Abrahamson announced their "Auto-Assembly" projects for the Signal Corps, which invested $5 million over three years in tube assembly. However, projects conceived during this period were limited by their use of tubes.

With the advent of the transistor, "Tinkertoy" was replaced by RCA's "micro-module" developed in 1960 for the Signal Corps. Each component was fixed on a ceramic plate, a process developed by J.W. Lathrop and James Nall, and one of the first applications of photolithography in electronics. Over a six-year period, some $26 million was chanelled into this research; it was, nevertheless, a failure; the most decisive setback occurred in 1959 when Texas Instruments patented the integrated circuit.

Integrated Circuits

In the race for miniaturization, reliability, and higher transmission speeds, advances in technology developed for junction transistors contributed to a new breakthrough: the integrated circuit. The first of these discoveries was the development at Bell labs of the Mesa transistor for which different methods were developed to allow a large number of transistors to be made on the same silicon wafer. A protective silicon oxide coating could be selectively removed by hydrofluoric acid, impurities diffused from gas or vapor, and components located using a photolithographic process. Without these very important bases, none of the subsequent discoveries would have been possible. Also important were the invention of Jean Hoerni's planar transistor at Fairchild Semiconductor in 1958, which made it possible to make a transistor on a flat surface and methods developed for epitaxial growth.

In June 1960, Bell developed the epitaxial method for making thin layers of single crystals. Gaseous epitaxy occurs in the following way: a semiconductor, such as a silicon semiconductor, is heated in a hydrogen atmosphere to a temperature below melting point. It is then placed in silicon tetrachloride to cool. When silicon tetrachloride decomposes, it leaves on the surface of the silicon a layer known as the "epitaxial" layer composed of silicon atoms in perfect crystalline order; because of its purity this layer is highly resistive. The components are thus integrated without any contact with the substrate. This method made it possible to construct a transistor in the epitaxial layer with characteristics different from those of the substrate material. During the same year, Hoerni, who explored the

methods of passivation on the transistor surface had the idea of diffusing the base layer inside the collector instead of assembling it on the collector substrate, which meant that he could obtain interconnections by evaporation and did not have to use wires attached by hand. His Planar transistor allowed him to amplify very high frequencies and obtain high power levels. In addition, as its name indicates, it was flat in shape, which favored miniaturization and mass production. Components are thus deposited without direct contact with the substrate.

A Planar transistor is constructed by taking a small silicon chip and covering it with an insulating layer of silicon dioxide. Windows are then formed in the oxide layer by a chemical process. P-type impurities (boron, for example) are then injected into the epitaxial layer through the windows. This process is used to construct the *base* of the Planar transistor. To construct the emitter, the whole unit is covered with a new insulating layer of quartz in which an even smaller window is made. N-type impurities (such as phosphorus) are then injected through this second, smaller window. Finally, to construct the *collector,* another insulating layer of quartz is added and pierced with two holes, one going to the emitter, the other to the base. The pin connections are made by evaporation of aluminum or gold. Finally, a conductive layer is blanketed over the lower surface of the collector.

Various different laboratories had demonstrated the possibility of constructing resistances with silicon and junction capacities with silicon oxide. Jack Kilby at Texas Instruments succeeded in gathering these properties together in one piece of silicon in order to make an "integrated circuit."

One of his colleagues, Harvey Cragon, recalls: *He had just been hired and was working in the semiconductor division. He didn't go on holiday that first summer: in two weeks he thought up the idea of integrated circuits all by himself in his corner.*

That is not completely accurate, of course. One of the first people to seriously envisage what the future integrated circuit might be was G. W. A. Dummer at the Royal Radar Establishment in England. In 1952 he wrote an article explaining that through development of the transistor and work on semiconductors, it would henceforth be possible to envisage electronic equipment designed in a solid block without connecting wires. This block, he said, could consist of layers of insulating materials, conductors, rectifiers, and amplifiers, while the electrical functions would be connected directly by cutting out zones in the various layers. This was the first time that the idea had been so clearly worked out, but no one was yet capable of actually applying it. The Royal Radar Establishment was working on this problem with the English company Plessey, and in 1957 presented a model fairly close to Kilby's. However, the English were at that time far less interested in miniaturization than were the Americans.

In America, Kilby believed that miniaturization could be achieved solely by the use of semiconductors and that resistances and capacitors could be designed into the same material as the active components. At the same time he realized that if all the components were made from the same material, they could be produced *in situ* and connected together in the form of a complete circuit. He very quickly set about designing a prototype and in October 1958 succeeded in building an integrated circuit with germanium mesa transistors, resistors, and capacitors on the same wafer.

He took out a patent in February 1959, but it was immediately contested, since a few months after Kilby's invention Robert Noyce showed that the components could be much more easily connected if the planar diffusion process was used instead: Noyce's method led directly to commercial production of integrated circuits. It was in fact Noyce who united and developed all the processes used today to manufacture integrated circuits: masks and photolithography (diffusion), batch processing, passivation of oxide on the surface of silicon (discovered at Bell but made possible by invention of the Planar) and metal evaporation over an oxide to make interconnections and resistors, which was Noyce's own original idea.

To construct an MOS (metal-oxide semiconductor) integrated circuit, a silicon chip is oxidized; then it is coated with a photosensitive lacquer over which is placed a mask representing the circuit to be etched. When it is in position, the lacquer surfaces not covered by the opaque parts of the mask are polymerized. The lacquer is then dissolved in the absence of light and the oxide is attached to form the "windows," thus exposing the "p" silicon. The impurities which dope the drain and the source are then dispersed. The remaining oxide is replaced by a uniform layer of oxide that undergoes the same photoengraving process and is only retained at the ends of the transistor. Another thinner oxide deposit, followed by a third photoengraving operation, acts to insulate the control electrode. Aluminum is then applied to form the connection and the control electrode, the gate.

The initial silicon slice, which has undergone the processing described above, is now covered with thousands of tiny printed circuits, each a few millimeters square, that are cut out into individual integrated circuits after the assembly process. To create the resistances, a given number of impurities are introduced into the circuit either by alloying or dispersion. Alternatively, the semiconductor can be coated with a very fine insulating layer (quartz, for example) and a resistant substance.

Integrated circuits were developed in response to immediate technical requirements and in order to mitigate the crisis at the beginning of the 1960s in the transistor market (prices were falling because of the large number of manufacturers) rather than from theoretical scientific research. Technical research then began to concentrate on obtaining the maximum integration within the minimum space. In 1971 Texas Instruments presented the first calculator chip. In November of the same year, Intel developed the world's first microprocessor, the 4004 (PMOS). In April 1974 Intel developed the first NMOS and in September of the same year, the first bipolar microcomputer. We should point out here that the microprocessor is just one element of a microcomputer, although it is often confused with it.

The microprocessor is the central processing unit (arithmetical and logic) in a microcomputer and controls other ancillary circuits. It is a chip that integrates several groups of elementary functions and acts as the "brain" of the data processing system. A microcomputer is constructed by bringing together some hundred or so of these microprocessor chips to form synchronization functions, memories (programs or data banks), input and output signal interface systems, etc. Microprocessors range from the chip designed for one precise function to multiprocessor systems as used in automation, with intermediary modules, small calculating systems, or development systems.

Since the invention of the integrated circuit, the cost per function has gone down by a factor of 10 every five years, as has the surface area of silicon occupied, while the number of components on a chip

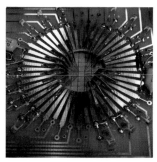

3

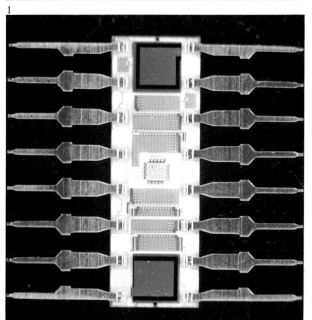

2

1. Oscillator integrated in gallium arsenide.
2. Microprocessor.
3. A test system capable of checking in a fraction of a second the on and off load characteristics of many thousands of different functions built into a chip.

have been increasing by a factor of 30 every five years.

LSI - VLSI

Integrated circuits were first of all developed and produced for logic, that is digital uses, mainly in calculators. Several observations were then made: for example, it was possible to increase the complexity of a basic circuit; several circuits could be interconnected by using suitable metalization patterns, and finally production figures were favorably affected since less silicon was used. In 1970 scientists invented a new category of transistor, the MOS (metal-oxide semiconductor), which was smaller and theoretically easier to make. The race towards density, speed, and lower consumption continued headlong. Technicians talked of degrees of integration* and distinguished between:

– SSI (small scale integration): approximately 10 gates (between 300 and 800 transistors) in which construction of a logic unit still necessitates external wiring.
– MSI (medium scale integration): approximately 100 gates (between 300 and 800 transistors) which allows use of a large number of TTL circuits.
– LSI (large scale integration): approximately 1,000 gates (between 3,000 and 8,000 transistors) which uses MOS rather than TTL circuits.

* The integration level is defined by the number of basic units per surface unit.

– VLSI (very large scale integration) is the latest development. The number of gates may exceed 100,000. In the thirty years between the first transistor and VLSI, the surface area has decreased 10,000 times.

In 1969, Harvey Cragon began work on the first LSI computer destined for work in geophysics. *Right up to integrated circuits,* says Cragon, *the military were behind most electronics discoveries. But the marketing of integrated circuits has pushed progress even further in both public and military spheres. Now a kind of reverse phenomenon is occurring: commercial systems are beginning to be applied to military systems. The intervention of the civilian world is basically due to telecommunications and civilian data processing.*

The larger Japanese, American, and European firms are developing new design and production methods. The Americans, for example, are developing a parallel program to VLSI, VHSI, very high speed integration. VHSI uses gallium arsenide, a new material that has already been discussed in the section on semiconductors. The silicon integrated circuit family tree is made up of two great "families" : the bipolars, whose operation is based essentially on currents circulating in the junctions between materials of different characteristics and in which both electrons and holes contribute *simultaneously* to electrical conduction; and MOS (metal-oxide semiconductors) which operate on the basis of the flow of a single type of electrical charge (electrons *or* holes) in a semiconductor material. The electrical charges run in the "channel" between source and drain controlled by the voltage applied to the gate, which acts on the channel via a thin layer of silicon oxide. Historically, the first integrated circuits made were bipolar. The first generation (Resistor Transistor Logic, Diode Transistor Logic) have almost entirely disappeared in favor of TTL (Transistor Transistor Logic) which incorporates a *multiple emitter transistor,* a transistor with two or three emitters sharing a common base and collector. This multiple emitter transistor controls a switching transistor which in its turn controls a group of three output transistors.

Two variations of the TTL are the ECL (Emitter Coupled Logic) and the I²L (Integrated Injection Logic).* Within the TTL family, the most commonly used type is the Schottky, where a metal-to-semiconductor junction is used to replace the p-n junction. A different distribution of free electrons in the semiconductors and metal produces a carrier-free zone in the semiconductors. The theory behind this effect was established by Walter Schottky in 1939, and this diode can be used at very high frequencies. Schottky diodes are used in conjunction with transistors to eliminate current saturation problems through a transistor.

The ECL, put on the market in 1976, has made it possible to produce the Motorola 10,800 microprocessor, which is the fastest integrated circuit ever built. The I²L, developed in 1975 by Philips, and baptized MTL (Merged Transistor Logic) by IBM, is competitive in terms of energy consumption but not nearly as fast. Bipolar integrated circuits are extremely competitive in terms both of speed and consumption, but they are far more difficult to make and integrate, with the possible exception of the I²L.

MOS transistors allow extensive integration. Depending on the doping of the semiconductor material making up the channel, technicians talk about two different types: pMOS and nMOS (nMOS type are preferred for microprocessors and semiconductor memories), that will allow much more dynamic operation than either TTL or ECL

and can memorize data over short periods. Other types of MOS circuits also exist: CMOS (Complementary MOS), which combine the properties of pMOS and nMOS circuits; the HMOS, which is the high-performance version of the nMOS; MOS/SOS or MOS, in which the substrate is an insulate and not silicon; and VMOS or Vertical Engraved MOS, which is used for extra high frequencies or power systems.

Charge transfer circuits – an extension of MOS enrichment transistors (CTD or Charge Transfer Devices) – are available in two alternative models: CCD (Charge Coupled Devices) and BBD (Bucket Brigade Devices); they enable a very high integration density to be obtained at very low prices.

CCD's were discovered in 1969 by two Bell Laboratory scientists, Williard S. Boyle and George E. Smith. The coupled charge signifies the accurate control of groups or "packets" of electrical charges in silicon. Boyle and Smith believed that the principle of bubble memories could be applied to semiconductors to create a new type of electronic system with both the inherent potential of bubble memories and the economic profile of integrated circuits. A CCD is a three-layer semiconductor device, one layer of metallic electrodes, another of silicon crystal, and separated by an insulating silicon dioxide layer. CCD's collect and transfer the information in the form of "packets" of electrical charges similar to the magnetic fields of bubble systems. Their greatest application has been in image detectors, such as the cigarette pack-sized heart of a solid-state television camera.

In 1971 Bell produced the world's first black-and-white CCD camera; in 1972 the first CCD color camera. These cameras went on the market in 1975 and are used by the Army for aerial reconnaissance and by astronomists for high-powered telescopes. They can produce very detailed images of distant planets. Their second major application is, as we have seen, for memories. A third CCD application is in the field of signal handling in communication systems. BBD's were in fact specifically designed by F. L. J. Sangster of the Philips Laboratories for this purpose.

We should also mention some other important developments:
– Tunnel diode conduction is assured by a quantum tunnel effect. If a potential barrier is sufficiently narrow, the electrons can penetrate it directly by the tunnel effect instead of having to cross it by increasing their potential.**
– Avalanche Transit Time Diodes (ATTD). This is also called an IMPATT diode and was discovered in 1964 by two Bell scientists, R. L. Johnston and B. C. de Loach. If the bias voltage at the p-n junction is high enough, conduction avalanche will take place in the high field region. The current rises to a high maximum level because of the ionization impact. By establishing a correlation between the passage time and the resonating frequency, it is possible to generate a few tens of watts in a pulsed power configuration.
– Trappat Diodes (Trapped Plasma Avalanche Triggered Transit). These are IMPATT diodes with a very high intensity electrical field.
– Gunn effect systems (discovered in 1963 by J.B. Gunn). If the voltage is increased, the electron speed increases also. When the field is raised beyond a certain threshold level, an increasing number of electrons are able to transfer to a higher conduction band which becomes an extremely populated "high field domain." This "domain" of "hot electrons" then moves through the semiconductor toward the

* The I²L were discovered at the same time (1972) by IBM, Germany and Philips, Holland.

** The quantum mechanical tunnel effect, which a number of different scientists had inferred as early as 1929 and for tunneling between valence and conduction bands in a semiconductor in an electric field by Clarence Zener in 1933, was discovered to occur in p-n junctions biased to large voltages in the reverse direction in 1958 by Leo Esaki (Sony and later IBM). Esaki received the Nobel Prize for his work. The term "tunnel effect" was chosen because it is linked to a quantum mechanics probability whereby electrons because of their wave property are able to pass through an energy barrier even though they do not have sufficient energy to go over the top of the barrier by classical motion.

positive end of the crystal. When the electrons reach the positive end the process starts again, and another "domain" is formed at the negative end. This in turn increases in size and propagates, whereupon the cycle begins again.

– Varactors. These are used as variable capacity devices and for frequency multiplication, whereas PIN diodes are used for switching.

Electronics experts are beginning to talk about submicronics systems in which the size of components will be smaller than one micron. As technicians continue working toward integrating more than a million transistors on a single chip, they are coming up against an increasing number of technical difficulties, such as the impossibility of verifying computer analyses of component characteristics whose electrical properties have been worked out according to a mathematical model. In fact, with such tiny dimensions, physical phenomena that are mostly negligible in physically larger systems have created entirely new problems. Calculations for these high density chips must be based on an amalgamation of probability and mathematical formulas that together provide more of a real idea of circuit performance. However, this work is still at the theoretical stage.

Ionic implantation methods of doping, allowing an even greater degree of accuracy, were developed at the end of the seventies, as were techniques for improving the insulation of bipolar transistors by means of oxide "wells" (Polyplanar, Isoplanar, V-Ate, etc.) together with different relief distribution procedures across the surface of transistors (LOCOS LOCMOS, PLANOX, etc.) using silicon oxidization techniques. The scale on which systems are now being integrated requires an accuracy of approximately 0.25 microns, which corresponds more or less to the wavelength of ultraviolet rays. But technicians are now trying to improve circuit integration even further by using electronic and x-ray beams whose wavelengths do not exceed 0.001 microns. From now on, computers will be used to produce the overall design of these mathematical models; to trace the photomasks that implant the microstructures on the silicon; to supervise the production process; and to carry out quality control tests on the chips, rejecting any that are found to be substandard. One of the crucial points for binary or pseudobinary semiconductors in particular is the depositing of the thin layers of materials on the substrate. Three essential epitaxial methods are used: liquid, vapor, and molecular spray (still at the research stage, but giving some promising laboratory results in the production of diodes, field effect transistors, and semiconductor lasers.)

Results obtained with liquid epitaxy techniques for exemple in semiconductor lasers are extremely good; they give layers whose optical and electrical qualities are excellent, with thicknesses ranging from a few tenths of a micron to a few microns. However, precautions must be taken to prevent the accumulation of impurities, particularly residual gases such as oxygen in the crucible and the oven. Gaseous epitaxy techniques are used where large substrates are required. Molecular spray epitaxy is particularly interesting in integration applications for the laying of very thin layers; it consists of separately evaporating the elements under ultravacuum conditions. This guarantees purity and enables the required layer composition to be obtained. Parameter control is by mass spectrometry techniques; however this method is still at the research stage.

The production of integrated circuits is now being oriented toward highlevel automization with the accent on machines, control software, and main- tenance. Production problems are still numerous: there is a great deal of waste after diffusion, oxidization, and cleaning processes as it is very difficult to detect and rectify system abnormalities earlier while the process is actually being carried out. At a later stage it is too late to correct these faults; and defective wafers must be rejected. Competition is fierce, and it is difficult to amortize production equipment; there is a tendency to discard still-new machinery for machines with better performance records. Finally, it is difficult to foresee and control the rate of growth and the changing tastes of consumers, as was shown for example by the extraordinary market performance of small pocket calculators. Thus user requirements and short-term (even medium-term) perspectives must be explored to provide control guidelines for a market whose inherent difficulties and pitfalls are only too evident. This means that parallel with the development of increasingly delicate techniques, the industry must also learn to anticipate the kinds of equipment the market will demand in the future.

The properties of some materials not yet in common use are now the subject of laboratory research; this group of new materials includes supernetworks made up of very thin layers (a few tens of Angstroms) alternating with two different materials with different forbidden energy gaps. However, there are difficulties involved in integrating the different components, and this technique is still far from being competitive with microelectronic techniques. In Bell Laboratories, at IBM in the United States, and also in Japan, technicians are working toward even larger-scale integration and increased miniaturization, as well as on improving the operation speed of integrated circuits.

VLSI:
The next stage of microelectronics

by Prof. Dr. Ing. Claus Reuber, Member of the Physics Department of the Berlin Technical University, Research Scientist at Fritz-Haber-Institute of MPG.

From the Planar transistors introduced in 1959 to the very high integration densities achieved today, all forecasts have been overtaken by reality. Where will miniaturization end? What is meant by SSI, MSI, LSI, VLSI and even ULSI? What materials and techniques will be harnessed to create an integration level of one million transistors per chip?

The Beginnings

The foundation of all modern integrated circuits was the introduction of Planar technology for transistors in 1959. Expectations expressed prior to 1960 of being able to combine several or even many transistors on one chip were then considered by many skeptical fabrication specialists as pure utopia. They did not believe that it could be possible to produce integrated circuits at all.

A transistor fabrication yield of 80 percent would, in the case of an IC with ten transistors, give a yield of only 10 percent. Anybody wanting to integrate thirty transistors would, with a 90 percent individual yield wind up with only 4 percent for the IC. For 300 transistors on one chip it would take a yield as high as 98 percent for the individual transistor, in order to obtain the ridiculously small and utterly unacceptable yield of 0.2 percent for the integrated circuit.

However, the skeptics had to learn differently and at a technical meeting in 1968, F. G. B. Casimir was able to declare: *If fabrication defects of the individual elements belonging to an integrated circuit were statistically independent, integrated circuits would hardly be possible. However, experience has shown that such an independence does not exist and that circuits with several dozen elements can be produced with good yield.*

Depending on the complexity of the integrated circuit, yields of 25 to 90 percent today are quite common. In the case of complicated new IC's, larger runs of serial production have been started even with yields as low as 10 percent, but then the 25 percent should be reached rather quickly.

Yet at that time Casimir continued by saying that in the case of large-scale integration, it would be somewhat doubtful how far one could go without redundance and subsequent selection of elements. He added that a miniaturization by a factor of 10 would probably be the extreme and, for the time being, still unattainable limit. And the optimists told us in the second half of the 1960s that in the not too distant future it would be cheaper to produce IC's with fifty or more standard single components than one transistor, and later on every household would have "a little calculator of its own to check the grocery bills."

From SSI to ULSI

All these expectations have been greatly exceeded by reality. Semiconductor electronics has progressed from small scale integration (SSI) via medium-scale integration (MSI) to large scale integration (LSI), and is at the beginning of very large-scale integration (VLSI). Some already speak of ultralarge-scale integration (ULSI).

The technological boundary lines in this case are not very sharp; they are generally defined according to the number of gates per chip. The SSI of the early period went as far as three to thirty gates, the MSI of the second half of the '60s, for example, went as far as 300. With this, memory units with 256 bits could be realized by MOS technology. The first half of the '70s brought LSI circuits for RAM's with capacities of from 1 to 16 Kbit and 3,000 gates on the chip. Still larger capabilities are shown by the VLSI circuits such as have become common since the end of the '70s in the form of the 64-Kbit-MOSRAMs. Today, those who think of ULSI put the upper limit for the VLSI at or above 30,000 gates and speak of ULSI in the order of a magnitude of more than 100,000 gates.

Thus far, the number of elements per chip has doubled each year; and the complexity of the circuits has, therefore, increased in one decade by a factor of 1,000. The future development will proceed more slowly, but doubling of the component density should, for the time being, not take more than three years. This then still adds up to a factor of at least 10 for the decade. At the same time, the cost per integrated transistor declined annually by 60 percent, that is, by a factor of 1,000 in approximately fifteen years.

1,000,000,000 Transistors on One Chip

The predictable physical limits of this development have not been reached by a long shot. Today, we are still far away, by a factor of about 500, from ten times the typical structural dimension of the crystal lattices. Hardly anybody dares prophesy how far miniaturization can actually be pushed. However, the rapidity of development is determined today by the capabilities of the development teams. There is a shortage of really good electronic engineers and physicists. In further developments in microelectronics, this technology must help itself with its own computers.

Although today the "one million transistors per chip" is not taken for granted, it is nevertheless within reach. After 1990, 10^7 transistors on one chip and in the next century, as many as 10^8 or even 10^9 transistors on a single chip may be well within the realm of possibility. The magnitude of these numbers is almost beyond comprehension. Today, the entire central unit of a large computer does not require more than the equivalent of 1 million transistors. Yet the human brain contains 10^{10} to 10^{12} nerve cells (neurons) with, perhaps, 1,000 synapses, to name an average figure. If, notwithstanding all the differences, we compare their effect with the elements of electronic logic, we arrive at a figure of 10^{14}. Even ULSI is left far behind by the human brain!

However, the miniaturization of the structures and of the elements in the integrated circuits does not stand alone, nor can it be realized as easily as these impressive numbers might suggest. Future structures will no longer be produced by means of conventional optical lithography. Electron beam, x-ray, and ion beam lithography are in the development stage and are in some cases already applied in practice. For smaller structures, the thicknesses of the insulating layers, the doping concentration, and the operating voltages must be matched so that the electric field strength and the current density do not rise to impermissible levels.

But why must integrated circuits remain essentially two-dimensional? An IC in the form of a cube instead of in the surface layer of a chip would offer possibilities still hardly dreamt of today. VMOS components are, in this respect, not even a first step, but they, at least, utilize the depth of the chip.

Important: the Functional Throughput Rate

The future of the VLSI revolves not merely around the question of more and more elements per chip, but also around the processing speed. A good indicator for this is the product of equivalent gates times the maximum pulse frequency. This product if often referred to as the functional throughput rate (FTR). Today, FTR values between 10^4 and 2×10^5 gate x MHz are common. Finer structures in the range of 0.5 to $0.8 \mu m$ should be able to increase this limit to 10^7 gate x MHz without exceeding the chip-load of 2 W, which is considered permissible.

Further improvements of the functional throughput rate are possible by using larger chips, more ingenious use of the individual elements, and finer structures. Presumably not very much can be gained by enlarging the chip, but it is particularly on this point that the predictions of the experts are very far apart. The same is true with regard to the utilization of the elements in the circuits, although a four-digit logic in place of today's binary logic may offer some advantages.

It may well be, therefore, that the refining of the structures from slightly less than $2 \mu m$ to $0.5 \mu m$ (and perhaps) less) by the end of the 1980s promises the largest gain. If the linear dimensions of a MOS structure are reduced to 1/k with k > 1, then the area is reduced to $1/k^2$. In order to keep the load per unit area constant, the voltage and the current must be proportionally reduced to 1/k. However, since with such a structure, the oxide-capacitance is at the same time also reduced to 1/k, the delay time is likewise reduced.

This means that according to this estimate, the FTR can be raised by a factor of k^3. This consideration shows, particularly, the possible gain in the case of the MOS structures, because there the voltage of 5 V, now used most commonly, may well be reduced to slightly below 1 V. Compared to that, the development prospects with regard to I^2L circuits are not as dramatic since here we are already operating at voltages that cannot be reduced further to any significant degree.

Gallium Arsenide for Later

Anybody who is thinking of very large-scale integration must at the same time also try to increase the processing velocity. For this purpose, gallium arsenide, with its distinctly higher charge-carrier mobility, is already waiting to follow the silicon age. In addition to that, the specific resistance of a gallium arsenide substrate is higher than in the case of silicon; this reduces the unavoidable capacitances.

Starting from a permissible chip load of 2 W, we can calculate the still acceptable dynamic switching energies for different degrees of integration and pulse frequencies. An SSI circuit, with ten gates per chip, and a pulse frequency of 1 MHz can operate with dynamic switching energies of the order of 10^5 pJ. In the case of a ULSI circuit with its approximate 100,000 gates per chip, a switching energy of only 10 pJ is permissible at this same pulse frequency. However, 1 MHz is much too slow. The step from 1 MHz to 1 GHz reduces the permissible switching energy by the same factor, that is, in the case of the SSI to 100 pJ and of the ULSI to 0.01 pJ or 10 fJ.* With the altogether desirable step to 10 GHz, the dynamic switching energy available in the case of the ULSI is then only 1 fJ.

Furthermore, the developments in the switching velocity have thus far not nearly been as dramatic as in miniaturization of the components. Gate circuits with 4 ns step delays were known as long ago as 1965, and bipolar ECL circuits with 1 ns were already reported in 1968. The future ultra-integration should, therefore, use above all GaAs as the basic material.

Here, MESFETS produced by means of electron-beam lithography present themselves; these are field effect transistors whose gates act by way of a Schottky contact. In this connection, there is also much talk today about VHSI or very high speed integration. Tests with such components have, at any rate, already led to switching times of between 30 and 300 ps. The corresponding dynamic switching energies would then amount to from 1.4 pJ down to 30 fJ. Switching times of 20 ps should be attainable.

With this, the GaAs-MESFET-logic then becomes approximately as fast as the logic with Josephson contacts.** The dynamic switching energy in the GaAs logic of this type is, by the way, smaller than that in the Josephson contact logic, if in the latter case we properly include in the calculation the energy expenditure for cooling with liquid helium. Only when this energy expenditure is disregarded does the Josephson logic rank first.

If we think of the cooling of computers, we also have to consider other III/V compounds besides GaAs. Thus, at 77 K it would be quite possible to think of logic circuits made of indium arsenide and indium antimonide with their still higher mobilities. However, for developments of this nature, nobody today would venture to name any date.

Lithography with Electron Beams

Lithography forms part of the integrated circuit technology. Optical lithography, which has been in common use, is limited in its power of resolution and, therefore, in the fineness of the structures by the wavelength of the light used. At and below $1 \mu m$, light-beam lithography must give way to electron beam, x-ray, or ion beam lithography. Work on electron beam lithography with acceleration voltages of between 10 and 20 kV has been going on for nearly fifteen years. Today it is used mainly for the preparation of extremely precise and fine masks.

However, with the electron beam the structures can be recorded directly on the wafer, a step further in fineness and precision with automatic adjustment by way of secondary electron emission. Electron beam lithography is, however, slower than lithography with visible light.

X-ray lithography has been under development since the beginning of the 1970s. Here, the shadowgram of a mask on the semiconductor surface with its resist is made use of. Future plans include the use of x-rays from synchrotrons with wavelengths in the range of 1 to 4 nm. With these methods, we have to consider that the resolution is no longer determined by the wavelength of radiation, but by the reach of the secondary electrons.

Ion beam lithography is a process that has been under discussion and development only for about the last five years.

Here, the chief problem is the ion optics required. Thoughts have been turning around lithography processes involving helium or argon ions accelerated by voltages between 30 and 150 kV. However, with each miniaturization step possible interference effects must be guarded against. Excessively narrow aluminum connections on the chips may be susceptible to electromigration and reduce the reliability.

VLSI and the Equipment

VLSI components are not produced for their own sake; they serve as parts of equipment and systems. How important microelectronics has become can be described by a single figure: today's worldwide output of semiconductor components is worth more than $10 billion.

The SSI components led to high-performance computers which W.T. Runge once called "utter idiots with an astonishing ability to do mental arithmetic." The MSI technology created the minicomputer. LSI circuits have resulted in the microcomputer. What true VLSI technology will bring requires some prophecy. What we can be sure of are faster microcomputers, for 16 and 32 bit words and maybe ten times faster than today's models, and also microcomputers with a genuine multiplication on the chip. Also the processing of digital audio signals will pay off. This will be urgently needed for the PCM audiotechnology of the mid-1980s, but also for a more perfect speech synthesis than today's. Digital processing of video signals and important functions in radar technology, as well as image processing at high resolution, also lie within the range of capabilities of VLSI logic.

Here, we should always bear in mind that in the next three to four years, the functional throughput rate will, with VLSI, grow by a factor of 20, which simply means that microelectronics will be improved by this same factor because it offers correspondingly more possibilities. However, in order to make use of these possibilities, it is necessary to create an adequate software to go with the hardware, and for this, an ever increasing number of good specialists will be needed.

VLSI logic by itself is nothing; it must be able to communicate with its environment. To this end, matching sensors and actuators for the controls are as necessary as the input organs and display systems for the interaction with the human being. All this is in the process of rapid development. In this case, silicon-based sensors have a particularly good chance because of their compatibility with semiconductor electronics.

The piezo-effect of silicon can be utilized for force and pressure sensors, its temperature sensitivity for the temperature, and for other process variables that manifest themselves in the form of temperature changes. Magnetic position sensors can be constructed with the use of Hall generators or similar components comprising either III/V materials or silicon. Sensors for moisture, flow, and motion which are compatible with microelectronics are known or in the process of development.

Thus, the automobile computer – today still a special feature of exclusive cars – may soon become commonplace. Aside from simple information such as time, distance traveled, velocity, tank contents and outside temperature, such a computer will be able to show the momentary and the average gasoline consumption in liters per 100 km, the actual distance to be covered with the existing tank contents, the average velocity, momentary distance from the destination and the anticipated time of arrival. It will, of course, warn the driver against exceeding a specified maximum velocity, against ice on the road; it will warn him, in time, that the fuel supply is nearing exhaustion. It will also, of course, include a burglar alarm system with a coded key for the ignition that, with the wrong input, will trigger an alarm.

VLSI Electronics for the "Day after Tomorrow"

Any discussion going beyond what microelectronics offers us today in new beginnings, and will perform routinely tomorrow, will lead to still other and quite different ideas. They may now still sound somewhat utopian, but the background from which they spring is real. Thus, the question may be raised whether the highly successful compact cassette and the Compact Disc PCM ready for 1983 will still be able to compete against microelectronics after the year 2000. Such a forecast is not too far-fetched.

Audio PCM is operated today with a scanning frequency of around 44 kHz and with a resolution of at most 16 bits. This results in a data flow of 704×10^3 bits/sec or 2.5 Gbits/hour. If we think of 10^9 transistors per chip after the year 2000, we cannot exclude the possibility that, by then, music from a semiconductor memory is feasible.

Or how about a camera completely without film? A semiconductor image-sensor coupled with a semiconductor memory for its digitalized signals. The CCD image sensor is already on the market. When the additional memory is full, the signals are processed at home completely without developing and fixing: perhaps rerecorded on the digital video recorder in the form of transparencies, or delivered for paper prints to the color printer then available for video texts and recently demonstrated in connection with projects for magnetic photography.

However, not everything that appears physically possible will also be realized in practice. It remains to be seen whether such extreme possibilities of microelectronics can be exploited economically. However, as a general rule, the realities of the future have always exceeded the beginnings and the prophecies of the given present.

* fj stands for femtojoule, or 10-15 J.

** The Josephson contact, a component used in-low temperature electronics, utilized the tunnel effect of the Cooper electron pairs typical for superconduction.

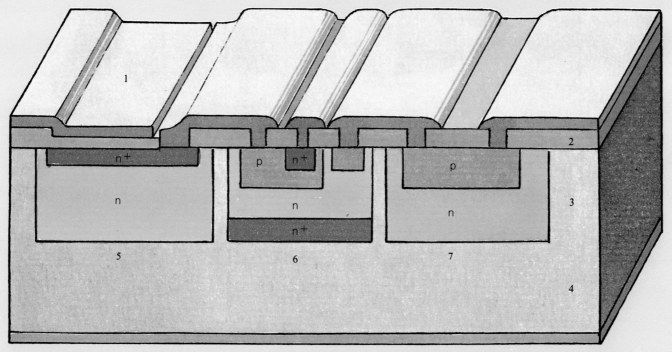

Circuit integration
1. Aluminum for the interconnections
2. Silica
3. Epitaxial layer between 10 and 20 μm thick
4. Initial monocrystal support (p), approximately
   100 μm thick
5. Capacitor
6. Transistor
7. Resistor

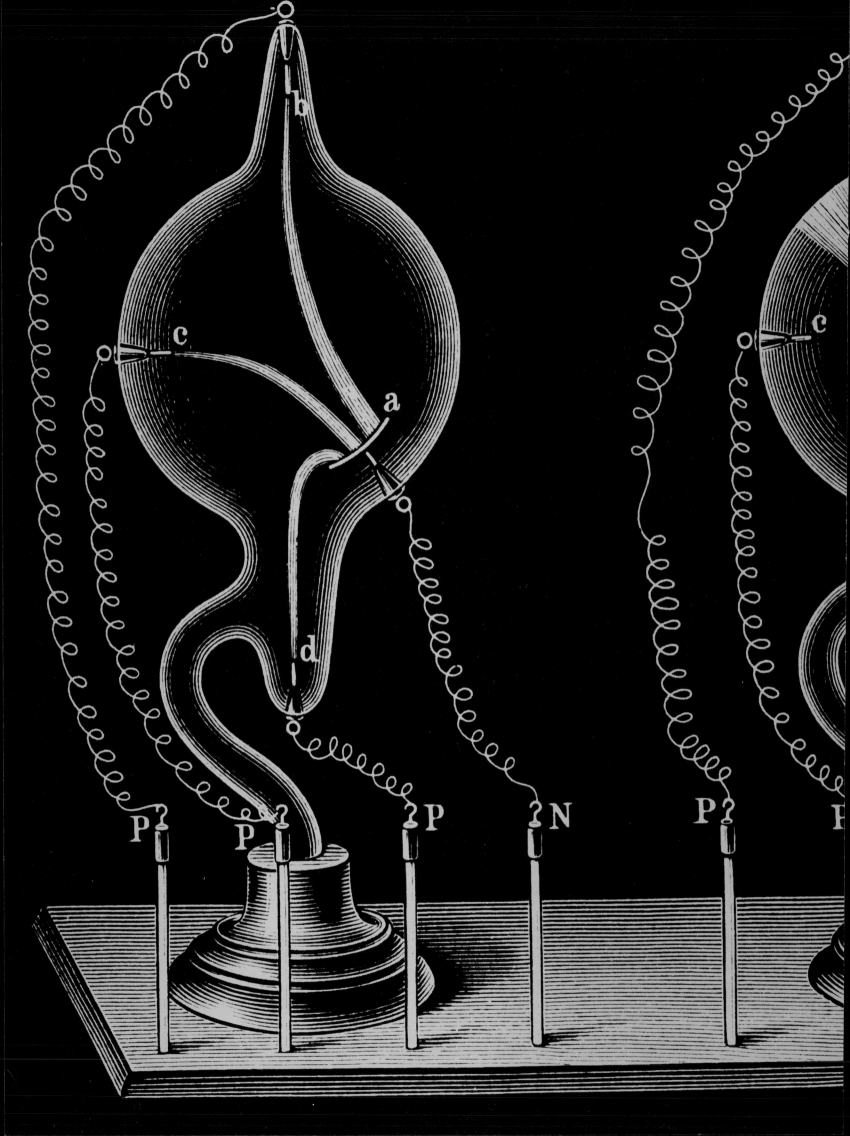

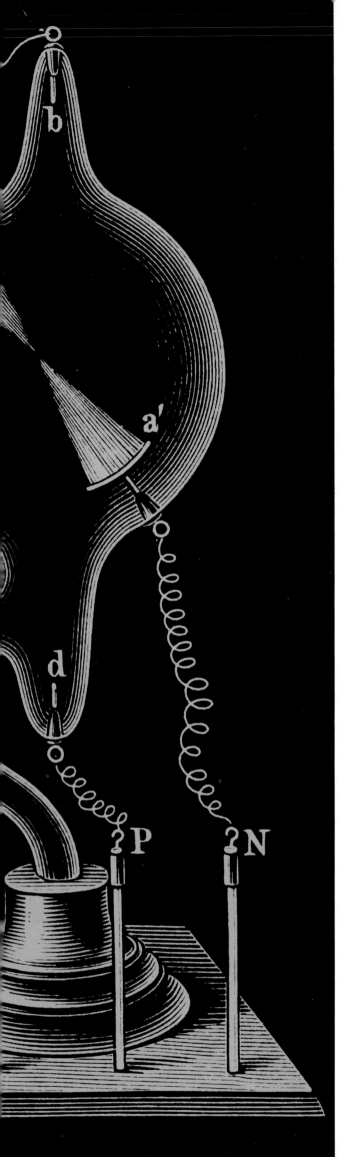

PART THREE

In the beginning there were tubes...

Once the arrow has left the bow, it pursues its course, and only a greater force can force it to deviate; but its original direction is determined solely by the person who aims it, and without an intelligent being capable of taking aim behind it the arrow would never fly. Therefore it seems to me that it is a useful thing to teach youth not to underestimate spiritual values.
Werner Heisenberg

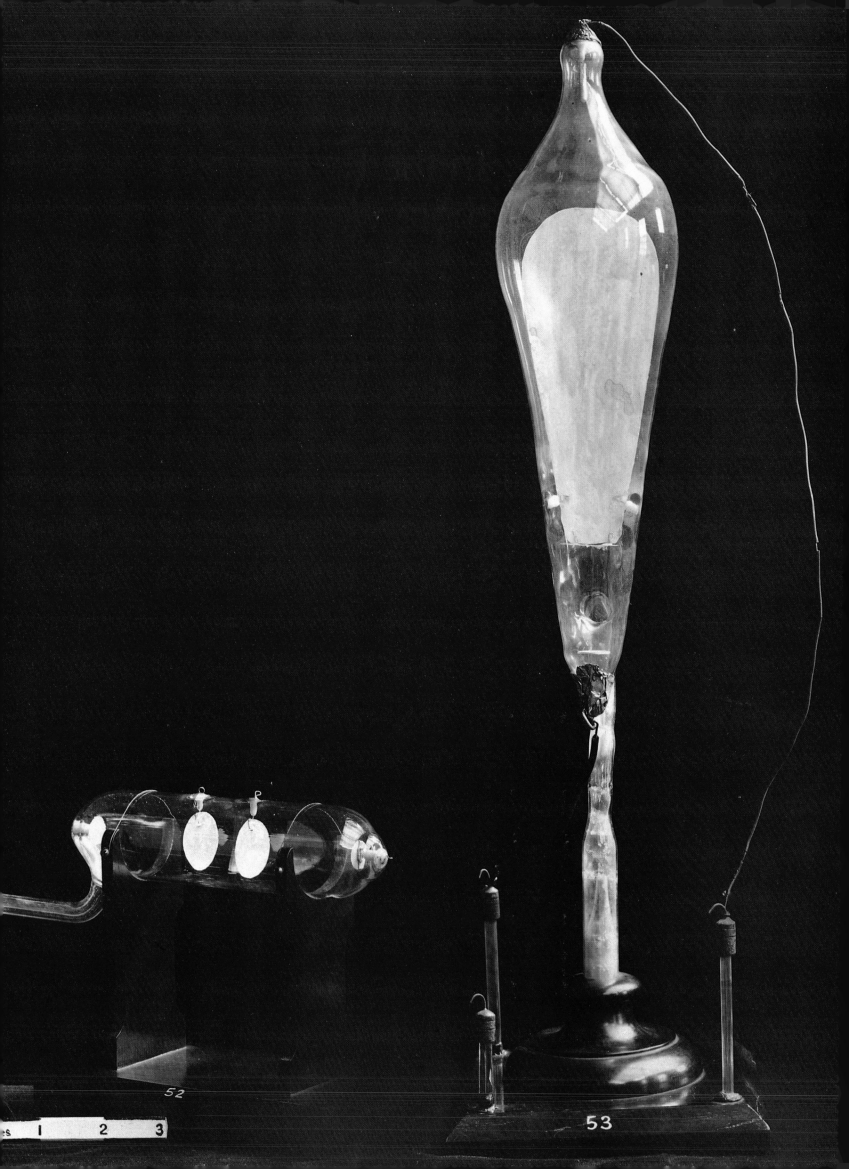

11

Tubes and cathode ray oscilloscopes: Crookes, Braun and Company

We began our description with integrated circuits because they represent the future of electronics and because they are a part of every application with which we are familiar. But most of the great electronic systems were first built on the development of tubes, of which the first and most important was the cathode ray tube.

In 1859 the German mathematician Julius Plücker studied rays emitted by the cathode of a vacuum tube. In his first experiments he used a heated negative electrode (the cathode) and studied the variations of light produced in the valve for different levels of air rarefaction; he noted a greenish fluorescence of the glass. He then suggested that the cathode must emit electrical rays, because if he interposed an object he could obtain its shadow. One of his students, Johann Hittorf, established that the rays could be made to deviate in the presence of a magnetic field. These experiments were the base for another discovery, that of the secondary emission of electrons where a large part of the energy was transformed into heat if it struck an obstacle. It is possible to extract new so-called secondary electrons if there is sufficient kinetic energy. This phenomenon was to be used very frequently in the new science of electronics. The English physicist William Crookes became interested in these experiments while investigating the light effects of electrical discharges in gaseous atmospheres. He reduced the quantity of gas to a few molecules, and also noted that when the light disappeared a greenish glow bathed the walls. In 1879 Crookes mounted a propeller in the tube to try to make the cathode rays revolve; he believed that these rays were molecules, and that he had discovered a fourth state of matter, the "radiant" state (the others being gaseous, liquid, and solid.) At the same time he developed the "Crookes tube," which emitted rays that Wilhelm Röntgen was later to call x-rays because he was unsure of their true nature. In the Crookes tube these cathode rays struck a high voltage anode and generated x-rays. Crookes never identified the true nature of the rays emitted by his equipment. An anecdote recounts that he was working on his tube on a table in the drawer of which another scientist had left some photographic plates. The x-rays clouded over the plates, but when the second scientist complained, Crookes merely replied, "You should have put them somewhere else," without bothering to enquire into the phenomenon. It is interesting to note that Becquerel discovered radioactivity in an equally accidental way. Nevertheless, Crookes tubes made Röntgen's discovery possible, and with the work done by J. J. Thomson, André Blondel, Jonathan Zenneck, and Ferdinand Braun, it provided a base for the rapidly-growing science of electronics.

It was a Frenchman, Blondel, who in 1893 first put forward the idea of the electromagnetic oscilloscope, or oscillograph, to study alternating currents. Three years later, the German Braun developed his cathode tube, later used as a basic element in television picture tubes. In Strasbourg, Zenneck joined Mathias Cantor as assistant to his former teacher Braun. Röntgen had just discovered x-rays and Braun had decided to experiment with Röntgen's tube, although his main interest lay in following up some of Crookes's work. On February 15, 1887, he presented a new tube, the cold cathode gaseous atmosphere low-pressure tube. It brought together a number of already-discovered elements: cathode rays and the illumination of a phosphorescent screen by cathode rays, the introduction of an obstacle into the path of the rays, and the electromagnetic beam displacement system. In Braun's tube, the cathode was located on the left of the tube and the cathode rays were directed through a small hole in a metal disc placed in the middle of the tube. Thus, a small straight beam of rays was projected onto a screen on the right of the flared tube where it formed a luminous point. Outside the tube and in line with the metal disc he placed an electromagnet. At first, however, it was only possible to displace the luminous point along one axis. In 1899 Zenneck added two rectangular coils to Braun's device and obtained simultaneous vertical and horizontal beam displacement.

The next step was taken by another German scientist, Arthur Wehnelt. In 1903 Wehnelt was studying the effects of applying a layer of metal oxide, such as barium or calcium, to a hot metal cathode. In the same year he published an article describing a method of obtaining large numbers of

The Crookes tube is used to study the effects of electrical discharge through gases under very low pressure.

J.J. Thomson enunciated his hypothesis of "electron particles" after studying the cathode rays emitted by Crookes tubes.

Tubes and Cathode Ray Oscilloscopes: Crookes, Braun and Company

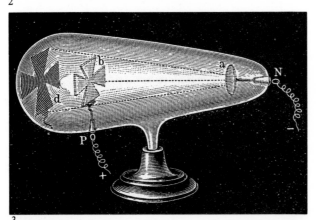

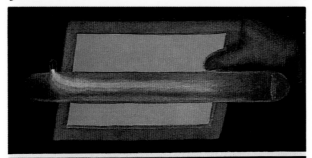

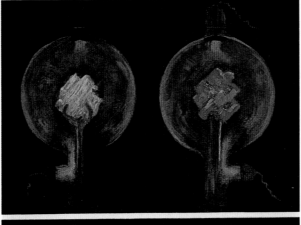

1. The house in which Ferdinand Braun was born in Fulda. Photograph taken in 1922.
2. German physicist Julius Plücker (1801-1868).
3. The famous Maltese Cross experiment with a Crookes tube.
4. "Electrical light phenomena" from a German engraving at the end of the 19th century: *1.* Illumination of a screen previously washed with barium platocyanide under the effect of cathode rays. *2.* Scintillation of different minerals, colorless in ordinary light, in Crookes tubes. *3.* These tubes, which were invented by German physicist Heinrich Geissler (1815-1879) are designed to study electrical discharges in rare gases.

"negative ions" by using incandescent metallic components, such as a platinum wire or strip covered with calcium or barium oxide. He found that electrons were emitted in large numbers when the cathode became red hot. In 1904 he took out his first patent for the AC to DC tube, which incorporated a hot high-emission cathode and a negative potential electrode. This "grid" was located near the cathode and controlled the density and fineness of the electron beam; it was a major development for the future of the oscilloscope. Wehnelt was also the first to develop an optical electronic device when he later added a focusing lens to accelerate the electron.

In France in 1917, Henri Abraham and Eugène Bloch developed the "multivibrator" oscillator which produced a very harmonic-rich 1,000 Hz signal. A signal of unknown frequency could be measured by comparing it with the harmonics generated by the multivibrator. In 1918 A. Dufour, another French scientist working at the Sorbonne University laboratory, succeeded in developing a method for the recording of electrical signals presented on a cold cathode oscilloscope by reflecting the electronic beam onto photographic film fixed on a drum that rotated inside the unit itself.

The oscillograph was perfected by the American Allen Du Mont and the German Manfred von Ardenne, and the first measuring oscillographs — the basic tools of any electronic laboratory — were largely invented and first sold in the 1930s. In 1939 William Hewlett and David Packard founded a company to develop oscilloscope technique for the purpose of studying fast electrical phenomena; it would prove to be of extreme importance in the fields of television and radar.

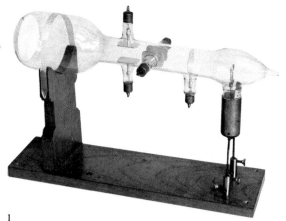

1

2

3

4

1. The hot cathode tube (1905) was first used to record electrical oscillations. It was invented by German physicist Arthur Wehnelt (1871-1944).

2. Sir William Crookes, English physicist and chemist (1832-1919), demonstrating his experiments to members of the Royal Academy.

3. Ferdinand Braun (center), in his laboratory.

4. Hewlett-Packard first began marketing electronic measuring devices or oscilloscopes in 1939. By 1981 Hewlett-Packard employed 60,000 people throughout the world, and David Packard (70 years) and William Hewlett (67 years) were both mentioned in Jacqueline Thomson's "Very Rich Book" with personal fortunes estimated at between 200 and 300 million dollars.

12

Marconi and the radio

The first application of the new science was the radio. At the end of the nineteenth century, scientists all over the world were fascinated by the idea of transmitting sound, thus abolishing frontiers and achieving "wireless" communication over long distances. One of the most famous pioneers of radio communication was Guglielmo Marconi, who was twenty-two years old when he took out his first patent, and not much older than that when he emigrated to England and founded the "Wireless Telegraph Company." He was twenty years old on the death of Hertz, the man who had first demonstrated the existence of the "waves" that would later be known as "radio waves." Marconi was already familiar with the work being done by most of his contemporaries. In 1890, Edouard Branly had brought his experiments with the metal filing "coherer or radioconductor tube" to a successful conclusion (this was the first time the term "radio" had been used). Using a Hertz oscillator, Branly demonstrated how signals could be received up to a range of thirty meters with an iron filing detector that became conductive when a spark was produced by the spark transmitter; this was the filing coherer.

Branly conducted further experiments into the antenna and its role in the transmission of these electromagnetic waves. He introduced rotatable coupling coils into both the conduction tube and spark emitter which resembled those used in the Hertz oscilloscope. The next step would have been wireless telegraphy, but Branly went no further; it did not occur to him that a variation of the same principles could be used for transmitting signals. During the same period, there was great public enthusiasm for the wireless and in the United States another character made his appearance on the scene. This was the romantic, appealing, and brilliant Nicolas Tesla, who began his experiments into long wave wireless telegraphy as early as 1890. Between 1897 and 1900, Tesla succeeded in transmitting signals over ranges up to 625 miles. J.P. Morgan sank enormous sums into Tesla's Colorado Springs Laboratory and his Wardensclyffe station, but the scientist died in solitude and poverty on the eve of World War II.

The young Marconi's ears really pricked up when Oliver Lodge gave a speech to the Royal Institute on June 1, 1894, on "the work of Hertz and some of his successors." Lodge had improved the coherer by coupling it with an automatic decoherer. His speech was published in *Nature* and *Engineering*. He went on to demonstrate the necessity of tuning the receiver to the transmitter and operating the stations on the same wavelength. In 1897 and 1898 Lodge took out three patents detailing the use of resonance, the tuning of the transmitter's antenna circuit and the receiver by means of an adjustable coil, and the inductive coupling of antenna to receiver. When Marconi wanted to take out a patent covering the same principle, the courts established in 1912 that Lodge had done it first and the Marconi Company had to buy back Lodge's patents. It took twelve years for the matter to be settled. This, as we shall see, was just one clash in the fierce battle of the patents that filled the courts at the beginning of this century. Lodge, like Branly, showed no interest in transmitting signals and left the invention of wireless telegraphy alone. He preferred to devote himself to another type of long-distance "wireless" transmission, spiritualism!

Marconi was to go further than all of his predecessors or rivals, further than his Russian rival, Alexandre Popov, or Lodge the Liverpool metaphysician; further than Tesla, the brilliant New York Croat and Branly the Paris Catholic Institute professor. He combined Branly's and Lodge's coherer, Popov's antenna and Tesla's inductive coupling winders and used them to transmit signals; it was in Marconi's laboratory that John Ambrose Fleming developed the first rectifier tube, the diode.

By 1895, all the pieces of the puzzle were there: the spark transmitter, the coherer, the decoherer, and the antenna. Marconi went to work on improving the systems known at the time and produced a prototype which, although rather crude, served convincingly in his later experiments to prove that electromagnetic radio waves did not move in straight lines but could be transmitted and received over hills and other obstacles. On June 2, 1896, Marconi, by then living in London, took out the first patent on his electrical pulse or signal transmission system and associated equipment that he had perfected. The rest is history; Marconi went to Italy in 1897, and

Marconi's first wireless telegraphy tests between France and England: a telegram is received at the Wimereux station near Boulogne-sur-Mer in 1899.

Marconi and the Radio

1

2

1. Thanks mostly to the coherer invented by French physicist Edouard Branly (1844-1940) wireless telegraphy became a possibility.
2. Guglielmo Marconi invited Benito Mussolini to visit his luxurious floating laboratory on the yacht Electra at the beginning of the 1930s.
3. Men of the RE Signals carrying short range wireless equipment (May 4, 1917).
4. Marconi (left) with Sir John Henniker Eaton, 1902, both wearing court dress for the Coronation of King Edward VII.

3

4

demonstrated his device to the King. He established radio contact between the La Spezzia dockyard and the battleship *San Martino*, a distance of 16 km. This was followed by his demonstrations in England at the Isle of Wight transmission center and also signal exchanges between the Dover lighthouse and the East Goodwin Sands lightship (20 km), and between Dover and Wimereux. The reporting of the America's Cup for the *New York Herald* and the great Atlantic transmission between Newfoundland and Cornwall in 1901 were Marconi's crowning glories.

Work was still proceeding on improving the detection of Hertz's waves. From 1890 a researcher at the Marconi Company, John Ambrose Fleming, later joined in 1901 by one of J. J. Thomson's students, O.W. Richardson, studied the "Edison effect," the main principle behind all electronic valves. What exactly is this Edison effect? When Thomas Alva Edison was working on his famous incandescent lamp, he noticed a secondary phenomemon: the bulbs often blackened and blew out prematurely. When he examined the phenomenon, he noted that a weak current traveled between the positive pole of the filament and a plate located inside the bulb. He noted this down in his logbook on February 13, 1880, and again on July 5, 1882. As the difference in dates indicates, he was really too busy to devote a lot of time to this secondary effect, but he did realize its importance. When questioned much later on the subject, he replied :

My theory was that the residual gases, coming in contact with the filament and part of the filament itself, became charged and were attracted by the glass and discharged themselves. As the polarity was unchanged, I thought this should give a constant current. The extra pole was put inside afterward to increase the current, as my first experiment was with only a piece of tin-foil pasted on the outside of the bulb. This gave a good deflection on the galvanometer. In fact the needle went off the scale. On putting wires and plates on the inside of the bulb the effect was greatly increased, so much so that at the Philadelphia Exposition I put a telegraph sounder in circuit and it worked well. As I was overworked at the time in connection with the introduction of my electric light system I did not have time to continue the experiment. (White: 1953-1955).

Edison also noticed the relatively quick change in the level of this electronic emission when the filament voltage was modified; he took out a patent on his observations in November 1883. By now he was much more interested in this discovery and demonstrated the system (using a special lamp with an electrode added to it) at the Philadelphia International Exhibition in 1884. Sir William Preece, who was visiting the exhibition, immediately recognized the full value of what he later baptised the "Edison effect." He took several of Edison's tubes to England with him for further analysis. At the same time, Professor Houston published the first article describing Edison's discovery. In 1903, after many experiments, Richardson succeeded in demonstrating that the level of electrical emission varied with the temperature and that the emission of electrons (he preferred the term "thermions" to that of electrons, which is why his demonstration took the name the *thermionic effect)* from a heated metal was analogous to the phenomenon of the evaporation of water. He also demonstrated that metals contained a large number of electrons that moved more or less freely inside the metal. The technological application of this discovery was not long in coming, for while Richardson was pursuing his research, John Fleming discovered the diode.

Fleming had studied with Maxwell two years before the older man's death and had worked at the Cavendish Laboratory. He was very familiar with the Edison effect since he had worked at the Edison Telephone Company and later at the Edison Electric Light Company. At this time radio waves were generated by means of damped high frequencies that were converted into audio frequencies by the receiver. The only device available to do this was the iron filings coherer — a difficult device to control. In November 1904, Fleming took out a patent on a signal detector based on a valve that let only the alternating high-frequency currents pass. In 1912 the physicist W.H. Eccles coined the word "diode" for Fleming's valve. The diode proved less efficient than the crystal detector of the same period, and like the triode could not be used commercially until after it had been perfected by Langmuir and Arnold. It would, at a later date, become an amplifier. However, it was the first tube to use the Edison effect and to open the way for a long series of electronic tubes. Everyone was keeping an eye on what was going on in England, and in particular, what Marconi was up to. In Germany, there were two opposing groups: the AEG group led by Adolf Slaby and his assistant Count Georg Arco, and the Siemens group headed by Ferdinand Braun. The German Emperor decided it would be better for the Germans to present a united front; and one fine day, in April 1903, fate helped him achieve this. While he was taking his daily exercise in the Berlin Zoo, he started a conversation with the rider of a black horse that crossed his path; the rider was none other than the director of the AEG. "Where are you going so fast, Mr. Rathenau? Are you trying to find shelter for the Wireless Telegraphy?" said the Emperor. "Why not, Your Majesty? replied Rathenau, "If Your Majesty so desires..." The Emperor replied, "Well, why don't you think a little about what you can take to be my formal wish. Competition in the field of wireless telegraphy in Germany weakens the German position with regard to the English Marconi company and gives it the opportunity to seize a monopoly. I don't approve of it because it is contrary to German interests." "I am grateful to Your Majesty for his suggestion. In an hour I will have the opportunity of explaining Your Majesty's point of view in the best possible place," replied Rathenau. "Does that mean you are going to take up your negotiations with Mr. Siemens about a merger again?" "Yes, Your Majesty," came the reply.

No sooner said than done: on May 15, 1903, the Telefunken Wireless Company was founded, amalgamating the two companies. The young couple's wedding gifts included the Slaby-Arco wireless system (Adolf Slaby, then professor at the Advanced Technical School in Charlottenburg, had succeeded in reaching a range of 21 km near Berlin with balloon-borne antennas in October 1903); the patent taken out by Braun in 1898 for inductive coupling of the transmitter to the antenna to decrease damping of oscillations; and of course Braun's famous cathode ray oscilloscope.

We should also mention the name of the Austrian scientist Robert von Lieben, who in 1905 was working on a mercury vapor tube to which he added a control grid in 1910. Siemens bought his patent and it was used by the German Navy as a radio-tube receiver at the beginning of World War II.

While von Lieben was working on the problem of telephone repeaters, he pursued his efforts towards developing a device that would be able to amplify signals without distorting the small variations of the incoming electrical current.

Between 1905 and 1910 he perfected his own amplifying tube. Von Lieben began by picking up where Wehnelt's experiments had left off. In 1906 he took out a patent on "cathode ray relays" which converted low-amplitude current variations at the input terminal into high-amplitude current variations at the output terminal, with no change being made to either the input signal's frequency or its wave form. He was not the only German scientist working on tubes during that period; in 1908 one of his most serious rivals, Otto von Baeyer at the University of Berlin, designed a three-electrode tube very similar to Lee De Forest's triode. On account of von Lieben's bad health, his assistant Eugene Reisz began to devote more of his time to the development of the tube and in 1910 the von Lieben-Reisz-Strauss tube with its control grid was patented. From now on it would be De Forest, whose work and discoveries were the most brilliant of his time, who would be chiefly responsible for improving upon the work done by his famous predecessors, despite competition from a number of brilliant rivals.

1. Von Lieben's tube and the patent which he took out.
2 and 3. Radio reception and transmission room at Sainte-Assise in the 1930's.
4. First airborne radio tests carried out by Captain Brenot.

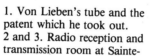

2

3

4

The diode: opening the door to electronics

by Pat Hawker,
Independent Broadcasting Authority

The first tube used to detect radio signals was John Ambrose Fleming's diode.
What were the exact circumstances surrounding its appearance?

The development and application of thermionic devices for the reception and generation of radio signals represented the birth of the electronics era. But the steps were not clear or sharply defined, and much controversy and patent litigation surrounded their development.

Even today the "invention" of the triode presents a curious paradox: for while there is no question of who was the first man to patent the concept of the triode as a means of amplifying weak electrical signals, there is considerable doubt whether, at the time, he appreciated either the true potential of the device, the way in which it worked, or how it could be used. Yet for many years the triode was the unique means of amplifying and generating electrical signals and superseded all earlier methods used in radio communication.

Early Work

The basis of the development of thermionic devices was the gradual recognition and investigation of the conduction of electricity in gases, and the interaction between heat, magnetism, electricity and atomic structure. Gilbert in 1600 noted that static charges on amber were enfeebled by heat; later that century Otto von Guericke invented the electrostatic generator "electrical machine" and the air pump (improved later by Robert Boyle). During the eighteenth century Charles du Fay and then Abbé Nollet studied the effects of the "leakage" of electrical charges from heated metals; in the United States, Franklin suggested "electric fire" as a common element in all substances. By the mid-nineteenth century Edmond Becquerel was launched on his highly significant investigation of electrical conductivity in gases, and developed for his experiments a form of thermionic diode. Becquerel concluded that a gas becomes a conductor of electricity at or above the temperature of red heat, but was unable to explain the phenoma that he observed experimentally. From Becquerel onwards, electrical conductivity of gases was investigated by a number of scientists, leading to the early development of a crude form of cathode-ray tube. However, the concept of a heated filament or cathode was not pursued, although Thomas A. Edison in 1880 observed the discoloration of the bulb of his incandescent lamps and utilized this effect to develop a voltage indicator ("electrical indicator" patent applied for November 1883).

The Fleming Diode

John Ambrose Fleming, in his work for the Edison Electric Light Company of London, became interested in the "Edison effect" from about 1882 onward. On March 27, 1896, in a paper presented to the Physical Society of London, he noted that even when the lamp filament was heated by an ac supply, the current flow to the separate electrode was unidirectional (dc); the device was thus a rectifier.

In 1899 Fleming became a technical adviser to the Marconi Wireless Telegraph Company that had been set up in London to exploit the use of Hertzian waves for communications. This work brought to his attention the requirement for improved techniques of "detecting" radio signals. Because he was suffering from progressive deafness, he wished in 1904 to use for his experiments a sensitive dc galvanometer rather than the then normal method of a coherer and bell.

To quote his own account:

I was pondering on the difficulties of the problem when my thoughts recurred to my experiments in connection with the Edison effect. 'Why not try the lamps?' I thought ... I went to a cabinet and brought out some lamps I had used in my previous investigations ... my assistant helped me construct an oscillatory circuit ... it was about five o'clock in the evening when the apparatus was completed ... to my delight I saw the needle of the galvanometer indicate a steady direct current ... we had in this peculiar kind of electric lamp a solution to the problem of rectifying high frequency wireless currents. The missing link in wireless was found – and it was an electric lamp. (Fleming: 1934).

He named his device an "oscillation valve" and the term "valve" remained in use for thermionic devices in Britain, in distinction to the alternative term "tube" adopted later in many other countries. A British patent application was filed in November 1904, and application was also made in Germany and the United States. Fleming, it will be noted, did not "invent" the device, but only its application to the detection (rectification) of radio signals. Nor were the detecting properties of the diode unique: coherers, magnetic detectors, electrolytic detectors, "crystal detectors" (an early utilization of semiconductors) – all preceded the "oscillation valve" and the Pederson Tikker and Tone Wheels followed after.

The Fleming diode, however, was a stable and efficient detector and was extensively used, although it was not capable of amplifying.

World War I transmitter.

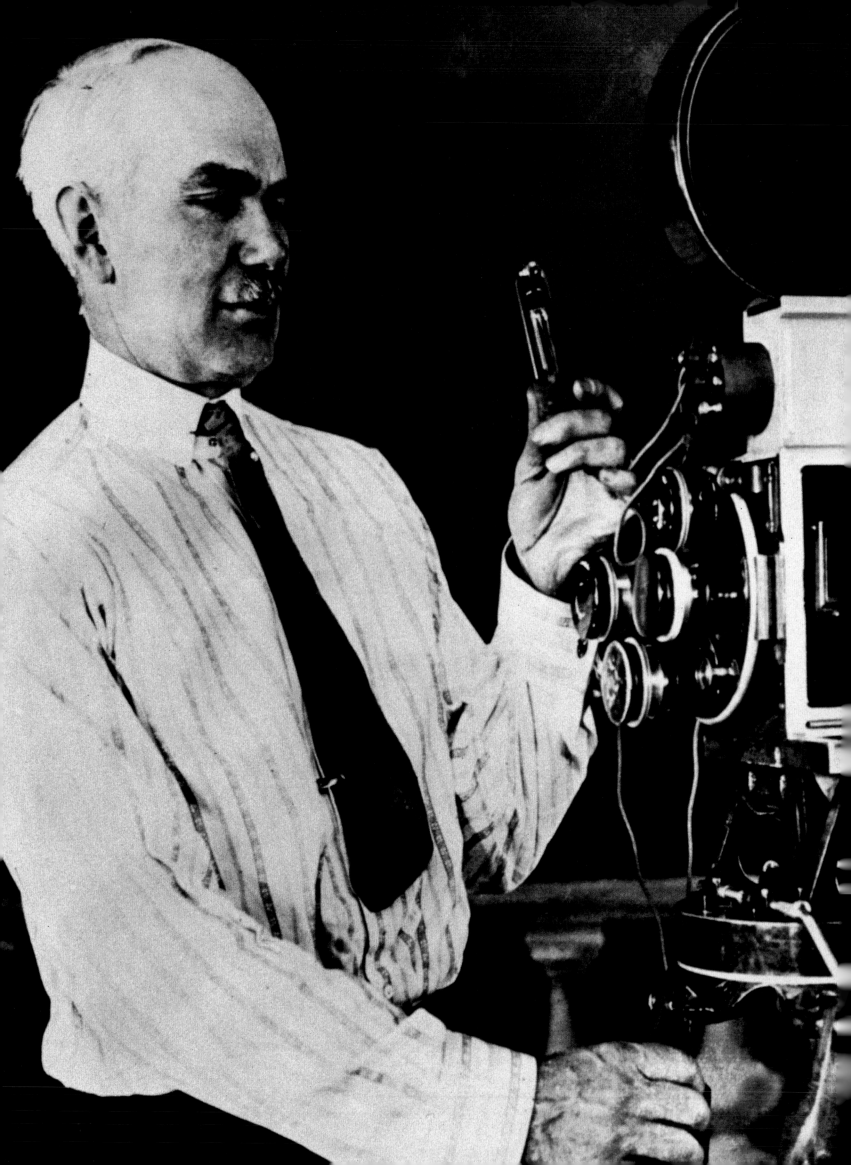

13

Lee De Forest's triode:
the golden age of tubes

In 1900, Lee De Forest was working as a telephone engineer for Western Electric in Chicago. He was passionately interested in wireless telegraphy and resigned from his job to devote his time to research. Borrowing an idea of Philip Lenard's – the addition of a control grid to study the movement of electrons liberated by the photoelectric effect – he invented the *audion*, which Eccles was to rename the *triode* in 1912. The addition of a grid to this tube made it possible to vary the intensity of the electrical current for a given operating frequency and enabled an increasingly large number of electrons to travel through the space located between the electrodes. Although Marconi and Fleming were both idolized in Europe – both men were eventually knighted – De Forest was practically ignored in his own country. He barely managed to get together the $15 necessary to patent his invention on December 31, 1906. In order to earn his living he took up another job, this time with Federal Telegraph in 1912.* During the same period he succeeded in establishing a radiotelephone link between two New York buildings using the audion valve, and in having the Navy install his devices on two of their ships. Encouraged by his success, he founded the Lee De Forest Telegraph Company, but the business folded fairly soon. This was not the end of De Forest's problems. In 1914 the Marconi Company attacked his patent, arguing that the triode was merely an improvement on the diode and thus fell within the scope of Fleming's patent. The courts betrayed their own understandable confusion when two years later they settled the matter by forbidding either side to manufacture the disputed devices! De Forest also lost the action brought against him by Armstrong over use of the audion as an oscillator. Nevertheless he did eventually manage to establish that he had been the first to invent the triode, although this took no less than three court decisions in 1924, 1928, and 1934. In January 1910, while waiting for the outcome of all these patent disputes, De Forest succeeded in transmitting Caruso's voice over a distance of some twelve miles, although he used the high-frequency arc generator designed by the Danish scientist Vladimir Poulsen rather than his own triode, then designed only for reception. In 1912, the hapless inventor had to face not only bank-

ruptcy but was also condemned by the courts for *abuse of trust on the basis of valueless patents, in particular a three-electrode lamp called an 'audion' which has been proved to be without any interest whatsoever!* Penniless once again, De Forest sold his patent for a mere pittance ($500,000) to the American Telephone and Telegraph Company for use as an amplifier in long distance telephone repeaters. In 1913 De Forest perfected the audion, which became the ultra-audion. In the meantime, however, Irving Langmuir at General Electric had begun work on De Forest's tube. Langmuir gave De Forest his due in the following words:

Lee De Forest, in discovering that an electric current in a vacuum tube can be controlled by means of an interposed grid, laid the foundation for an extension of man's senses and an increase in speed and in sensitivity of many millionfold. The revolution has been as great in its way as that which may now be envisioned in other fields through our new control of nuclear power. (Langmuir, 1950: 3).

In 1932 Langmuir was awarded the Nobel Prize for his chemical research into surface layers. He began to interest himself in Richardson's work and in the thermionic effect, which he wanted to develop further. In the December 1913 issue of the *Physical Review* he wrote:

It has generally been found that the saturation current is independent of the pressure of the gas and increases rapidly with increasing temperature of the filament. However, certain gases were found to have very marked effects; for example, traces of hydrogen were found to enormously increase the saturation current obtained from hot platinum. Recent investigations have shown that at least in some cases, the current is due to secondary chemical effects. Pring and Parker showed that the current obtained from incandescent carbon could be cut down to very small values by progressive purification of the carbon and improvement of the vacuum. The opinion seems to be gaining ground, especially in Germany, that the emission of electrons from incandescent solids is a secondary effect produced by chemical reactions, or at least is caused by the presence of gas.

Lee De Forest (1873-1961), "Father of Radio".

* The Federal Telegraph Company was founded by Cyril Elwell (its first name was the Poulsen Wireless Telegraph Company, as it was one of the first firms to market Poulsen arcs). Around this time De Forest discovered that the triodes he used for amplifiers could also act as oscillators. In 1925, the Federal Telegraph Company realized that vacuum tubes would soon replace arcs and began to intensify its experiments on the triode.

1. The audion or triode, one of a series of American stamps celebrating Progress in Electronics.
2. Extract from one of Lee de Forest's notebooks (August 28, 1912).
3. The first triodes: "these lamps which don't even light" as they were called at the time.
4. General Ferrié's radio receiver (1902).
5. The Eiffel Tower transmitting station in 1914.
6. General Gustav Ferrié (1868-1932) who gave France wireless transmissions in 1899.
7. Hans Bredow (left) and Paul Pichon (right) in 1913.

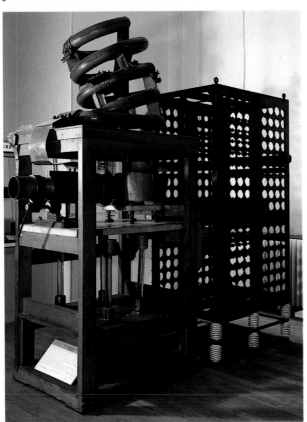

In November 1912, Langmuir proved that the thermionic effect could be demonstrated in a vacuum even in the absence of any trace of gas. He gave the name *pliotron* to these high-vacuum triode tubes.* In his notes, Langmuir wrote:

We found that these pliotrons possessed very great advantages over the audions that had previously been available. Because of the fact that we could work at anode voltages of 250 volts, it was possible to obtain a very much greater degree of amplification than De Forest had ever obtained and we were able to work with very much larger amounts of power. Furthermore, we found that our pliotrons had a remarkable constancy in their characteristics. We found that there was no critical anode voltage, such as De Forest had found, at which a maximum amount of amplification or sensitiveness was obtained. This meant that the tube could be adjusted once for all, and did not need constant regulation to operate satisfactorily.

During World War I, both the United States Navy and the Bureau of Standards used the pliotron as a transmitting valve. The De Forest Company was the first to produce tubes for the Signal Corps and the Navy, but only Western Electric and General Electric possessed the manufacturing facilities that could ensure regular production of the large quantities required by the American Army. Tube technology continued to progress, and in 1913 Langmuir and Rogers at General Electric began work on the thorium filament (marketed in 1921). At Bell Laboratories, Research Director Harold D. Arnold was one of the first scientists to take an interest in current research on the electronic high-vacuum tube that was eventually introduced into the first intercontinental telephone links. Thanks to improvements made in triode technology by Langmuir, and later by Arnold, transmitters became mobile devices, receiver sensitivity was increased, and it became possible to transmit spoken messages much more effectively. At the same time as he was doing this research, Arnold was working for ATT.

He was particularly anxious to satisfy ATT's

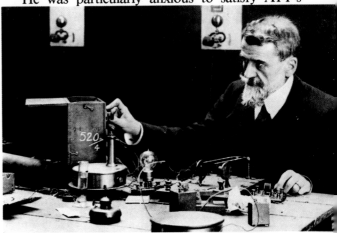

pressing commercial requirements. GE was already discussing immediate marketing and was closely following Langmuir's project. The two companies had very different objectives: ATT was continually trying to improve the highly standardized system developed by Bell Telephone, while GE preferred to work on diversified industrial applications and encourage new discoveries. The two companies had very different vocations as well: ATT was interested in low-current handling for telephone systems, whereas General Electric was mainly interested in the applications of lighting systems.

In 1913, when Langmuir published his article, there was still some doubt about the possibility of electron emission in high vacuum. Scientists' resistance to this hypothesis was so strong that in 1916, when Richardson published *The Emission of Electricity from Hot Bodies,* he devoted ten pages to discussing its probability!

We should emphasize that the first real application of the triode took place in France. It is rather a fantastic story. In 1908, with the consent of Gustave Ferrié, Lee De Forest came to Paris to try out his radio equipment at the Eiffel Tower radio station. His music broadcasts were picked up as far away as Marseille, although they were somewhat distorted. Ferrié took little notice of De Forest's work at the time, although in June 1914 he commissioned a Fessenden three-stage amplifier equipped with De Forest's tubes. The equipment later rather mysteriously disappeared from sight. It was thanks to a French deserter from the 1870 war, a gentleman by the name of Pichon, that De Forest's tubes made a second appearance. In Germany, Pichon had landed a job as French teacher to the children of Count Arco, head of the patents division at Telefunken.

As the storm clouds of war began to loom threateningly on the horizon, Pichon turned up in London at the offices of the managing director of Marconi, who advised him to return to France. Pichon was arrested in Calais shortly after landing, but he insisted on seeing the head of the military radio section to give him some highly important information. Confronted with Ferrié, he took one of De Forest's triodes out of his pocket and began to explain its virtues to Ferrié, who drafted him on the spot, and sent the tube to the Lyon radio laboratory for examination by the brilliant scientists working there: Henri Abraham, Eugène Bloch, and Leon Brillouin. Abraham, Peri, and Biguet, Bocuze's production manager, had a good look at the "new" invention and proceeded to make a number of improvements. In 1915 Biguet organized industrial production of the valve in collaboration with Grammont workshops despite the difficulties experienced at the time in obtaining the requisite high vacuum.

More than 100,000 TM (military telegraph) tubes were produced in 1916, and more than 300,000 in 1918. A new French industry — the electronic tube industry — had been born. Ferrié's role was not to cease at the end of the war, although from 1910 onwards, Emile Girardeau, founder of the SFR (French Radioelectric Company), and later Henri Damelet, head of Radio-Technique (1919), were to play an increasingly important role. It is interesting to note that despite the general enthusiasm for the newborn radio, it also aroused some criticism. H.G. Wells, for instance, who was one of the most prophetic spirits of the century, wrote in 1927 that radio had no future and that broadcasting stations were addressing themselves to a "phantom army of nonexistent listeners."

In 1918 the Russian Commissar for Posts and Telegraphs established a radio laboratory at Nijni-

1

2

3

4

1. Visit of the Russian delegation to the Nauen Telefunken Station during the Berlin International Conference on radio transmission in 1906.
2. The May 1910 visit of the Prince Tsaio (2), brother of the Chinese Emperor, to Nauen.
3. These ladies are taking tea in front of the latest technical marvel: their 4-valve radio receiver (1923).
4. Presentation of various types of transmission tubes in 1927 by female workers at Radiotechnique.

* extracts from letters from van der Pol to Appleton.

Novgorod. One of the laboratory engineers, M.A. Brontsch-Brujevitch, had heard of the French TM tube and produced a similar tube toward the end of 1919. For two years his PRI tube was the only one available in Russia. The anode was made of aluminum as it was the only suitable material to be found in the country at the time. The Nijni-Novgorod laboratory benefited from the direct protection of Lenin, who was most enthusiastic about this new means of propaganda. In January 1920, Russian scientists began research on the telephone.

Back in France, Henri Abraham threw himself not only into research, but also into what was then rather pejoratively known as "applied physics." He was commissioned by Ferrié to look into adaptation of the triode. During this period he also worked a great deal with Eugène Bloch; both men were to fall victim to the Nazi roundups and extermination of Jews during World War II. A former student of both men, Maurice Ponte, who became Chairman of the Compagnie sans Fils (French Wireless Company), tells the following story:

In 1923 I undertook my first research work at Eugène Bloch's laboratory at the Ecole Normale Superieure. Bloch and Abraham were both excellent teachers, although very different in temperament. Abraham was interested in the applications of science and particularly in the Great Adventure of the time, Wireless Telegraphy. Bloch was a theoretician who was extremely enthusiastic about quantum theory. After the 1914/1918 war the ENS laboratories were a kind of junkyard overflowing with all sorts of bits and pieces, including the famous triodes — "these lamps which don't even light" — as people used to sneer. During the war, the triode which Abraham and Peri had adapted was used only in receiving equipment, while the famous Sainte-Assise transmission station continued to use arc alternators as a source of power. Some incautious scientists even predicted that high frequencies could never be used for radio communication because they would be absorbed by the earth!

The Dutch company Philips was one of the first in the world to take an interest in triodes after the war. Like General Electric in the United States, the company first concentrated on achieving mastery of a similar technique, that of domestic lights. In particular, Klaus Posthumus who had worked with Ferrié in France, helped GE develop the 25-kilowatt rated triode, an extraordinary accomplishment considering the limited technology of the period. However, the most important contribution was made by the brilliant Dutch physicists Balthazar van der Pol who was made head of Philips' Radio Research Section. At that time the company was successfully manufacturing tubes and had also managed to develop the most powerful triodes available at the time.

21 December 1922: I was in London with Holst more or less as a "commercial traveller" in 25 Kilowatt triodes! We visited the Marconi Co., dear old Fleming and Eccles. They were developed here, as I told you, and now it happened that quite independently Langmuir did similar experients. Of course the whole thing is an application of a new joint between glass and metal. The external appearance of our triode is similar to Langmuir's but the material used is *much better* (i.e. our material is).

9 January 1923: I hear from Turner that Eccles was very much impressed by the high power valves. He thinks it is one of the biggest things yet done in wireless.

19 August 1923: In Carnavon I got 200 kilowatts

in 8 triodes at a high efficiency, but this is still a secret.*

However, van der Pol was now beginning to take an interest in another problem posed by the introduction of power triodes into a circuit. An electrical network is a system governed by a certain number of differential linear equations, and introducing triodes meant introducing not only negative resistances (to induce oscillations) but also nonlinear terms that got mixed up in the equations and complicated the different problems. Linear phenomena are those in which the magnitudes vary while staying proportionately the same among themselves. They are thus governed by algebraic relations, including only first degree terms which do not affect signals in an electronic system. In parametric amplification, the basis of which van der Pol had constructed in 1926, it was a question of using the phenomenon of non-linearity which up until then had been considered to be a defect as it caused distortion. The theory behind it was fully expanded and articulated in 1956 by two American researchers, J. M. Manley and H. E. Rowe.

Van der Pol was also doing research into cosmic rays and the ionosphere, whose existence was only proved in 1925 by wave reflection measurements. Along with H. Bremmer, he articulated a ground wave theory concerning the diffraction of waves emitted by a dipole antenna around a spherical body such as the earth, that is the wave propagated along the surface of the earth, which is particularly important during daytime over short distances and in particular for the ultra-short waves used in television, frequency modulation, and radar. To explain the "sky wave" propagated in the ionosphere, he went back to the work begun by Kennelly and Heaviside in 1902 and established that waves traveling long distances were propagated in a zigzag pattern by successive alternating reflection between the ionosphere and the earth's crust. Back in 1912, Eccles and Larmor had developed the theory of wave propagation in conductive gas, and this theory was now immediately applied to the ionosphere.

Van der Pol and his friend Appleton continued their experiments. Appleton's research led straight to the development of radar. In 1930 he gave a lecture at Eindhoven, where B. Tellegen had discovered one of the more remarkable effects of the ionosphere concerning the interaction of radio waves at different frequencies. Thanks to the linearity of Maxwell's equations, the principle of superimposition could be exactly applied to a vacuum, and the two wave systems could interpenetrate each other without interference on either side. But in the ionosphere, this interaction is of a partly nonlinear nature so that waves propagated by powerful transmitters through the ionosphere can cross-modulate depending on the frequencies of the transmitters. This was the effect that Tellegen observed when signals transmitted by the Beromunster station interfered with the powerful Luxembourg station signals. In 1929 J. Ballantine and H. Snow of RCA invented the variable slope tube which compensated for this so-called Luxembourg effect.

The end of the war saw the birth of another very important device in the history of wireless telegraphy and electronics (or radioelectricity, as it was known at the time.) This was the *superheterodyne*, a technique for reception and amplification for use in both wireless telegraph and telephone systems; it reduced the pass band in a receiver, thus reducing atmospheric interference. A number of inventors joined the race to claim credit for inventing the superheterodyne: Meissner in Germany in 1914, Laut in France in 1916, Lucien Levy (who took out two patents in France, on August 4, 1917, and October 31, 1918, respectively), and Edwin Armstrong (who applied for a patent on December 30, 1918). The Columbia courts judged in favor of Levy on December 3, 1918. However the controversy was still alive in December 1954 when an article written by Pierre Braillard for the magazine *L'Onde Electrique* on the life and work of the recently deceased Armstrong drew a reply from Lucien Levy:

That Armstrong brought to the development and the promotion of the superheterodyne the strength of his personality, his dynamism, and his enthusiasm cannot be denied; but to say that he designed or produced the first model is quite inexact. The American patent was delivered on November 5, 1929 to L. Levy and his assignee, the American Telephone and Telegraph Co. The German patent was awarded on October 1, 1931, to the Telefunken Company, L. Levy's German assignee.

Armstrong invented the superregenerative receiver in 1922; it was an extremely sensitive receiver which amplified very weak signals. In 1928 he predicted frequency modulation and in 1933 he took out patents for special wide band frequency modulated (FM) transmission and reception circuits. The reticence of most technicians to adopt his FM technique is explained by the high cost of frequency modulation detectors which had to be of extremely high quality to operate at the low transmission frequencies then used. In addition, the size of the spectrum required becomes even larger when we consider that the same sound requires 9 Hz when amplitude modulated and 150 Hz when frequency modulated. But with frequency modulation, reception quality proved to be incomparably superior, because of the elimination of distortion and interference and this is why it has been developed for radio communications in particular. And yet at one point it seemed as if this invention would meet with a tragic and premature death. In 1912 Armstrong succeeded in picking up very remote signals using De Forest's triodes, and went on to invent the regenerative receiver. He then demonstrated that the triode could behave as an oscillator and generate high frequency currents. De Forest brought the first action against him for infringement of patent. At the same time Armstrong missed an opportunity to collaborate with any of the large companies, since the American Marconi Company and ATT, who were both initially interested, finally decided not to buy his patent; he did, however, manage, to sell it for several thousands of dollars in 1914. Also around the same time, De Forest, who had taken out a patent on the "oscillon," took him to court. The case was delayed by the advent of war and dragged on for some twenty years. Not even Armstrong's influential backer Westinghouse could prevent him from losing.

In June 1918 another inventor, Walter Schottky, who had also worked on improving the triode, took out a patent on the superheterodyne, although he did recognize Levy's prior right. In 1915, two years after Langmuir had made the same modification, he added a grid to improve amplification: this was the *tetrode*, which would be followed by the pentode, and so on. In fact almost everybody was now working on adding some kind of grid or electrode during this period; in 1917 the SS-1 tube was developed in Germany by the Siemens-Schottky team.

Schottky's tetrode was a space charge grid, that is, one with a wide mesh grid placed between the cathode and the input (control) grid and maintained

1. Cat-whisker type crystal receiver and double headphone, made in Japan in 1925, were very popular in the initial stages of broadcasting.
2. Walter Schottky.
3. Bernard D.H. Tellegen.
4. Almost fifty years ago: manufacturing radio receiver tubes by hand.

* The mathematical theory behind counterregeneration was invented by two American scientists, Harry Nyquist and H. W. Bode.

UNE SEULE MANŒUVRE avec le
"SFER-20"
RADIOLA

ARCOLETTE 3W

Arcolette 3

1, 2, 3, 4. Posters from the 1930s.

at a positive voltage lower than the anode voltage. This led to a reduction of internal resistance and enabled the anode voltage to be reduced. In 1925 Schottky at the Berlin University Laboratory and Hull at GE simultaneously designed a screen-grid tetrode, with a tight-mesh grid inserted between the control grid and the anode and charged to a voltage similar to that of the anode. In this way, amplification was increased without increasing the internal resistance, thus avoiding the phenomenon of self-oscillation in the tube, which had proved a problem. However Schottky's tetrode was still not the perfect solution. In Holland, Bernard Tellegen had read about Schottky's modifications but was himself preoccupied with finding a solution to the problem of secondary emission of electrons from the anode, which prevented the new tube from functioning correctly. Between the screen-grid and the anode he introduced a third grid (the suppressor grid) which sent the secondary electrons back toward the anode from which they were liberated, thus eliminating secondary emission. This was the five-electrode tube, or pentode, which enjoyed unparalleled success and which is still used in radio transmitters almost to the exclusion of any other device.

Bernard Tellegen was almost the opposite of a man like van der Pol, who was not at all keen on teamwork or contact with others, and who carried out his research in his own ivory tower. As Tellegen said, "Van der Pol was a physicist, I was a graduate engineer from the Delft Technical University, and I was much more interested in the immediate application of an invention." Tellegen gave an interview in which he admitted the potential value of the tetrode:

W. Schottky had written papers on the screen-grid tube (or tetrode) which I found very interesting. I also read papers on the use of triodes as transmission tubes, and noticed that the triode must present a very low internal resistance to favor high outputs. Then in my mind I combined the two tubes and came to the conclusion that a screen-grid tube was very suitable for a transmission tube despite a high internal resistance. We have to remember that the power source for an anode was supplied from dry batteries and that we wanted to obtain the strongest possible transmission but also a saving in the number of tubes and less signal distortion, because the current in the loudspeaker would be proportional to the current in the control grid. But at that time the tetrode was perfectly suited to transmission.
(B. Tellegen, February 1979.)

In the triode the problem of secondary emission did not occur because the electrons returned to the electrode which emitted them, but in the tetrode the secondary electrons emitted by the anode were attracted toward the screen-grid when the anode potential fell below that of the screen. The invention of the pentode was as important for receivers as the triode had been for transmitters. The EF 50 pentode was finally completed in England and played an important role in the design of radar receiver radio frequency amplifiers. During the same period, Klaas Posthumus, who was still working for Philips, developed the principle of *counterregeneration*, which required a lower amplification factor and at the same time improved sensitivity and reliability. Harold Black at Bell Telephone came up with the same principle, which Bell used commercially for the telephone before Philips got around to using it in radio receivers, even though Posthumus's patent (September 19, 1928) was taken out some time before Black's

(August 13, 1929.)* According to an anecdote, the idea of this new form of amplification circuit came to the twenty-nine-year-old Black as he was taking the ferry across the Hudson River to his job at the laboratory. He drew a sketch on his newspaper along with a few mathematical formulae and when he arrived at the laboratory, pipe in one hand and newspaper in the other, the discovery was already made. This principle was used for stabilizing all systems using amplification. For superregenerative designs, a higher lever of amplification is needed in order for the stage to oscillate.

As a brief summary of the evolution of radio broadcasting in the first years of its existence, we can say that the different authorities only began to take an interest in it during World War I, when the possibility of exchanging messages over long distances was seen to have important military implications. The first organizations to take a real interest were the big American companies contributing to the war effort, the first individuals to become interested were army officers such as Ferrié and theorists of the Continuing Revolution such as Lenin. With the cessation of hostilities, scientists began to concentrate on improving techniques. The first regular radio transmissions had taken place in Laeken from March through July 1914. After the war, radio continued to be used as a tool of political propaganda, but people were not slow to recognize its cultural potential. On November 2, 1920, Westinghouse started up the Pittsburgh KDKA, the world's first regular radio broadcasting station, which was a continuation of the experimental station founded in 1916. Other stations followed: on September 1, 1922, the Russian Comintern station began transmitting in Moscow; on November 6, 1922, Radiola, later to become Radio-Paris (1932) began broadcasting in Paris; and the British Broadcasting Company was founded on November 14, 1922.

Japanese interest in wireless telegraphy research began in 1897. The Ministries for Communications and the Navy both followed Marconi's experiments with great attention; in fact, it was a message transmitted from a ship that launched the news of the advance of the Russian fleet towards Kyushu, thus playing a decisive role in the Tsushima battle. In 1911, Tsunetaro Kujirai succeeded in broadcasting messages by means of arc equipment. In 1912, TYK coordinated research into radio telephone systems with Uichi Torigata, Eitaro Yokoyama, and Seiji Kitamura. In 1917 Tokyo Denki, one of the two companies which merged to create Toshiba, experimented with the first vacuum tubes (triode type); and in 1926 NHK (Nippon Hoso Kyokai) opened three radio stations in Tokyo, Osaka, and Nagoya. It took twenty years from Marconi's and Popov's first transmissions for radio to become a public service: twenty years for successful development of amplification and amplitude modulation techniques capable of transmitting regular broadcasts. Wells's "phantom army of non-existent listeners" lost no time in assembling, and radio became a familiar element in everyday life.

But let's "touch base" again with tubes: while technicians were working on developing the various different "-odes," the "-trons" also made their appearance on the electronic scene. These "trons," which took their name from the last syllable of electron, would ultimately be applied to the amplification of ultrashort wavelengths.

Triode, tetrode, pentode

by B.J. van Westreenen, Philips

Most historians agree that the triode marks the true beginning of electronics. It was in fact the first tube capable of amplifying radio signals. The addition of an extra grid (the tetrode) compensated for some of the triode's defects. But the best amplifying tube was the pentode.

The Triode as an Amplifier Tube

The modern triode where modern means, say, from the thirties onward comprises a tubular cathode, a thin wire, the grid (as a wide pitch helix at a small distance around the cathode), and a cylinder-shaped anode surrounding the grid at a somewhat larger distance. This structure is placed in a glass or metal envelope, which is then evacuated. Within the cathode a heater is provided by which it can be heated to the temperature of about 900 °C, the temperature necessary to effect sufficient emission of electrons.

The usual electric setting of a common type triode has, with the cathode potential taken as 0 V, an anode voltage V_a between 100 V and 300 V, and the grid negatively biased to V_g between -10 V and -2 V. The grid then draws no electrons (that is there is no grid current). On the contrary, it pushes electrons back, either to the cathode or to the "cloud" of space charge around the cathode. However, depending on its negative bias voltage, the grid does allow part of the electrons to pass between its windings and to escape to the positive anode. In this way an anode current I_a is drawn (of, say, some milliamperes) that is strongly dependent on the value of V_g and also, to a much lesser extent, on the anode voltage V_a. In fact, the ratio μ of these dependencies is approximately equal to the ratio of the electrical capacitances of grid-to-cathode and anode-to-cathode. This quantity μ is called the *amplification factor* of the triode, for reasons that will become clear further on. Its value usually lies between 15 and 100. The influence of the grid voltage on the current, for constant V_a, is normally expressed by the transconductance S, the change ΔI_a in I_a due to a change ΔV_g in V_g, or $S = \Delta I_a/\Delta V_g$. Normal values for S are between 1.5 mA/V and 12 mA/V. On the other hand, at constant V_g, I_a may be varied by ΔI_a by varying V_a by an amount ΔV_a. The ratio $\Delta V_a/\Delta I_a$ is called the internal resistance R_i of the tube. A low value of R_i implies a big influence of V_a on I_a. Normal values for S are between 1.5 mA/V and 12 mA/V. On the other hand, at constant V_g I_a may be varied by ΔI_a by varying V_a by an amount ΔV_a. The ratio $\Delta V_a/\Delta I_a$ is called the internal resistance R_i of the tube. A low value of R_i implies a big influence of V_a on I_a. Normal values for R_i lie between 5 kΩ and 50 kΩ.

Mathematically the following relation between the three tube parameters applies: $\mu = SR_i$ (Barkhausen's relation). So a triode with $S = 3$ mA/V and $R_i = 20$ kΩ would have $\mu = 60$.

The amplifier function of the triode can best be explained by means of an example. For a tube with the parameters as in the preceding paragraph the operating point settings may be $V_g = -3$ V, $V_a = 200$ V, $I_a = 4$ mA. On these set point values alternating voltages and currents of not too large amplitudes may be superimposed. These will be denoted by lowercase letters. For an input voltage $V_g = 1$ V the current will be $I_a = SV_g = 3$ mA. If a resistor R_a of 1 kΩ is inserted between anode and anode voltage supply, I_a will develop across R_a an output voltage of 3 V. So in this circuit the tube produces an amplification or voltage gain of 3.

It seems easy to obtain more gain by choosing a larger R_a. With $R_a = 10$ kΩ one might expect an output of 30 V, so a gain of 30. However, one would find only 20 V. This is a consequence of the fact that the voltage drop V_a across R_a represents also an appreciable variation in V_a, and due to the R_i of the tube this counteracts the primary variation I_a. If R_a is not small compared to R_i the effect of the internal resistance can no longer be neglected. As a consequence it appears that the attainable voltage gain in limited to μ, a value which is only approximated when R_a can be made large compared to R_i. This explains the name given to μ.

A high-voltage gain is not always the required property of a triode. For various other applications, such as in oscillators, transmitters, power amplifiers, wide band amplifiers, or multivibrators, other properties are of more importance. Thus triodes were designed not aimed at high μ or high R_i, while for high voltage gain they were superseded later by pentodes.

The Tetrode

With the first triodes the fundamental possibility to amplify electric signals had been established. Subsequent development of the tube led to important progress in its applications, but also brought its practical limitations to light: the low internal resistance of the triode restricts the attainable voltage gain, and the anode-to-grid capacitance gives cause for instability when high-frequency amplification is aimed at.

A low internal resistance means that in a triode the anode voltage has an appreciable influence on the anode current. This is a direct consequence of the construction of the tube. The grid surrounds the cathode at a short distance, and therefore its voltage has a large influence on the current, as desired. The anode is further away from the cathode but electric field lines from the anode still reach through the open structure of the grid to within the cathode region, so that the anode

The introduction of a grid between the anode
and the cathode provides a means of
controlling electron flow
1. Anode
2. Grid
3. Cathode
4, 5, and 6. Batteries

voltage still influences the current, albeit a factor μ less than the grid voltage.

Another consequence of the construction is that the anode, which surrounds the grid, naturally has a certain capacitance to the grid (C_{ag}). This implies that any alternating voltage that appears at the anode will induce also a voltage on the grid. But a voltage on the grid will lead, in an amplifier circuit, to an amplified voltage in antiphase, at the anode. So this is then a case where a feedback loop exists. For low frequencies, say, audio frequencies, the phase relations are generally simple enough to cause no harm. At higher frequencies and in particular when the load impedances are no longer resistors but resonant circuits, there is every chance that there is at least one frequency where things go wrong. As soon as the gain around the loop is at least unity and the phase shifts in the loop are such as to add up to just 360 degrees, oscillations start. Slight changes in circuit parameters easily lead to changs in frequency and a very unstable situation results.

A remedy against both troubles at the same time is found in the provision of a second grid in the tube, between the first, or control-grid, and the anode. Thus the tube has four electrodes and is called a tetrode.

The second grid is more tightly wound than the first, so that it screens off the cathode region rather effectively from the anode. It is called the screen-grid and is normally kept at a constant DC voltage, somewhat below the anode voltage. From the total cathode current that passes the control-grid, some 15 percent is intercepted by the screen-grid.

The insertion of the screen-grid was, historically, primarily aimed at reducing the anode-to-grid capacitance C_{ag}. The target was amply reached: a reduction by at least two orders of magnitude proved possible. And by the same means the influence of the anode voltage on the current from the cathode was drastically reduced: the internal resistance of the tube could be increased by about a factor of 100, to a few 100 k Ω or more. The corresponding amplification factor μ was raised to between 2,000 and 5,000.

The Pentode

If the tetrode is a triode with just a grid added, the *pentode* is a tetrode with just one more grid. And here the series stops: the pentode proved to be, for many years how, the perfect amplifier tube.*

The tetrode excellently remedies the two weak points of the triode — its rather large-anode-to-grid capacitance and its small internal resistance and related low attainable voltage gain. But the insertion of the screen-grid with a voltage about as high as that of the anode, introduces an undesirable side effect. This side effect is due to secondary emission at the anode and its consequence can best be explained by means of the characteristic curves drawn in the figure 1.

These curves show the total cathode current I_c, the anode current I_a, and the current to the screen-grid I_s, of a tetrode, all as functions of the anode voltage V_a and at fixed voltages of the control-grid V_g and screen-grid V_s.

I_c, which is equal to the sum of I_a and I_s, is nearly independent of V_a, which reflects the high internal resistance of the tube ($\Delta V_a / \Delta I_a$). At zero or low V_a, most of the current goes to the screen-grid. At values of V_a some 20 V or more above V_s, a large, constant portion of I_k reaches the anode, the rest goes to the screen-grid, as would be expected. At anode voltages in between, the effects of secondary emission become apparent. An electron accelerated by more than 10 V may give rise to the emission of one or more electrons by the electrode it strikes. So, an electron arriving at the anode at 50 V may liberate there a few others. These secondary electrons leave the anode with about zero energy and will be drawn to a nearby electrode with a higher potential, thus to the screen grid. Consequently, a current of electrons will flow from the anode to the screen-grid. This is the explanation for the hump in the curve for I_s and the corresponding dip, down to negative values, in the I_a curve. As soon as the anode voltage is sufficiently increased, secondary electrons emitted at the anode will find no electrode at higher potential and fall back to the anode.

In a tetrode used as an amplifier at low signal levels the consequences of secondary emission are hardly noticeable. It is only at higher alternating voltage levels that the anode voltage will drop momentarily to below V_s. At these moments the cathode current goes to the screen-grid. As a result the shape of the anode voltage is no longer a good replica of the input voltage at the control-grid. In other words, the amplifier gives a distorted signal.

In a pentode it is the function of a third grid to suppress the deleterious effects of secondary emission. This suppressor grid is located between screen-grid and anode, and is kept at cathode potential — in most tubes it is directly connected to the cathode. In this way a minimum in potential is obtained between screen-grid and anode, through which no secondary electrons — emitted at low energy — can pass. So wherever secondary electrons are emitted, at the anode or at the screen grid, they cannot but fall back to where they originated. The resulting characteristic curves are shown in broken lines.

With its structure of three grids between cathode and anode and the voltages to which these are adjusted, the pentode exhibits a performance that fulfills nearly all amplifying needs. It can be optimized for an l.f. power amplifier, for an h.f. amplifier, or for a controllable i.f. amplifier. The only place that it had to leave, later on, to the triode was that of those extremely high frequencies that are commonly reckoned to belong to the microwaves.

* of course, engineers went on adding more grids — hexode, heptode, octode — but these tubes served other purposes.

Anode current of a tetrode has a peculiar behavior as a function of anode voltage, due to secondary emission at the anode. Insertion of a suppressor grid at 0 V would cause I_a and I_s to follow the broken lines.

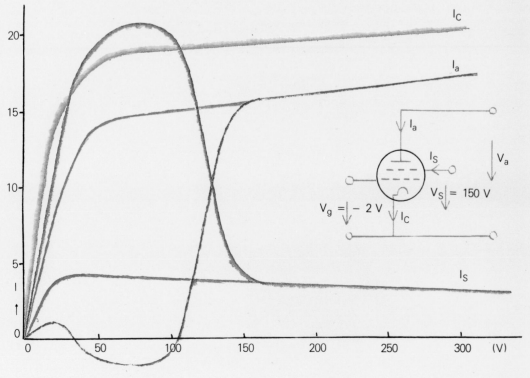

14
The "tron" forest: magnetron and klystron

Langmuir's pliotron was the first in a long series of "trons." On November 19, 1913, Langmuir noted: *Last Sunday we got together with Professor Bennet to think up some names for electronic tubes. We thought about it for a bit and finally came up with pliotron, from the Greek* pleion *meaning more and the suffix* tron, *signifying an instrument for giving more, an amplifier, a relay.* At first this suffix was only used for General Electric's tubes – magnetron, klystron, axiotron – which led De Forest to comment that they were "Graeco-Schenectady" products. (General Electric's main center was in Schenectady, New York.) Then "tron" began to be used more generally to designate power tubes of all makes. The two most famous "trons" produced in the period immediately preceding World War II were the magnetron (high-frequency vacuum tube oscillator) used mainly for radar and the klystron (from the Greek *kluzein* meaning "to unfurl") which was a high-vacuum tube used as an amplifier and oscillator in the UHF band up to several million Hertz. The term magnetron was first used in 1921 in an article written by A. W. Hull for the *Physical Review*. The term described a cylindrical anode tube with a cathode filament to concentrate thermionic emission. The passage of electrons in this tube was governed not by a control grid but by a magnetic field. In 1924 Hull produced his magnetron for medium frequency amplifiers and oscillators.

Albert Wallace Hull started work at General Electric in 1913, where he carried out electronics research with Langmuir and Coolidge. He had already invented the thyratron, which could generate medium output power levels and which served as a time base for oscilloscopes. His magnetron, which used a magnetic field to control the passage of electrons, was suitable for use as a regenerative oscillator, as was the pliotron, but was subject to the same limitations as regards high-frequency operation. During this period a great deal of effort was devoted to expanding the radio tube market. Two articles that appeared at the same period concerning the generation of oscillations in a constant magnetic field went unnoticed by other scientists at the time. Dr. A. Zacek of Prague wrote about the possibility of generating oscillations at a wavelength of 29 cm,

and a second article, written by Erich Habann of the University of Iéna, touched on the same subject. During 1927 and 1928 work continued both in Germany and Japan, where Kinjiro Okabé of the Imperial University of Sendai published an article on his work in this field. In February 1, 1928, Hidetsugu Yagi gave a lecture in New York. The Americans could not help being interested and General Electric invited Yagi to Schenectady to discuss buying the rights to one of his patents. In May the Army realized the potential of Yagi's pulse magnetron (which could operate at higher frequencies than Hull's), and General McArthur gave the order to start research along these lines. As in the time of Ferrié, it was once again an officer from the armed forces, Major E.S. Darlington, who was among the first to become really interested in this new technique. He established contact with Yagi, and after Yagi's Schenectady visit continued his studies at the College of Engineering at the Tohoku Imperial University in Sendai. In his letters he writes:

How did Dr. Hidetsugu Yagi first learn about the existence of Dr. A. W. Hull's magnetron and how did he become interested in the magnetron and then develop the first microwave power magnetron which today is the most important component of radar and similar electronic devices?

I learned the answer to the above questions directly from Dr. Yagi himself in Tokyo, Japan, on 26 November 1951 at his residence.

As we sat around the charcoal fire in his room, Dr. Yagi first told me about his former home and books. In April 1945 a fire bomb set his house afire and it burned to the ground. All his notes, books and research data covering over 30 years of intensive work in the field of electronics and physics were completely destroyed. Dr. Yagi was always a keen student and follower of Dr. J. A. Fleming and his work on the diode during the latter's early research work. In 1916, Dr. Yagi left Fleming's laboratory in London and went to Cruft Laboratory at Harvard University, where he further studied Fleming's contributions to the field of electronics. Upon Dr. Yagi's return to the College of Engineering, Tohoku Imperial University at Sendai, Japan, he instituted a

Manufacture of klystrons, power tubes used mainly in television and radar applications.

135

The 'tron' Forest: Magnetron and Klystron

1. Irving Langmuir (1881-1957), American engineer, chemist and physicist, who was awarded the Nobel Prize for Chemistry in 1932.
2. The pliotron.
3 and 4. H. Yagi and Professor Okabe.

1

* In fact, Professor Yagi gave the name magnetron to a tube that was very different from Hull's magnetron.

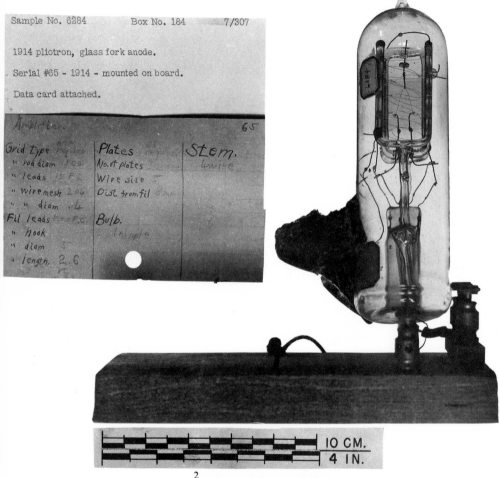

Sample No. 6284   Box No. 184   7/307

1914 pliotron, glass fork anode.

Serial #65 - 1914 - mounted on board.

Data card attached.

2

3

course at the University in the functioning of the diode and added classroom experiments to be undertaken by the students.

Professor Okabé also became much interested in the functioning of the diode and in an investigation of its characteristics, since teaching the subject brought up questions to which no clear-cut answers could be given. One of the most important questions discussed by these two professors was why did not the diode influenced by an axial magnetic field show a shart cut-off of the plate current as was to be theoretically expected at a definite magnetic field strength. This led to the suggestion, noted later in the letter from Professor Okabé, that minute oscillations might possibly be the cause of this lack of sharp plate cut-off.

Sometime prior to 1924, a Japanese Naval Officer returning from an eastern college in the United States had reported to Dr. Yagi that he had heard of a new tube that was called a "magnetron." All that the Japanese officer could report to Dr. Yagi was that the "magnetron" was a large vacuum bottle, encircled with a large magnetic coil.

This was Dr. Yagi's first knowledge that the name "magnetron" existed and was the introduction of this word into Japan. The Japanese Naval Officer had been referring to Dr. Hull's magnetron, which was described by Dr. Hull in a 1921 publication.* (White: 1952).

Darlington wrote to Okabé, who had in the meantime become a researcher at the Physics Department of the University of Osaka. Okabé replied:

As the oscillations obtained with single-anode magnetrons (Dr. Hull's type magnetrons) were weak in intensity, I tried to increase the output power, as well as to shorten the wavelength, by improving the electrode configuration and invented the split-anode magnetron after experiments with more than fifty magnetrons of different electrode configuration. All of these tubes were manufactured by myself or by my two young assistants. The above invention was patented in Japan, U.S.A. and Canada. I assigned the U.S.A. and Canada patents to the General Electric company.

In this device, inverse potential variations across the two anodes modified the electron speed, and the current developed in the two symmetrical oscillator sections of the tube would follow these potential variations.

Okabé continues:

The first one who took an interest in my work was Dr. Yagi. He gave me valuable suggestions in the course of the development of my invention. I learned later that Dr. Zacek succeeded in detecting oscillations from a single-anode magnetron before I did. At that time there was no information available to me on Dr. Zacek's work, as it was written in [a] foreign language. Therefore I am not the first discoverer of the fact that a magnetron could produce short-wave oscillations. However I may be the inventor of the multi-split-anode magnetron, and also the first one who succeeded in showing experimentally and theoretically that intense oscillations of very short wavelengths can be produced with magnetrons.

By an irony of history, the Japanese invention was taken to the next stage by the Americans and it helped to perfect the radar used by Japan's future enemies in World War II. In France, it was Maurice

Ponte who developed the multiple-segment magnetron that used tuned cavities located in the anode section as a means of frequency control. (A circular movement of the electrons is created and their orbital time governs the operating frequency.) One of Ponte's assistants was Henri Gutton, son of Camille Gutton, Professor of the Science Faculty at Nancy. In 1927 Camille Gutton and Pierret had undertaken experiments in the 16-cm band using a braking field tube. Research into this device and into increasing the power of magnetrons, which at the time worked well up to 430 MHz (70 cm wavelength) continued at the same time. From the beginning of work on radar, Maurice Ponte had been trying to work with higher frequencies and wave guiding designs. In 1934 Henri Gutton, inspired by his father's work, submitted to him the idea of using decimetric frequencies. Ponte recalls that when his team was testing the first magnetrons in a courtyard in Levallois, they were astonished to find that the returning signals were being somehow modulated between transmission and reception.

When we examined the problem more closely, we found out that the source of these changes in waveform was a bicycle wheel. This was how we first realized that instead of going straight, the beam was propagated perpendicularly towards a nearby bicycle shed. We finally decided that it was a reflective effect and we later went on to develop the magnetron for higher frequencies.

But to limit ourselves to the magnetron itself, the CSF team developed the resonating segment magnetron in 1936. It produced a 10 W continuous carrier wave, and could go down to wavelengths as low as 8 cm. In this tube the cathode was surrounded by a set of oscillating circuits. In June 1939, CSF brought out a new thorium filament magnetron, five times as powerful as the earlier model. In his *Memoirs,* Emile Girardeau is quick to point out that credit for the invention of radar does not completely belong to American or English scientists. He recalls a phrase spoken by Alfred Loomis, director of MIT's Microwave Committee:

The possibility of using ultra-short waves for radiodetection was a problem that seemed speculative in the extreme, since we had no convenient source of energy for such frequencies at the time. (Girardeau, 1968: 218.)

The Americans were extremely happy with the present that the head of the British mission brought with him in September 1940: the resonating cavity magnetron. This magnetron was at least in part of French origin. On May 8, 1940, under the terms of Anglo-French cooperation agreements, Maurice Ponte arrived at the Wembley Laboratory with his resonating segment magnetrons and the oxide-coated cathode that made it possible to generate pulses at high power levels. The Randall-Boot team at the University of Birmingham, under the direction of Marc Oliphant, recognized that Ponte's contribution had saved them six months in the development of multicavity magnetrons and had "a considerable effect on the second half of the war," since its UHF radar applications could be more rapidly perfected. The resonating cavity magnetron is made up of an anode in the shape of a thick copper cylinder in which cylindrical cavities have been drilled out. These cavities are linked to the interior space by means of fissures and in the middle of this center section there is a pure tungsten cathode. These fissured

The 'tron' Forest: Magnetron and Klystron

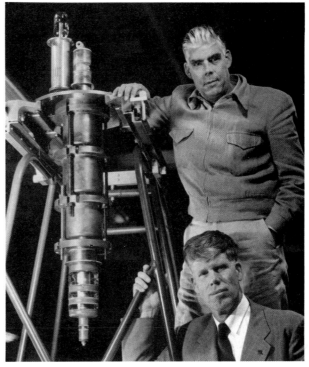

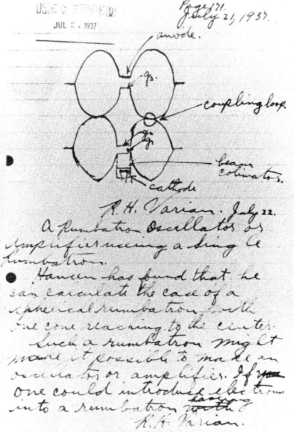

1. Randall and Boot's magnetron.
2. The Varian brothers: Sigurd (top) and Russell (bottom).
3. The cavity magnetron.
4. Notes made by Russell Varian.

The 'tron' Forest: Magnetron and Klystron

cavities, which resonate at a predetermined frequency, represent the heart of the oscillator in which the electrons circulate, their orbital time governing the oscillator's frequency. High-energy frequency is picked up by means of a loop and probe inserted into one of the cavities. In 1938, William Hansen in the United States drew the attention of scientists to the numerous benefits that could be drawn from the resonating properties of electromagnetic cavities when used as high-power UHF oscillators. These cavities are completely closed and empty metallic enclosures whose basic operating frequency corresponds to the wavelength of the enclosure.

For example a hollow sphere 10 cm in diameter resonates on a wavelength of 11.5 cm.

Hansen worked in Stanford in close collaboration with the Varian brothers who did not take long to develop their own very high power tube, the *klystron*. The invention of the klystron was made at the same time as research in the field of experimental physics and the first experiments on television. From 1930 to 1933 Russel Varian worked on television with Phil Farnsworth in San Francisco and with the Philco Corporation in Philadelphia. Like most engineers at that time, he was trying to produce power tubes. People were just beginning to understand the effects of electron transit time between the electrodes of positive or negative grid triodes operating at very high frequencies. From 1933, Varian began working with William Hansen on research started prior to his arrival in the field of high voltage power supplies of around 2 million volts for x-rays. The third decisive event was the threat Hitler posed for Europe and that led to the Americans accelerating their radar research. Hansen worked on resonators. Following an interview with him, Russel Varian decided to classify all projects on power tubes in order to study them systematically.

While he was working on this classification, he suddenly thought of the principle of "velocity modulation." The principle of velocity modulation had nothing to do with any of these classification projects that Varian had invented and the idea must have come to him because he was unconsciously trying to test the value of his classification system. He then began to think of an exception which in time became the fundamental principle behind the klystron.

In 1936, G. F. Metcalf and W. C. Hahn at General Electric had developed a klystron that they showed to the Navy and the Signals Corps and, in the following year, to RCA. On January 30, 1939, the *New York Times* published an article on the klystron written by Webster and the Varian brothers, Russel and Sigurd. Fifteen days later *IRE Proceedings* published another article on the work of the General Electric team. The English were the first to realize the potential of klystrons in the field of radar and the first to produce a working system. American klystron radars were based on English models. In their book *Electronic Control Tubes Using Velocity Modulation* Robert Warnecke and Pierre Guénard of CSF give a brief outline of the history of the klystron:

Witt and Clavier should undoubtedly share the credit of making the first precise and accurate suggestions, Witt for the diode in 1932 and Clavier the following year for the operation of the spatial oscillation triode, on the method of generating an electron group in the heart of an electronic flux resulting from velocity differences induced by a rapidly alternating electrical field. An earlier study by Moller (1930) on the operation of the Barkhausen lamp and J. Muller's analysis of the working mechanism of the diode at very high frequencies, as well as certain remarks made by H. Gutton on the operation of Pierret's oscillator, can all be seen as making some sort of contribution to the field. (Warnecke and Guénard, 1951: 1 and 2).

One of the ancestors of the klystron was no doubt the device described by German scientist Oskar Heil and his wife Arsenjava in 1934. For the first time this description talks about the systematic use of localized velocity variations producing "electron" packets. Warnecke and Guénard continue:

In velocity modulation tubes, the control signal is not used to produce a variation in the number of electrons governing the alternating component of the electronic current but mainly to vary their velocities.

The tube presented by Hahn and Metcalf is an elongated tube along the center of which straight, narrow beams of electrons emitted by the high-emission cathode are directed through three grids, of which the last two are placed at either end of an electrode shaped like a rather long pipe. If a weak high-frequency voltage is applied between the two first grids, the electrons are either slowed down or accelerated, which leads to a modulation in the density of the electronic current. A fourth grid is added to induce oscillations. The Varian brothers' klystron is a device in which the current flow is governed by constant variation of electron velocity (that is the velocity modulation level) and upon the use of a new design electrical resonator. At the end of each tube are placed fissured toroidal chambers called *rhumbatrons*. These act as resonating cavities. The first cavity has grids on either side of the slit, and it is possible to obtain high-frequency voltages and velocity-modulate the electrons.

The principle of the klystron is to localize electron packets and play on the electron's kinetic energy. The input grid is charged and the velocity of the electrons modulated in accordance with the field applied to the electrode. The speeded-up electrons catch up with the slower ones preceding it and thus form packet of electrons in the interelectrode area or "drift space." A feedback loop between the two rhumbatrons concentrates the energy into a single rhumbatron. Owing to velocity modulation, the electrons are now bunched, and oscillation is sustained because they are reflected in phase. The klystron was to be extremely useful for radar, television, linear accelerators, and various radio applications. Klystron research was undertaken systematically in France after 1939 by Warnecke (CSF) and Clavier (LMT).* In 1940, the klystron reflex, also known as the braking field tube, was designed by Robert Sutton for the Wembley Admiralty Signal Establishment. This was a klystron with only one cavity which functions at UHF frequencies in a similar way to the positive grid diode, rather than at decimetric wavelengths as did the Varian klystron.

After the war, in 1947, Robert Warnecke produced the first traveling wave magnetron amplifier. Three year later came C. Beurtheret's discovery of an original technique for cooling transmission tubes, the *vapotron,* which used vaporized water whose specific heat is higher than that of water in its liquid state. But most importantly, Bernard Epzstein (CSF) invented the *carcinotron oscillator* in 1951. This tube made it possible to produce extremely high-frequency waves and is now used for fundamental research into the area between the electromagnetic spectrum and infrared regions. It is the only existing

The first programmable magnetron.

* The completed klystron theory was constructed during the same period by Jean Barnier.

generator that can operate at these frequencies in the 300 GHz region. A Bell researcher, Rudolf Kompfner, was responsible for the conception of this device but Epzstein was the first to take out a patent for it. Klystrons were much improved over the next few years, and in 1952, R. Zwobada (Radio-Industrie) designed the first millimetric magnetrons and klystrons. From 1955 to 1958. Metivier developed a 30 MW klystron for C.F.T.H. which developed 20 kW of power. In 1956 P. Guenard developed wide band klystrons after having first produced power klystrons.

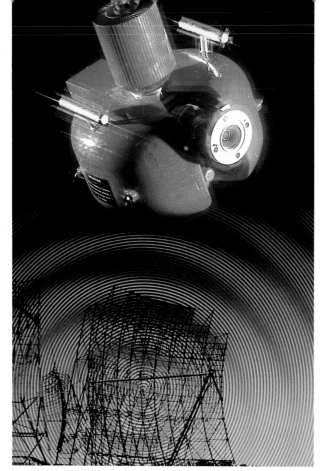

1 and 2. Klystrons: transmission tubes which can be used as oscillators or amplifiers in various applications: from radar to HF communications, microwave ovens to particle accelerators.

1

2

15

The electron microscope

The roots of electronics are firmly planted in electricity, but at times electronics almost seems to be related to poetry. L. Marton was aware of this when he wrote *optical electronics began before the discovery of the electron, when the first man gazed in admiration at the northern lights of the aurora borealis.* (L. Marton, 1968: 2). It was a Norwegian specialist in this phenomenon, Fredrik Carl Mulerzt Stormen, who in 1907 first established a theory of electron optics, an optical technique using beams of electrons instead of the luminous radiations of traditional optics. A few years earlier, in 1895, Swedish scientist Kristian Birkeland discovered that a single magnetic pole had an effect on cathode rays similar to that exercised by a lens on light rays; that is, it concentrated them into one focal point. He noted however, that the cathode rays could be made to bend by magnetism; and he could displace the deformed image of a Maltese cross situated in front of the cathode along the glass walls of the electronic tube, thereby showing that the size of the shadow produced by an ordinary light source at the same distance was larger. Despite these early discoveries, the true birthdate of electron optics occurred much later, in 1926, when the German Hans Walter Hugo Bush calculated the trajectories of electrons in a symmetrically revolving magnetic field and showed that they behaved exactly as light rays in symmetrically revolving optical systems. In particular he showed that it was possible to focus these electrons and to design lenses similar to those used in photonic optic. This research was continued by the Dantzig Physics Institute.

The first practical opportunity to widen experimentation into optical electronics was the creation in Berlin of the AEG-Telefunken research center. Bush wrote: *Electron optics was really created from the fundamental research done by Knoll and Ruska on the one hand and on the other by their colleagues at the AEG research institute.*

All these efforts paid off in 1931, the year that saw the birth of the first electron microscope. On June 4, 1931, in a lecture given at the Berlin Technical School, Max Knoll described the experiments he was then doing with his student Ernest Ruska. Basing their work on experiments done between 1928 and 1930 on magnetic lenses, they built a two-lens microscope that operated at voltages in the region of 1,000 volts. A beam of electrons emitted from a heated tungsten filament is accelerated by a potential difference and is propagated across a vacuum. The accelerated electrons are then guided across one or two condensors by the magnetic lenses. These lenses can be used to obtain magnifications 350 times larger than is possible with optical microscopes. Although the first electron microscopes operated at voltages of 1,000 volts, the classic microscopes from the 1950s operated at between 50 and 100 kV. At AEG in 1930, Ernst Brüche introduced the idea of an index of fictive refraction, thus establishing an analogy between electron optics and geometric optics. In August 1931, J. Johannson used a unit designed and built by Brüche to obtain the first electronic image using a heated cathode. It did not take long for this new German discovery to spread to Europe and the United States. As far back as 1931 two Bell Laboratory physicists, Davisson and Calbick, gave a report on their work on electronic lenses to the American Physics Society meeting in Pasadena. In 1933 Brüche and Knecht in Germany constructed the first emission microscope. Burgers and Ploos were doing similar work at Philips during the same period. In 1935 an English team made up of Martin, Whelpton, and Parnum built a magnetic two-stage microscope for the Metropolitan Vickers Company. During the war Zworykin, Hillier, and Snyder in the United States developed the first scanning microscope and obtained a resolution of 500 angstroms,* while in France Pierre Grivet designed the first French microscope with electrostatic lenses.

Work in Japan and the USSR, on the contrary, did not begin until after the war.

Later, in an attempt to increase the effective penetration of electrons into matter, research was oriented towards the development of instruments using increasingly higher voltages. Just after the war, two attempts were made in the United States and in Europe to construct two such microscopes, a 300 kV model and another 450 kV model. The work of Gaston Dupouy in France proved decisive. He decided to build an electron microscope that could operate at 1 million volts. In December 1960, he gave the first

In the electron microscope, the light rays of the traditional optical microscope are replaced by an electron flow.

* 1 angstrom represents 10^{-7} of a millimeter. One nanometer = 10^{-6} and 1 micron = 10^{-3} mm.

The Electron Microscope

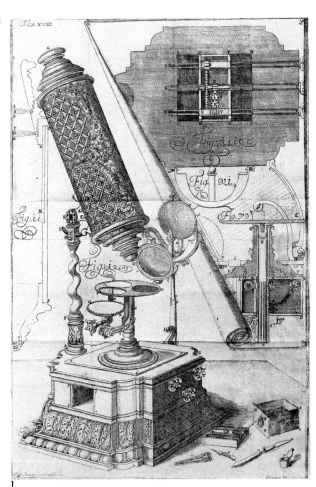

1. Hertel's compound microscope, 1716.
2. Professor Ernst Ruska.
3. Bodo von Borries.
4. Borries and Ruska's microscope.
5. Vladimir Zworykin (standing) and James Hillier worked on the first electron microscope marketed by RCA.

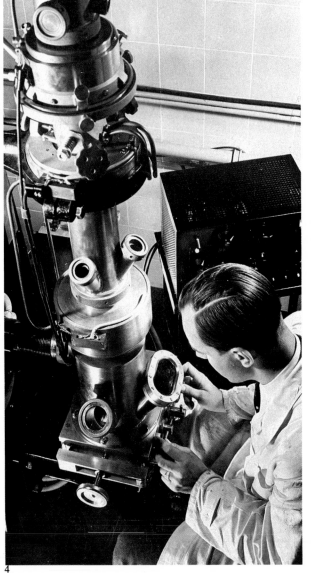

demonstrations to the Academy of Science of the 1 MV microscope, which he had built in Toulouse with the help of Frantz Perrier, and installed in a spherical chamber. In 1962 English researchers at Cambridge, under the direction of Cosslett, developed a 750 kV microscope.

In 1966 the Hitachi firm in Japan succeeded in building a 1 million volt unit. Gaston Dupouy decided to go even further and succeeded in building a 3 MV version that was imitated soon afterwards by the Japanese. Even more powerful equipment is now being envisaged, although no such microscopes have yet been commissioned. Most of the electron microscopes in use throughout the world operate at a voltage of 100 kV. Recently, research centers specializing in this kind of equipment have begun to develop instruments of between 200 and 300 kV. STEM (Scanning Transmission Electron Microscope) made it possible to observe atoms for the first time. In 1970 Crewe became the first scientist to distinguish uranium atoms on a layer of carbon.

The electron microscope which at the beginning of its useful life seemed to be best adapted to giving topographical or crystalline information now seems to be headed toward a more quantitative and analytical role in the same areas of spatial resolution. Finally, we would point out that studies in the field of biology are proceeding more slowly than expected because of the difficulties in obtaining specimens as close as possible to their natural state, and the necessity for reducing the doses of radiation used for observation in order to reduce the effects of deterioration under the microscope's beam.

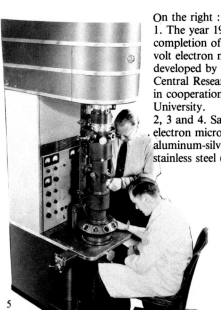

On the right :
1. The year 1970 saw the completion of the 3-million volt electron microscope developed by the Hitachi Central Research Laboratory in cooperation with Osaka University.
2, 3 and 4. Samples under the electron microscope: aluminum-silver alloy (2), stainless steel (3 and 4).

1

2

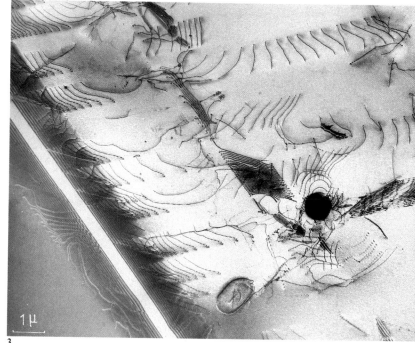

3

4

The properties of the modern electron microscope

by Christian Colliex, Maître de recherches au Centre National de la Recherche Scientifique (C.N.R.S.)

Today we distinguish between two types of microscope, the CTEM (Conventional Transmission Electron Microscope) and the STEM (Scanning Transmission Electron Microscope). What are the characteristics of the two types?

The main components of the electron microscope are the lenses, which are of two types: electrostatic lenses and magnetic lenses. The first are the simplest to design: they use the forces exerted by electric fields on the electric charges. A set of circular electrodes centered along a common axis and carrying different voltages constitutes the simplest system. Unfortunately, it is hardly suitable to use them at very high voltage (100 kV) due to the risks of insulation problems and breakdown, as well as relatively significant aberration defects. They are hardly used anymore today, except in the gun of the microscope where a potential difference is necessary for electron acceleration. If the first microscopes operated at a voltage of several thousand volts, the standard microscopes from the 1950s operate at a voltage of 50 to 100 kV.

In fact, magnetic lenses are used to focus electrons along the entire column of electron microscopes. Contrary to the case of the electrostatic lens, where the force experienced by the electrons is colinear to the electric field, for a magnetic lens the Lorentz law

$$\vec{F} = \frac{e}{c} \vec{V} \wedge \vec{B}$$

provides for a force F whose direction is perpendicular to the plane formed by the magnetic field B and the speed V of the electrons. Thus, an axial magnetic field imposes a helical motion on the electron paths along this axis. It is the leakage field at the ends of the polar components that is responsible for the focusing properties of magnetic lenses. Let us first consider the case of a transmission electron microscope, designed to examine specimens prepared in the form of thin films, the thickness of which varies between 10 angstroms and several microns. The electrons pass through the specimen and the interactions that take place during their passage (deceleration, angular deviation) constitute the information utilized to reconstruct the properties of the specimen. Today, there are two types of microscopes singled out by their design:

— The CTEM (Conventional Transmission Electron Microscope), is the direct descendant of the devices built in the 1930s and made commercially available by a good twenty or so companies over the last forty years. Let us cite the most well-known of these: RCA in the United States, AEI in Great Britain, Siemens in Germany, Philips in the Netherlands, Hitachi and Jeol in Japan. The total number of microscopes built borders on 10,000. In this type of instrument, the analogy with the optical microscope is obvious: two groups of magnetic lenses provide, accordingly, the illumination of the specimen (these are the condensers) and formation of the image between the object and the observation screen or the photographic plate (objective lens, intermediate lenses, projector lens). There is a biunequivocal correspondence between the various points in the object plan and the corresponding points in the image plane, at the level of which detection is effected simultaneously. The main lens is the objective whose properties define the resolving power of the microscope, that is, the minimum distance between two points of the object that it is possible to discern. The other intermediate lenses allows us to alter the magnification of the image at screen level. Finally, in the case of conventional microscopes, the condensers are adjusted so as to illuminate uniformly a relatively wide area of the specimen (several tens of microns).

— This is one of the basic differences with the STEM (Scanning Transmission Electron Microscope), the other type we will consider. Here the condensers are excited so as to focus the electron beam in a very fine probe at specimen level. Around 1970, Crewe thought of associating this concept with the field-effect electron gun, whose intrinsic luminance properties permit the realization of electron probes of 2 or 3 angstroms on the surface of the specimen. In this case, the various signals due to the interaction of the beam with the specimen are manipulated electronically to modulate the brilliance of a television system in which scanning is provided in synchronism with the scanning of the primary beam over the specimen. Here we are therefore concerned with a sequential mode of image formation, where the specimen is scanned point by point, with a resolution determined by the size of the probe. This new generation of microscopes, accompanied by a relatively costly electronic infrastructure, is gaining a foothold on the market and approximately twenty instruments manufactured by the British company Vacuum Generators are currently in operation.

The STEM, designed for the observation and high-resolution characterization of thin specimens, represents the most recent version of the SEM (Scanning Electron Microscope) commercially available for over fifteen years for the observation of thick specimens. Following a notion put forward by von Ardenne in 1938, it was Nixon and Oatley who, toward the end of the 1950s, developed the first instrument in which an electron beam (with a diameter of several hundred angstroms) was moved over the surface of a thick specimen. We thus observe reflected electrons that can be either back-scattered with the same energy as the incident electron, or emitted secondarily in the case of electrons pulled away from the target with a relatively low energy (10 to 100 volts). In the latter case, we first scan a layer 10 angstroms thick below the surface, and the results are very sensitive to the degree of cleanliness of the latter.

From the de Broglie relationship

$\lambda = \dfrac{h}{m\,v}$, a wavelength λ is

assigned to electrons having a speed v, for electrons accelerated at 100 kV and having a speed $v = 0.548\,c$ (c being the speed of light), a wavelength of 0.037 angstroms. It is this very small value for the wavelength of electrons used in electron microscopy, as well as their high probability of interaction with the material, that makes it a preferred tool for the study of solids with a resolution always limited by the properties of the objective lens (in the CTEM) or the objective-condenser lens (in the STEM) to a value on the order of 2 angstroms.

On this scale, after the work of Crewe and his collaborators, it was possible to observe individual atoms on very thin support films and even to follow their movement due to thermal agitation and the irradiation of incident electrons. Highly significant possible applications in molecular biology can be foreseen as soon as we master the techniques of selective tagging by heavy atoms on preferred sites in molecular biology. In the case of highly crystallized specimens, the diffraction of electrons through a network constitutes the main point. According to Bragg's law, verified in 1927 by Davisson and Germer, $\lambda = 2\,d\sin\theta$: if we know the energy (or the wavelength) of the electron, if we measure its angular deviation θ as its leaves the crystal, we can deduce the nature and the spacing of crystal stacking. This effect has been used for nearly twenty years by all metallurgists to visualize structural defects (twins, dislocations, precipitates) in a periodic stacking. More recently, by reconstructing the interference images of diffracted beams on groups of crystal planes, we learned how to obtain diagrams of lattices showing the regular succession of atomic planes. These images, of very high resolution, enable us to study the nature of the deformations experienced by a crystal lattice on a very small scale (less than 1 angstrom) around an isolated defect like the center of a dislocation.

These possible applications of the electron microscope will increase still more when it is possible to obtain information on the chemical composition of specimens, even on their electronic properties. To this effect, we

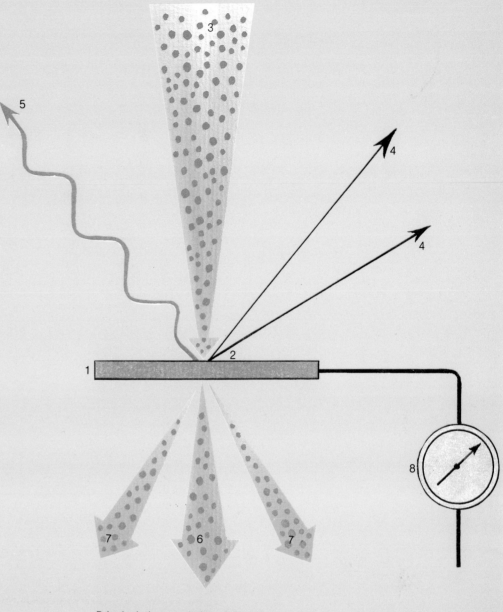

Principal characteristics of the electron microscope (Jed JSMT 200)
1. Sample
2. Probe
3. Incident beam
4. Reflected or secondary electrons
5. Secondary photon emission (visible UV,X)
6. Transmitted beam
7. Beams diffused elastically (with no energy loss) or inelastically (with energy loss).
8. Measuring the current received by the sample.

145

can profit by the elementary atomic specificity of certain events that occur during the interaction between high-energy primary electrons and the target. This is why there exists a significant probability of pursing certain electrons located initially along the underlying energy of atoms in the solid. This results in a slow electron and an ejected electron that leaves behind an empty hole on an atomic level. Three basic methods that use these data have been developed for the purpose of chemical analysis:

— First, in the 1950s, Raymond Castaing developed x-ray microanalysis, which consists in measuring the wavelength of rays contained in the x-ray beam emitted following deexcitation of the primary hole. By concentrating the beam in probe form of reduced size on the specimen, we are able to effect a point-by-point study of

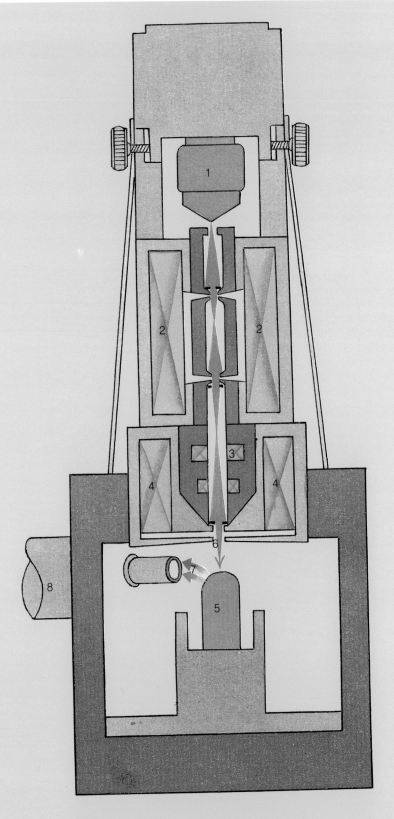

Column of a reflex scanning electron microscope
1. Gun
2. Condensing lens
3. Scan-coil
4. Lens system
5. Sample
6. Primary beam
7. Secondary or reflected beam
8. Pumping system

the chemical composition of the specimen with a resolution on the order of a micron when a solid is involved. This is an indispensable complement to electron microscopy on a solid specimen, as described in the preceding paragraph.

    — The second method uses the Auger effect. A high-energy electron (for example, belonging to the valence band) can fall into the hole of the inside orbit while imparting the equivalent energy to another ejected electron whose kinetic energy has to be measured. Because of the weak energies involved, the depth that can be analyzed is very reduced (less than 10 angstroms) and we are therefore dealing with a method of surface analysis.

    — Finally, the spectroscopy of energy losses consists in measuring the energy loss of the incident electron which has been slowed down on first impact. This method, proposed by Hillier and Baker from the United States, was developed during the 1970s in Orsay, Cambridge, and Chicago. At the present time, it seems to propose the most important intrinsic possibilities for exploring the chemical nature of masses as small as 10^{-21} gram.

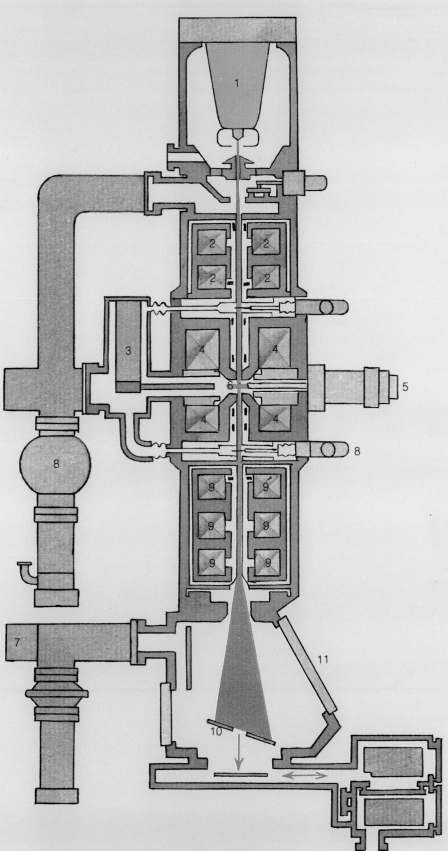

Column of a conventional transmission type electron microscope (Siemens CT 150)
1. Electron gun
2. Illuminating lenses
3. Anti-contamination device
4. Lens system
5. Introduction chamber
6. Sample (thin slice)
7. Pumping system
8. Contrast diaphragm
9. Projection lenses
10. Fluorescent screen
11. Viewing window

16
The eye: television

The term "television" was first used by Captain Constantin Perski of the St. Petersburg Artillery School, a delegate at the 1901 International Electronic Congress in Paris. However, the idea of remote transmission of images was already twenty years old. Traces of it can be found in the books of Jules Verne and in 1880 the Russian biologist P.N. Bakhmenev mentioned such a concept in his writings. It also appears in a satiricial futurist novel, *The Twentieth Century: The Story of a Frenchwoman of the Future*, written by French humorist Albert Robida in 1884. From 1873, as we have already seen, the American researcher Carey had the idea of using the photoelectrical properties of selenium cells. His idea was to project an image onto a screen made up of millions of these cells, each of which was connected to a light bulb on a remote screen that would reproduce the image. Following the work done by the Reverend Caselli, a French notary named Constantin Senlecq published his version of the "telectroscope" in 1881 in Paris, London, and New York. He described a process of image exploration point by point across an ebonite plate pierced with selenium-filled holes, with sequential transmission by electro-mechanical switching. At the same time another Frenchman, Maurice Leblanc, described another method of scanning and image exploration. A major break-through came in 1884, when a young German student from Berlin, Paul Nipkow, invented the scanning disc that broke down the image and converted it into electrical pulses. For the first time the image had been cut into lines. E. Belin used this procedure to transmit documents, while Nipkow's spiral perforated disc, in front of which was placed a photoelectric cell (the "electric telescope"), was to be used by the pioneers of television in many different countries. The first electronic image analyzer was developed by another German, Jonathan Zenneck, using the principle of electromagnetic deviation of electrons in a Braun tube. Subsequently, Wehnelt's "electron cannon" made it possible to eliminate the gas from the tube and to use vacuum tubes instead.

A. A. Campbell-Swinton also envisaged using electronic tubes for image scanning and drew up a sketch using a very fine cathode beam to trace the outline of an image projected onto a fluorescent screen. However his research remained theoritical, and it was left to a Russian researcher to put Campbell-Swinton's theories into practice. In 1907 Boris Rosing (at the St. Petersburg Institute of Technology) used a two right-angle drum system for transmission and one turning drum system for reception. This so-called cathoscope could produce several pictures per second. Another pioneer was the Hungarian Denes von Mihaly, who in 1914 undertook the construction of a television set with the support of the Hungarian Ministry of War and the Telefonfabrik company in Budapest. In July 1919 his "Telehor" succeeded in obtaining moving pictures of scissors, pincers, and letters over distances of 4 to 5 kilometers.

The main principles governing television were already known by the 1920s. The potential for integrating applied electronics into this new system was already obvious and, in the United States, Zworykin was already working on his kinescope, the first completely electronic picture tube. But the first breakthroughs were in mechanical television. In 1923, in the United States, Charles Francis Jenkins succeeded in transmitting television pictures of President Harding from Washington to Philadelphia. In 1925 John Logie Baird in England and August Karolus in Germany gave public demonstrations of their respective systems, followed by Herbert E. Ives's demonstration in the United States in 1927. Ives's career was heavily influenced by his father who had a passion for photography and optics. The son first became interested in photoengraving and painting, but later switched to electronics. His team at Bell Telephone used a mechanical scanning system of polygonal rotating mirrors or perforated discs. As the object was scanned point by point, a photoelectric cell recorded the different light intensity levels of the target. In the receiver the image was reproduced on the screen using a rotating perforated disc, synchronized to the scanner's disc and a neon light source that was modulated by the incoming signal. The picture was still rather foggy. Much later, scanning by electronic beam would eliminate the moving parts of the mechanical system and increase the sensitivity by providing a larger concentration of information at the camera and on the TV screen. However this was

With the advent of television the world became what American sociologist Marshall MacLuhan called "the global village".

149

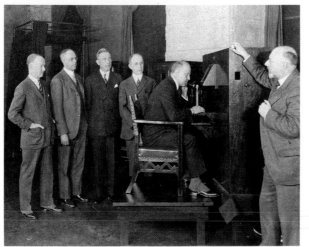

1. The multiplex telegraphoscope receiver built by E. Belin in 1907.
2. Baird's mechanical television system.
3. A.A. Campbell-Swinton.
4. Paul Nipkow (black suit) examining some of the first television equipment.
5. In the Bell Labs Auditorium in Washington D.C. Chairman Walter S. Gifford (ATT) watches the Honorable Herbert Hoover, Secretary of Commerce, as they chat during the first public demonstration of television on April 7, 1927. Dr Herbert Ives is on the right.

still in the future. The Prince of Wales agreed to be filmed by Baird, and in May 1925 Baird founded Television Ltd., the world's first television company. Remote reproduction of moving images relies on two optical limitations of the human eye: the limited power of resolution (the eye's capacity to separate is limited and we can only see a maximum of 20,000 points on a drawing 10 cm square); and the persistence of light impressions (the slower the speed of a flash, the clearer our sensation of alternating light and darkness.) If the changing speed increases, this variation becomes more and more difficult to perceive. The unpleasant flicker induced disappears at speeds above sixteen flashes per second.

The BBC was not very impressed by the thirty quite badly-defined line images and Baird had to work very hard to be taken seriously. His main adversary at the BBC was its Chairman, John Reith.

1928 was an important year for English television, with the first appearance of family receivers and the standardization of transmission equipment. In February Baird succeeded in making the first long distance transmission between London and New York and between London and the steamboat *Berengaria* the Atlantic. The same year saw the inauguration of a television department in the French Compagnie des Compteurs in Montrouge. The new division was directed by René Barthelemy with the collaboration of Dimitri Strelkoff. It was also in 1928 that Nipkow discovered the practical application of his patent using the Mihaly-Karolus system and demonstrated it at the fifth German radio exhibition in Berlin.

In 1929 the war between Baird and his collaborators on the one hand and the BBC on the other intensified. Baird was also receiving television transmissions from Germany at his English studios. In September the BBC transmitted its first televised broadcast. In 1930, when the BBC transmitted its first TV play (Pirandello's *The Man with the Flower in his Mouth*), Baird was working on improving the image and developing a bigger TV screen. In 1931 the BBC finally admitted defeat and Baird began regular broadcasts. This was also the year of the first outside broadcast, the Epsom Derby. Between 1926 and 1932 Baird attempted to develop new techniques: infrared television, radar and disc research, color television, long distance and transatlantic television. In the meantime, HMV and Columbia had merged in 1931 to form EMI. The new company's first objective was to build an entirely electrical television system based on the work done by the Scots scientist Campbell-Swinton. In 1934 EMI merged with Marconi. Isaac Schönberg, Director of Research for the new company, was friendly with two other Russian-born scientists, David Sarnoff and Vladimir Zworykin, who had developed a similar system in the United States. In France, another of Zworykin's friends, Dimitri Strelkoff, was working with Barthelemy on a technique (later demonstrated by Barthelemy) that could produce sixteen images a second. The first public demonstration to be held in France took place on April 14, 1931. The thirty-line transmission equipment was set up at the Montrouge Laboratory of the Compagnie des Compteurs and the single television receiver in the amphitheater of the Malakoff Advanced College of Electricity. Barthelemy had previously worked on radio with Ferrié. In 1929 the Chairman of the Compagnie des Compteurs made a trip to England to watch Baird's first experiments; this visit inspired him to create a television department of his own, with Barthelemy to head it up. In November 1935, Georges Mandel, then Minister of the PTT, commissioned 180-line television equipment, with

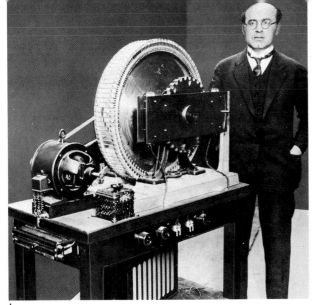

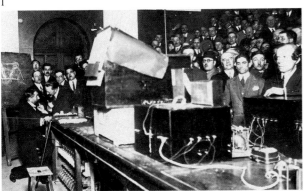

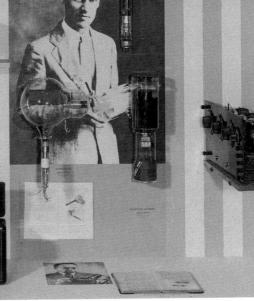

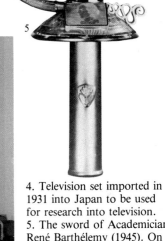

1. August Karolus (1893-1972), German television pioneer, with his mechanical television device (1928).
2. Fritz Schröter began his experiments in 1924 in the Telefunken Laboratories.
3. The first French demonstration of television at the Ecole Polytechnique in Paris carried out by Professor René Barthélemy.
4. Television set imported in 1931 into Japan to be used for research into television.
5. The sword of Academician René Barthélemy (1945). On the hand-guard is the Eiffel Tower, intended as a hommage to pioneers of wireless transmissions; on the pommel is the mirror drum; on the hilt, the cable specially designed for television which finishes in scroll-like ribbons of film.
6. Farnsworth and his image dissector.
7. Douglas Birkinshaw, the BBC's first television engineer, with one of the original Marconi EMI cameras at Alexandra Palace.
8. Alan Blumlein, who was killed in a wartime air crash while working on airborne radar.
9. King George VI and Queen Elizabeth visit Hayes in 1940. The Queen is talking to Schönberg.

151

1. The supericonoscope.
2. In 1929, Dr. V.K. Zworykin demonstrated the first electronic television receiver with the kinescope or picture tube, which he had developed along with the iconoscope or "eye" of the camera.
3. The Berlin Olympics in 1936 were filmed for television by Walter Bruch, later to invent the PAL TV system.

1

2

4. At left, John Logie Baird (1888-1946); at right, the Baron Manfred von Ardenne (born in 1907), pioneer of television and the electron microscope.
5. A cathode ray tube in 1939: a real monster!
6. *Charade,* the Japanese pioneer TV quizz program,, was broadcast regularly from February 20, 1953 to March 27, 1967.

3

4

5

6

SFR commissioned to supply high frequency equipment (transmitter, feeder, antenna). Louis Lumière and Ferrié took part in the first tests but were skeptical as to the potential of the new system. Barthelemy who, like Baird, was a man of rather delicate health, was assisted in his work by Strelkoff.

In 1933 Baird met Farnsworth, the American inventor (who, as we shall see later, had already felt the muscle of the large companies backing Zworykin) and decided to develop Farnsworth's idea of the picture director. But the bad luck that haunted Farnsworth was transferred to Baird. A fire at Crystal Palace in November 1936 destroyed all his equipment. In November 1936 the BBC installed studios and a television transmitter at Alexandra Palace, alternately using the Baird system and EMI's (Marconi) electronic system (which used the Emitron camera, and was based on the principle of the iconoscope). It eventually opted for the Marconi system. Nevertheless, Baird's name almost always crops up in any discussion of the early years of television, often to the exclusion of other researchers, particularly the EMI team led by Schönberg and Alan D. Blumlein, who were the true inventors of electronic television in England.

This is partly because in Europe there is a tendency to promote individual achievement at the expense of research teams, unlike the United States where Farnsworth was almost entirely neglected in favor of Zworykin and his team, and partly because Schönberg himself was reluctant to take advantage of any publicity. The victory of the Schönberg team sounded the deathknell of mechanical television and the defeat of Baird, who had always believed that he would obtain better definition with his system than would be possible with electronic television. His health, which had always been bad, began to deteriorate even further, but he continued his research with great courage and enthusiasm and began to concentrate more specifically on color television. When he finally turned to electronics just after the war, he founded a new television company; but he died in 1946 before he was able to produce an elctronic color system. If it is true that Baird's life was devoted to giving an idea of what electromechanical television could have been, Zworykin, on the other hand, quite adequately symbolizes the destiny of electronic television.

Zworykin was Boris Rosing's assistant in 1911 when the Russian obtained the first image of four white bars on a black background on a fluorescent cathode ray screen. In 1917 he set up an electronics laboratory (in those days the term "radioelectric" was still in use), first at the headquarters of the Russian Marconi company in St. Petersburg, subsequently in Moscow. The first demonstration of television in Russia would not take place until 1931, however, partly because Zworykin's work came to an end with the Russian Revolution. While Baird believed all his life in the potential of mechanical television, Zworykin had inherited from Rosing the view that there was only one future for television and that was electronic. In 1912 Zworykin had worked on x-rays with Paul Langevin at the College de France. When war broke out he enrolled in the Signals Corps. But in 1917, when Russia was overtaken by the maelstrom of revolutionary confusion, Zworykin hid and then rejoined the Allied Armies at Archangelsk, where he buttonholed an American diplomat and spoke to him about his concept of television. The diplomat was taken aback but allowed himself to be convinced and arranged a visa to the United States. In 1918, when he arrived in America, Zworykin found a job at Westinghouse and began once again to talk about his chief obsession: television. No one seemed very convinced, but Zworykin persisted and on December 19, 1923, he took out a patent on the first entirely electronic picture tube, the *kinescope* through which he succeeded in obtaining the picture of a cross. He was also working very hard on another tube at the same time, the revolutionary *iconoscope*. His demonstration of the cross transmission aroused little enthusiasm. The chairman of Westinghouse found it amusing but advised him to work on something a little more useful, more commercially oriented! Fortunately for Zworykin, he had great success in Rochester in 1929 when he showed his system to the Institute of Radio Engineers.

He met David Sarnoff, Managing Director and Chairman of RCA, which had been founded ten years earlier. Sarnoff was very impressed with Zworykin's ideas, and hired him to direct RCA's electronics research laboratory. In 1931 Zworykin built the first all-electronic camera, thanks to his iconoscope.

Strelkoff admits that "the iconoscope revolutionized television, but it did have one defect — it produced a secondary interference effect, a kind of black spot on the image which we couldn't correct with any of the different circuits we tried. But without the iconoscope it is unlikely that the other television tubes would have been developed as fast as they were."

During the war, Zworykin developed the "sniperscope" or "snooperscope" for the Allies. It was a device capable of converting invisible infrared rays into visible light and was used to spy on enemy camps. During this same period an American farmer by the name of Phil Farnsworth, working almost alone and without any kind of backing, developed an original electronic device called the *image dissector*.

Farnsworth's system employed a photocathode which statically registered changes in image intensity, and the image was made to move electrically across the photocathode by the use of magnetic fields instead of an electron beam scanning across a fixed image as had hitherto been current practice. But although the dissector corrected some of the iconoscope's defects, the system proved too complicated; this was one of the reasons for its failure as a commercial product. Farnsworth's backyard inventiveness was no match for the giants of the industry. Between 1931 and 1939, RCA unleashed its might against the tiny Farnsworth company, spending $2 million on patents and $7 million on research into television development. Notwithstanding the competition, Farnsworth's equipment and patents were extended as far as England (by Baird) and Germany, and in July 1936 the Fernseh AG company used Farnsworth's cameras to televise the Berlin Olympic Games alongside Telefunken's iconoscope cameras. But in 1940, discouraged, though definitely not beaten, Farnsworth retired.

After the war, Zworykin worked with John Von Neumann at Princeton on electronic devices for meteorological forecasts and cyclone control. At the beginning of the 1940s he worked on the electron microscope with a group led by James Hillier. 1936 was a very important year for electronic television. Zworykin's system was used in New York at the recently-opened RCA station in the Empire State Building, and during the same year F. Schröter demonstrated his device in Germany. Manfred von Ardenne had been working for many years on the transformation of the electrical image into a light image and in 1931 he and Siegfried Loewe gave the first tests of electronic television in Berlin-

Lichtenfelde. In France, after a first public demonstration given by Barthelemy at the Malakoff Advanced College of Electricity, the French research effort started to take off, and the first all-electronic equipment was built. During the period from 1929 to 1935 the equipment in use had been almost entirely electromechanical, the only electronic components being photoelectric cells, amplifier tubes, and the light source in the receiver. In Paris in 1928 the Minister of the PTT inaugurated the new 25 kW Eiffel Tower television transmitter built by the LMT company to Emile Labin's design. In 1939 A. Rose and Lams invented the orthicon which eliminated the iconoscope's main defect — secondary electron emission.

The stage was set for television's fantastic postwar success. Regular transmissions started in the U.S. on April 30, 1939, and in the USSR in December 1945. They started much later in Japan (1953) and in Peking, where the first experimental station was not set up until 1958. In 1927, Kenjiro Takayanagi developed a forty-line TV system; this represented Japan's first real success in the field of TV. Research continued at the University of Waseda in 1928, and in 1930 NHK also entered the fray. The Japanese were to inaugurate their television system at almost the exact moment that the Pacific War began (December 1941). After the war, television had to wait in the wings until the signature of the 1951 San Francisco Treaty brought the American occupation of Japan to an end.

The electronic picture tube consisted of a cathode ray which swept back and forth across the TV screen — 525 lines 60 times a second in the United States, and 625 lines 50 times a second in Europe. These standards were not chosen at random; the scanning speed was dependent on multivibrators which were in turn governed by the mains frequency cycle (50 Hz in Europe, 60 Hz in the United States.) In 1947 and 1948 cathode television tubes made their debut in France (TTC, Lampes Fotos) along with other specialized tubes (SFR, Radio industrie). Over the next few years, improvements were made in picture analyzers, with the iconoscope being used first to establish the light/current image transformation system based on rapid electron scanning (1,000 to 1,500 volts). The related defect, as we have seen, is due to the emission of secondary electrons produced in the scanning process, which are then randomly redistributed across the target, producing spurious marks. The orthicon abolished this disadvantage by using a slow electroscanning system whereby the electrons emitted by the cathode were first of all accelerated to a level corresponding to approximately 250 volts and then slowed down, so that in the absence of light on the photosensitive plate, this speed would be practically nil and even negative in the vicinity of the target plate. The target plate was illuminated on one side by the light image and scanned on the other side by the electron scanning beam.

After 1940, Zworykin developed and marketed the supericonoscope which gave very good definition. The orthicon picture which in 1946 was claimed to be "as sensitive as the human eye," was now very large and used for black and white photography. All these analyzer tubes are constructed on the principle of photoemission. In 1950 RCA presented a tube that was also based on photoconduction using sensitive layers on which the luminous image was screened. This was the vidicon, smaller but less sensitive than the orthicon. The latest of these picture analyzers, used more particularly for color television, is the plumbicon. The plumbicon was developed by a Philips team including E.F. de Haan, A. van der Drift

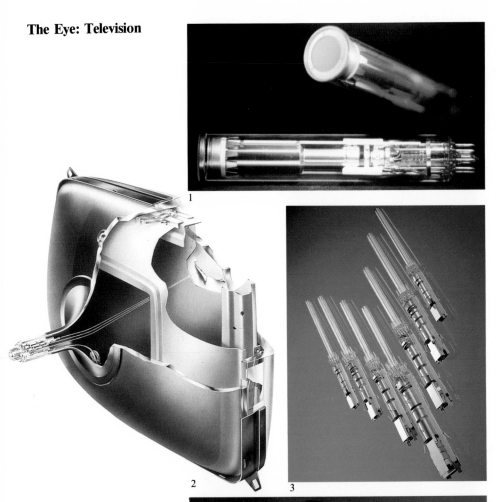

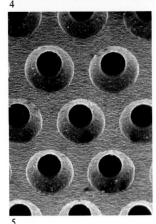

1. The vidicon.
2. Exploded view of an auto-convergent color picture tube.
3. Trinitron electron gun.
4. Internal view of a color television from the 1975 generation. Each year the introduction of new high-performance components leads to a reduction in the number of components required, and a resultant simplification of assembly procedures.
5. Hole mask of a color tube: each ray (blue, green and red) is modulated differently as it passes through approximately 0.5 million holes making up a total of 1.5 million points. To respect the homogeneity of light and the purity of the curves it is essential for the shapes of the holes and the distances between them to be carefully controlled.
6. Morphological structure of the light sensitive layer of the color tube (here zinc sulphide is used).

and P. P. M. Schompers. Up until then, the iconoscope or the orthicon had been used for live transmissions of moving scenes, because they were the only two types of tube with sufficient resolution power and response speed for transmitting detailed moving pictures in TV production work. The vidicon was particularly appreciated for its simplicity and ease of use but it produced an uneven picture and had a very slow response time in weak light. It could therefore only be used when the ambient light was fairly strong. The plumbicon had the same advantages as the other tubes, combined with greater sensitivity, faster response time and better resolution. Like the vidicon it was smaller and less complex than photoemission tubes, an important factor in the case of color television where three tubes were needed (blue, red, and green.) But whereas the orthicon picture distorted picture contrasts (a factor not particularly important in black-and-white television but obviously serious when used in color television as it tended to fade one color into another) the plumbicon eliminated this distortion. Work on the new tube began around 1954. Like the vidicon it was a photoconduction tube, but it used lead oxide, in preference to antimony. This compound is very unstable and its operating conditions need to be closely controlled. In the plumbicon tube, a sheet of glass is coated with a transparent conductive layer of tin dioxide upon which is placed another fine layer of a photoconductive material (lead oxide). The scene to be transmitted is projected through the glass and the tin dioxide onto the lead oxide. A beam of slow electrons hits the other side of the lead oxide layer. The plumbicon can thus be seen as a kind of vidicon with an essential difference. The photoconductive layer is not only constituted from a different photoconductor (antimony for the vidicon and lead oxide for the plumbicon), but what is even more important, the lead oxide layer joined to the tin dioxide layer forms a consistent three-layer unit each constituting a different type of conductor: a layer of almost pure lead oxide (intrinsic semiconductor), a layer of tin dioxide (type n), and the contact layer between the first two layers which can also provide an n-type layer in the lead oxide.

Today the plumbicon is the best professional quality picture analyzer. It has one rival on the general market, the trinitron, or single-tube camera, developed by Sony in Japan. Ibuka officially announced the new color system in April 1968. In 1961 LEP in France designed an analyzer tube for infrared pictures and in 1963 a cathode tube capable of visualizing transitory phenomena traveling at 1/10th of a nanosecond. In the field of picture tubes, cathode ray tubes have been improved, and the screen has become flat rather than convex while the angle of beam deviation has gone from 60° to 110°. In 1939 Fisher developed a special procedure which gave birth in 1950 to the Eidophore. This device consisted of utilizing a beam of light reflected from the surface of a film of oil, which in turn was modulated by the electrical charges produced by the scanning beam. As early as June 12, 1902, a German researcher, Otto von Bronk, had taken out a patent for color television. In 1925 Zworykin took out his color tubes patent. In 1928 Baird had developed a color technique that Ives was to use a year later in the United States. "Shadow-Mask," the first successful color tube, was developed by David Sarnoff of RCA, and in 1953 that company standardized the NTSC (National Television System Committee) developed by the Hazeltine Laboratories and its team of engineers led by Charles Hirsch. In 1959 Henri de France invented the Secam (Sequential

Memory System). In 1962 a German researcher, Walter Bruch, combined Secam's delay line procedure and the NSTC to come up with PAL (Phase Alternative Line).

1

2

1. Assembly line of color video projection system.
2 and 3. Television screens in a production line.

3

Television

by Dimitri Strelkoff, retired engineer.

*Dimitri Strelkoff was, with René
Barthélémy, one of the pioneers of
French television. In this article he
explains the basic principles of the
television image and describes the tubes
used and the techniques of color
television.*

Analysis and Reconstitution of an Image

In order to transmit an animated
image over a distance, the following
operations are generally performed
during the entire transmission period:
a) Successive breakdown of the optical
image to be transmitted into elementary
points.
b) Conversion of elementary light
impulses thus obtained into electric
impulses.
c) Amplification of electric impulses.
d) Transmission of electric impulses over
a distance.
e) Inverse conversion on reception of
electric impulses into light impulses and
successive reconstitution-synthesis of the
image.

In order for the image to be
reconstituted faithfully on reception, it is
essential for each elementary point of
the image analyzed on transmission
having the coordinates X and Y to be
reproduced on reception with the
coordinates X1 and Y1 such as
$X1 = Ax$ and $Y1 = Ay$, where A is
the proportionality coefficient defined
by the ratio of image dimensions on
reception. The rates of displacement of
the analysis "spot" on transmission and
that of synthesis on reception should
have the relationship

$$\frac{dX}{dt} = \frac{Adx}{dt} \text{ and}$$

$$\frac{dy}{dt} = \frac{Ady}{dt}, \text{ i.e., analysis and synthesis}$$

are effected according to the two
simultaneous displacements parallel to
the X axis and Y axis. In other words,
the image is scanned and reconstituted
point by point and by parallel lines, like
reading a book, but a book in which
you go abruptly from the end of the last
line to the beginning of the first.

This implies that the two
simultaneous movements of the analysis
"spot" should be synchronized with
those of synthesis and that the law of
their displacement is the same.

In the case of black-and-white
television, the width of the frequency
band to be transmitted is proportional
to the number of elementary data per
second characterizing the picture, that
is, at the repetition rate of the frames,
to the number of points per frame and,
for each point, to the number of half-
tones distinguished.

Naturally, all the reasons, whether
technical, economic, or the number of
simultaneous transmissions possible in a
given frequency range, tell in favor of
the transmission of a picture of given
quality by means of a frequency band
as narrow as possible.

If the number of points per picture
had been since the beginning of
television essentially limited by the
maximum performance of transmission
analysis devices, the determination of
the desirable frame speed had to make
allowance for an unforeseen
phenomenon related to the physiological
properties of the eye. Whereas for
cinematographic projection the
flicker effect disappears before
16 frames/second and is consequently
not visible, even at strong illumination
with the current standard of 24 frames/
second, the phenomenon is still very
pronounced when the picture is
described by the very bright mobile spot
of a cathode-ray tube, even if one tries
to attenuate it by using slightly
remanent screens.

In order to avoid this "flicker"
effect it was necessary to increase the
frame speed in television as compared
with cinematography. To facilitate
synchronization, it was initially chosen
equal to that of the electric distribution
network: 50 in Europe, 60 in North
America. This led to the doubling in
Europe and multiplication by 2.5 in the
United States of the frequency band
necessary for the transmission of movie
film without improving the "definition,"
the fineness of detail visible in the
picture. One device permits us to avoid
this "wasting" of frequencies: interlaced
scanning.

For this, the complete cycle is
broken down into two periods:
the scanning of even lines is effected at
1/50th or 1/60th of a second, followed
by the scanning of odd lines during the
next scanning sequence. We thus
transmit the total number of points in a
picture in 1/25th or 1/30th of a second
without changing the bandwidth, and
the "flicker" is eliminated.

Once the rate and scanning model
were defined, the efforts to improve
picture definition could only have been
crowned by the advancements made in
the basic components of the complex
chain of devices used between the shot
and the screen of the receiver: relaxation
oscillators, pulse amplifiers, electron
optics. It was the lack of technical and
even scientific know-how in these fields
that limited the early experiments in
1920 to 1930, despite the ingenuity of
the precursors.

Electromagnetic Analysis and Synthesis Techniques

Since 1920, when the electronic
amplifier and sensitive photoelectric cells
first appeared, electromechanical systems
made it possible to embark on the study
of applied television. In 1923, large
companies in all countries, foreseeing
commercial possiblities, opened
television research laboratories while
entrusting the organization and
management to internationally renowned
pioneers in applied television: John
Baird (England), René Barthélémy,
Henri de France and Marc Chauvière

(France), Jenkins and Zworykin (USA), F. Schröter and A. Karolus (Germany). All experimented with the following devices on transmission and reception: the Nipkow disk with one spiral, the Nipkow disk with several spirals, the Brillouin disk with lenses, the Weiller mirror drum. For transmission, they worked with simple photoelectronic cells and photomultipliers, as well as cells with several internal amplification stages. For reception, the devices included a large-surface neon lamp, a "crater" neon lamp and the Kerr cell; the cathode ray tube has been used exclusively since 1930.

Going through the succession 30, 60, 120, and 240 lines the results obtained were.encouraging but practically inadequate to satisfy our vision requirement: the resolving power of the eye. The slightest attempt to improve definition by increasing the number of lines created growing and insurmountable difficulties. This is why, since 1935, all television laboratories abandoned electromechanical systems to replace them with a special cathode ray tube proposed by Zworykin — the iconoscope.

Electronic Television

1. High-Speed Electron Analyzer Tubes
a) Iconoscope

This is an analyzer tube consisting mainly of a cathode ray oscilloscope whose screen is composed of a multitude of individual photoemissive cells placed on a mica plate, following a special procedure, supported by a conducting electrode called the "signal plate." The picture to be transmitted is projected onto this mosaic. The electron beam sweeps across the entire surface of the mosaic according to the scanning law.

b) Supericonoscope

To increase sensitivity even more, Zworykin proposed to combine an electronic image intensifier and an iconoscope, that is, the optical image is no longer directly formed on the photoelectric mosaic of the iconoscope but on an auxiliary transparent photocathode, the electron transmission from which will form, due to suitable electron optics – a homothetic (scaled-down) high-speed electron picture on the target of the iconoscope. This target is made from a material selected for its high secondary emission factor in order to obtain a relief of charges on the elementary condensers amplified by a factor of 3 to 4.

c) Disadvantages

The impact of the electron beam with the elements of the mosaic occurs at a high speed on the order of 1,000 volts (high-speed electrons). When the beam reaches the first cell of the first line, it discharges from the elementary condenser only the quantity of electricity due to the illumination of this cell. However, the emission of secondary electrons produced by this first discharge will more or less charge all the adjoining cells. When the beam reaches the second cell it will have, on discharge, not only a quantity of electricity produced by illumination but also that of all the secondary electrons from the first cell. The number of secondary electrons gradually increases as the cathode ray beam moves along the line. The same phenomenon occurs when the beam goes from one line to another.

This emission gives rise to the appearance of signals called "spots" that are superimposed on the picture signals, properly speaking, and their amplitude can reach a value several times higher than that of the picture. The parasitic secondary emission is stronger as the density of the electron beam increases. In practice, this defect is corrected by an entire set of circuits with linear and parabolic characteristics. Despite the defects mentioned, high-speed electron analyzers have for many years occupied an important position in television technology because of their high definition, with an illumination on the order of 1,000 to 2,000 lux.

2. Slow-Electron Analyzer Tubes
a) Image orthicon

It was the orientation of research toward the reduction of the speed of electrons in a beam that led to a system which R.C.A. called "superorthicon" or "image-orthicon."

The picture is formed on a semi-transparent photoemissive surface. The electrons from this surface form a second picture on a target composed of a glass plate, which supports a grid with a very fine mesh. The secondary electrons emitted by the glass are collected on the grid, which is scanned by the cathode ray of the tube. A decelerating electrode, located near the grid, is at negative potential in relation to the cylinder of the tube. Under these conditions, in the absence of light on the photoemissive surface, the electrons in the beam are repulsed by the grid and turn to the cathode of the tube cylinder. For a strongly and uniformly illuminated photoemissive surface, the grid potential becomes positive with respect to the electrode, and the electrons in the beam are fully used to bring the grid potential back to the initial state without return electrons. Under these conditions, if the photoemissive surface is illuminated by an optical image to be transmitted, the electron density of the return beam is proportional at all the light points of the picture to be transmitted, thereby forming the electric signals of the picture. By means of suitable focusing devices, these signals are delivered to the input of a multistage electron multiplier, making it possible to obtain a sensitivity exceeding that of the photographic emulsion.

b) Vidicon

This is a photoconductive-effect tube similar to the plumbicon, reduced to the basic elements of an electron gun and a target, composed of a photoconductive layer deposited on a transparent metal plate, called the signal plate. The image to be transmitted is projected through the signal plate onto the photosensitive layer. The inside wall of the tube is raised to approximately 300 volts, and the signal plate to about 20 volts. In the absence of light on the target, the conductivity of the photosentivite layer is very low (very high resistance) and the output current to the signal plate under beam action is almost zero. For maximum illumination of the target, the internal resistance is minimal under beam action, and the output current through the signal plate is maximal; that is, the output current along the signal plate is proportional to the illumination at each point of the image analyzed. The vidicon tube, of very simple design, has a high photoelectric sensitivity, very high resolution, and is at present commonly used in both black-and-white and color television.

Color Television

1. Physical Principle of Color Television

The physical principle of color television is based on the use of an additive mixture of colors. After a very large number of experiments, the C.I.E. (Commission Internationale d'Eclairage) selected three primary colors of light as international standards: red(R), green (G), blue (B). By varying the ratios of light intensities of the three primary colors R, G, and B, we can obtain all the colors of the visible spectrum. In all color television systems, black is obtained automatically in the absence of light.

2. Sequential Color Transmission System

In a standard black-and-white television set, a disk containing filters for the primary colors R, G, and B is positioned between the lens and the analyzer tube. On reception, a disk with three filters R, G, B is mounted in front of the white-screen picture tube, whose speed and phase are synchronized with that of the photographic camera or movie camera.

In a rotation speed of two disks high enough to avoid "flicker" due to the successive color change, we see a color picture of the scene on reception.

The system was studied and perfected by C.B.S. In 1950 this same system was approved by the F.C.C. (Federal Communication Commission) for public broadcasting on a provisional basis.

Disadvantages of the System:

The successive transmission of each color requires a frequency band equal to the black-and-white system. The successive transmission of three colors requires a frequency band three times wider than that of black-and-white television. This solution is unacceptable. The use of electromechanical components renders the system inelegant, especially as the filter disk of the receiver, taking into account the dimensions of the picture tube, can have cumbersome dimensions. This is why, in 1951, the F.C.C. proposed that different companies combine their effort under the name N.T.S.C. (National Television Systems Committee) to find a satisfactory solution for color television.

3. Simultaneous Color Transmission System

In order to make wide commercial development possible, the N.T.S.C. imposed two major restrictions:

– Obligatory use in the color system of the same frequency band as for black-and-white

– Possibility for black-and-white receivers to receive color transmission (compatibility of the system)

These would be scientifically unsolvable if the physiological properties of the eye did not again allow technicians to cheat without much harm for the final subjective appreciation of the television viewer as regards the apparent quality of the color picture.

Actually, the resolution of the eye is much weaker at the two ends of the spectrum (blue and red) than in the middle (green and yellow). This justifies the use of the following procedure.

By means of suitable filters, the picture to be transmitted is divided into three pictures, R, G, B, characterized by two factors: *luminance* (brightness of the picture) and *chrominance* (color of the picture). The three analyzer tubes in each camera simultaneously convert the three color pictures into three video-frequency channels – R, G, and B. The G channel, occupying nearly the entire width of the frequency band, is authorized to fulfill two roles: chrominance for color television and luminance for black-and-white sets. The R and B channels, at a reduced frequency band, provide chrominance. The width of the frequency band is used for a whole series of additional techniques such as coding, decoding, carrier current, subcarrier current, amplitude modulation, frequency modulation, and electronic switching to obtain three video-frequency channels R, G, and B, at the output of the three amplifiers in the receivers.

The mask-type color television tube, proposed by R.C.A. (Zworykin), comprises three adjoining cylinders, RGB, arranged in a triangle. The three beams are deflected simultaneously in accordance with the scanning law by a single set of magnetic coils.

The screen is composed of a mosaic of three fluorescent powders RGB, forming a regular network of small triangles in which the colored phosphor dots RGB are located at their vertices.

At a short distance behind the screen there is a metal mask containing as many holes as elementary triangles so that the three electron beams RGB converge at mask level and then separate, each one falling on the phosphor dot corresponding to each of the colors RGB as it passes across a hole.

A color television set with a simultaneous color transmission system should necessarily contain:

a) For transmission, a color (filter) discrimination lens, three analyzer tubes, a horizontal scanner, vertical scanner, and focusing scanner, three video-frequency amplifiers, and a frequency distribution coder.

b) For reception, a frequency distribution decoder, three video-frequency amplifiers, a horizontal scanner, vertical scanner, and focusing scanner, and a mask-type picture tube.

Despite the serious problems in developing the basic system, it is now in generalized use. There are three versions of it:

– The American N.T.S.C. system of 525 lines and 30 frames/second

– The French SECAM system of 625 lines and 25 frames/second

– The German PAL system of 625 lines and 25 frames/second

The difference between the three systems only concerns the coding and decoding (frequency discrimination) devices that require the use of special matrices to go from one system to the other.

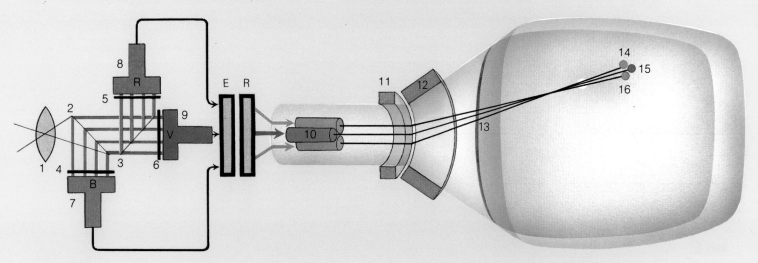

Mask tube:
1. Lens
2, 3, 4, 5 and 6. Color discriminator
T. Transmitter, R. Receiver
7, 8, 9. Video frequency analyzer tubes; RGB
Color video frequency outputs
10. Three electronic guns
11. Focusing
12. Deflection coil
13. Mask 14, 15 and 16. Elementary color triangles
17. Focusing

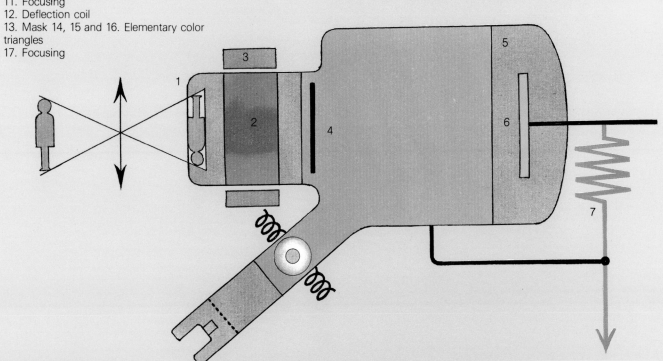

Supericonoscope:
1. Emitting photocathode
2. Focusing field
3. Focusing coil
4. Acceleration voltage
5. Photoemitting mosaic surface
6. Signal plate
7. Signal output

The ear:
tape recorder and record player

The great ancestor of sound reproduction was, of course, Edison's phonograph. A recording was made by positioning the singer or musicians under a large metallic funnel, the tip of which was directly connected to the recording needle. The needle vibrated in sympathy with the vibrations picked up by the funnel and traced out furrows on a rotating tin drum that slowly traveled along its axis, leaving a spiral pattern on the surface of the tin. To replay the recording, the drum was simply returned to its starting position and the needle retraced its path, thus reproducing the recorded signals from the drum (later replaced by the disc). Various different attempts to reproduce sound were made before Edison's phonograph. For example, Poulsen presented his "Telegraphon" at the 1900 Universal Exhibition. In 1886 Werner von Siemens had suggested a special listening system with huge loudspeakers linked to the phonograph. In 1887 the Frenchman Janet had written about the "transversal magnetization of a conductor," and he described the possibility of reproducing sound by magnetizing a thread of homogeneous steel. Then, in 1888, the first person to envisage all the technical problems associated with magnetic reproduction was Oberlin Smith, who wrote an article entitled "Some Possible Forms of Phonograph" for the magazine *Electrical World*. But the most important breakthrough occurred in 1898 when Waldemar Poulsen developed his Telegraphon, which consisted of a thread of steel wound in a spiral around a cylinder. It recorded magnetic signals that were then reconstituted by means of an electromagnetic reader, and that could also be erased. But the essential component – electronic amplification – was still missing.

Poulsen's invention, in 1900, was taken no further for another twenty years because of the lack of immediate results and the technical means necessary to develop it.

Toward 1918 Curt Stille decided to take another look at Poulsen's system, followed in 1922 by Mario Marchetti and Antonio Padiglione. For thirty years, the thread and the steel strip had seemed the best possible equipment for magnetic reproduction of sound and any subsequent research merely involved attempts to reduce the thickness of the components.

Independently of the work done on Poulsen's device, Carlson and Carpenter, working at the Naval Research Laboratory in the United States in 1921, installed magnetic detectors for recording and reconstituting signals. They found that it was possible to record weak magnetic signals if an alternating current of around 10,000 Hz was superimposed onto the signal. But they did not interpret the effect as they should have.

1932 the first recorded documentaries were broadcast, including King George V's Christmas Message transmitted by the BBC. Some progress had been made in the field of amplification. Poulsen had obtained a signal-to-noise ratio of 15 to 18 dB, whereas now 25 to 30 dB was considered normal. However, spool techniques had not yet been completely mastered and the wires broke easily and very frequently. Oberlin Smith was the first to suggest carrying sound by means of a flexible support sprinkled with very fine iron powder, but he did not succeed in developing his idea. In 1921 a Russian, Boris Nasarischvili, tried his hand. He had some degree of initial success with copper wire covered with nickel; he then put forward the idea of a recording material consisting of strips of nickel paper. The third attempt turned out to be the best. Fritz Pfleumer, a German researcher, took out a patent in 1928 for his magnetic tape. Pfleumer was a somewhat bizarre and flamboyant character. He had already been responsible for a number of inventions including crepe latex and improved cigarette tips. He had the idea of sprinkling bronze powder on the plastic film used at the time so that it would not discolor. Perhaps inspired by an article he had read on steel strip recording by Curt Stille, Pfleumer replaced the steel strip with a fine strip of plastic or paper material with a coating of steel powder. His first demonstration at AEG was not very convincing because the iron sprinkling technique was not uniform enough and the band was too thick. If iron or iron oxides are subjected to a magnetic field, they will retain a residual magnetism when the magnetic field is removed (hysteresis) and can then be used to record signals. To make a recording the tape has to be brought near an electrical circuit in which a current has been induced. Pfleumer was the first to build a unit that could

The compact stereo-disc music center for discs and cassettes.

The ear: tape recorder and record player

1

2

3

4

1. Edison's phonograph (1877).
2. Thomas Alva Edison (1847-1931) was a most extraordinary man, a self-taught genius responsible for a number of inventions including the 2-way telegraph, the microtelephone and the incandescent lamp.
It is less well-known that he set up on his own property the world's first movie film

5

6. Synthesizing chamber for magnetic powder: the reaction speed of the magnetic powder is controlled and made to react uniformly.

studio and that he founded the first film production company in the United States in 1898.
3. Vladimir Poulsen (1869-1942).
4. Poulsen's coil and motor telegraph (1898).
5. An Ampex F-44 audio-receiver (1963) next to John Mullin's German Magnetophon, manufactured around 1940. The Ampex F-44 is only intended as a consumer product, while Lyndsay's Model 200 and 300 were professional recorders.

162

6

reconstitute the recorded sound. His technical means were still quite primitive and the sound produced was distorted by noise, but he did manage nevertheless to play back signals through a loudspeaker.

In 1933 a German minister decided to try to improve Poulsen's unit. He succeeded in interesting BASF in his idea and F. Matthias was made research director. AEG then asked Eduard Schuller to devote himself to developing tape heads, and Schuller obligingly came up with the first real breakthrough. He developed a round tape head that produced a magnetic field, thanks to the inclusion of an electromagnetic coil. He designed the head in such a way that the recording or playback signal field could only leave or enter the head through a tiny, critically-positioned slot on the face of the device. The sensitized steel strip was held in contact with the coil on both sides by two pointed poles. In this system the tape head recorded, erased, and played back. For many years this was the basic principle behind all reading heads.

In 1935 another decisive breakthrough occurred when a much finer chemically produced magnetic powder replaced steel powder. Interference decreased and tapes became more supple and more adhesive. Their width was decreased to 6.5 mm and their speed reduced to 30 inches a second. In the same year T. Volk demonstrated his system, which used a three motor drive system to increase maneuverability and tape speed, which became ten times faster. The first "Magnetophon" produced industrially for cinema equipment was shown for the first time at the Wireless Telegraphy Exhibition in Berlin, in 1938. During the war some 500 researchers were set the task of improving this equipment, exhibited in 1941 at the Palast UFA in Berlin.

The main problem encountered was the elimination of interference, and W. Weber (Reich Radio Company) worked on this problem with H. J. von Braunmühl of AEG. To eliminate the distortion due to nonlinearity of the magnetic tape, they developed a signal linearization technique based on a polarization effect of a DC signal; the next step was to replace this DC signal by an ultrasonic oscillator. Von Braunmühl discovered a phenomenon similar to that discovered twenty years earlier by American researchers Carlson and Carpenter and rediscovered two or three years earlier by Nagai, Sasaki, and Endo in Japan. But von Braunmühl knew how to apply it. He had actually rediscovered the effect of high-frequency premagnetization, in which a high-frequency signal, oscillating outside the audio spectrum, is mixed with the recording head signal. This high frequency causes a demagnetization effect; it eliminates all the tape's magnetizations. Just like a completely nonmagnetized tape, it cannot induce any tension in the reading head and no noise can be heard where there is no signal (in the intervals between words or sounds). As soon as a sound, or magnetization, does occur, the noises due to the granular structure of the strip also reappear, but they are not audible as they are covered by louder sounds. AEG acquired the rights to the Weber-von Braunmühl patent and demonstrated the new device in June 1941.

Although the Germans continued to improve the tape recorder after the war, the best work was actually done in the United States where it contributed to the success of a number of companies including Ampex. Alexander Mathew Poniatoff, founder of Ampex, had retained from his Russian childhood a very strong love for violins and Tartar music. Legend has it that this passion was responsible for his interest in recording. Harold Lindsay, who had worked on the Manhattan atom bomb project, and who had an unashamed love of classical music, joined the team along with Jack Mullin, a former officer in the Signals Corps, and like the others, very keen on classical music. When Mullins had been posted to London, he heard German radio transmissions of concerts that impressed him with their high quality. He also noted that Hitler's speeches, as he moved around from one city to another, were retransmitted so quickly that only a flying carpet would have enabled them to be live transmissions. He began to wonder what recording method the Germans were using. Before the war, Mullin had worked on sound recording for a San Francisco cinema company that gave him leave to go to Germany to have a look at recording techniques. In Germany, Mullin discovered the tape recorder. Taking advantage of the post-wartime practive whereby returning soliders could take possession of confiscated equipment as war souvenirs, he made off with two tape recorders, took them apart, and sent them in pieces to San Francisco. During this period, Lindsay was discussing the possibility of producing high fidelity recording procedures with G. Forrest Smith. In July 1946, Lindsay attended a radio engineer's meeting (IRE) in San Francisco, where Mullin demonstrated the tape recorder he had brought back from Germany. But Mullin was keeping the nature of the recording head a secret.

In December 1946 Lindsay left the company he was working for and started designing magnetic heads for Ampex. He designed the company's first magnetic recorder, the 200. The most popular figure in show business at that time was Bing Crosby, who was particularly interested in recording quality. Mullin was invited to record the Crosby show on his famous tapes, despite the skepticism of ABC company technicians. Mullin, of course, obtained superior quality and Crosby, convinced, asked Mullin where he could buy advanced magnetic recording equipment designed by Ampex. In April 1948, the first machines were delivered to ABC. The 300 (which was much more economical and whose tape speed was slower) replaced the 200. A military model, the 500, was later developed for the Navy.

In 1953 a Dutchman, W. K. Westmijze, summarized the theory of magnetic recording. In the meantime, work had been done on improving heads, and Philips introduced ferrite tape heads. In the field of the microgroove disc, Columbia (Peter Goldmark) introduced the vinyl resin 33 1/3 disc in 1948 and in the following year RCA presented the 45. To record a disc, technicians had always used a self-centering steel needle located in the field of an electromagnet that was placed between two poles of a permanent magnet. When the electromagnet was subjected to an amplified microphonic current, it began to vibrate and the steel needle traced out grooves in the disc, which was covered with a wax layer and turned at a speed of 33 1/3 rpm. The recorded disc — and this was, in fact, where Goldmark's true originality lay — was then covered with a conductive layer of fine graphite powder and dipped in a copper sulphate bath. A copper plate was placed against it, and a continuous voltage applied between the disc connected to a negative pole, and the plate, connected to a positive pole. A galvanoplasty process then took place which caused copper to be transferred between the plate and disc. After a certain time the new master disc was removed, from which a number of copies could be made. The final disc was obtained by compressing polyvinyl chloride plates against the master disc at high temperatures.

Record player pick up heads were designed using the same basic principles as recording heads. A diamond or sapphire needle was fixed to the electrical

magnet placed between the two poles of a permanent magnet. The movements in the magnet field induced currents that were amplified and transferred to the loudspeakers. Piezoelectric devices also began to appear. In these devices the vibrations were transmitted by means of an electric coupling substance instead of by the electromagnetic procedure.

1. Recording studio control room.
2. From the classic 33 rpm record to the Compact Disc (with its player).
3. The latest generation of recording cassettes: the micro-cassette for dictating machines.

1

2   3

18

The Second World War: radar

Sonar was discovered some twenty years before radar. After the sinking of the Titanic, an Englishman suggested that the presence of submerged obstacles such as icebergs could be detected by a system which analyzed the echoes reflected by such objects. In 1914 there was only one way of obtaining short or supersonic waves, by using mica oscillators. In February 1915, a Russian engineer, Constantin Chilovski, discovered that the frequency of ultrasonic waves that could be used for underwater soundings corresponded to radiotelegraphic wavelengths. But the technique for transforming high-frequency electric oscillations into ultrasonic oscillations had yet to be discovered. Chilovski submitted his project for an electromagnetic vibrator made of laminated iron to the French, and the French Minister for the Navy then commissioned physicist Paul Langevin to look into the problem. In 1917 Langevin came up with the idea of producing ultrasonic waves by using the piezoelectric effect a method of producing electric voltage by mechanical pressure and vice versa discovered by Jacques and Pierre Curie in 1880. The 1917 conference on underwater detection brought together the greatest scientists of the time, Millikan, Langmuir, Arnold, Rutherford, and Abraham. Following this conference, the British Admiralty's experimental station in Harwich developed the Sonar System in 1918.

Before the invention of radar, airborne objects were located by radiogoniometry (or direction finding), a "system for navigation and position finding using angular bearings." A given object was located by plotting the angles on a map of relative bearings (gonios is the Greek word for "angle"). However the major disadvantage of radiogoniometry was that it could only be used in the location of "friendly" targets, those sending out an active signal. In addition, the distance between the base station and the mobile station could only be approximately reckoned. By 1914, Bellini and Tosi had already discovered the basic procedures (rotating frame, simulation of a rotating antenna). The discovery of the triode valve made it possible to increase receiver sensitivity. Subsequently, direct-reading instruments for radio direction finding became available; Watson-Watt's in 1926, Busignies's in 1927 and Marique's in 1931. Airbone radiogoniometry or radiocompass systems owe a great deal to the Frenchman Henri Busignies.

Busignies took out a patent for the "Huff Duff" radiocompass (a high-frequency direction finder) in 1926. It is estimated that his system cost Doetnitz forty submarines a month. The Huff Duff was eventually replaced by the Loran system, developed by Fernand Bac under Busignies's direction.

One of the first applications of the cathode tube was the instant-display cathode ray direction finder. The first such instrument was developed by the Englishman Watson-Watt in 1926, while he was working in the field of aviation on the location of storm centers that produce static. Radar uses a completely different process; it can be used to locate moving objects by calculating their polar coordinates (ray r and angle θ), and, in particular, moving objects emitting no signal (radioelectrically passive objects) by processing the returning echo reflected by any object hit by the radar beam. A radar "map" is built up rather like a television image; the cathode ray display is tuned to the frequency of the returning short wave echoes that have been reflected from any object coming into contact with the scanning radar beam. But it does not build up the image from left to right and from the top to the bottom, as in a television picture. The cathode ray sweeps from the center to the edge of the screen. The French, the English, and the Americans all claim credit for the invention of radar, but the work was actually accomplished in stages. Hertz had already proved that electromagnetic waves were reflected from conductive surfaces, but this electromagnetic echo only occurred with short waves. In 1904, Christian Hülsmeyer, a Düsseldorf engineer, noticed that radio wave propagation was disturbed by the passage of boats moving along the Rhine. He took out a patent for his "telemobiloscope," the ancestor of the radar. A wireless transmitter sent out waves which only reached the receiver if they were reflected by a metallic obstacle. When the reflected pulse arrived at the receiver it triggered an alarm bell. The disadvantage of this method was that it only allowed detection of objects located very close to the device.

In September 1922, two research workers at the

Meteorological radar at the top of Mount Fuji in Japan.

1

1. Sir Robert Watson-Watt (1892-1973), Vice-Controller of Communications Equipment .at the Ministry of Aircraft Production, Scientific Adviser on Telecommunications to the Air Ministry, Deputy Chairman of the Radio Committee of the War Cabinet.

2

Das *TELEMOBILOSKOP 1904*
von *CHRISTIAN HÜLSMEYER, DÜSSELDORF, D.R.P. 165546,169154*

Gerät zur Feststellung und Entfernungsbestimmung bewegter metallischer Gegenstände im Nebel (Schiffe, Wracks, Unterseeboote u.s.w.) durch hör- und sichtbare Signale.
DRP Nr. 165546 v. 30.4.1904, u. 169 154 v. 11.11.1904.

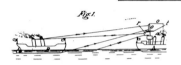

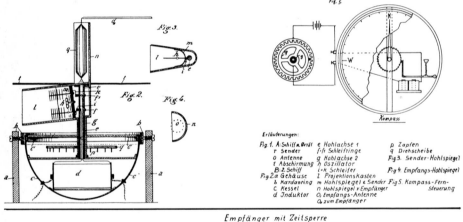

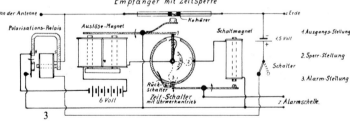

3

4

2 and 3. Christian Hülsmeyer (1881-1957) and his Telemobiloscope.
4. A.H. Taylor.
5. Trials on board the Normandy. 5

Anacostia Naval Aircraft Radio Laboratory in the United States, A. Hoyt Taylor and Leo C. Young, noticed that some radio signals were reflected by steel buildings and metal objects. They suggested beginning work on a system whereby destroyers located along a line a certain number of miles long could be immediately informed of the movement of an enemy warship despite fog, cover of night or smoke screens. According to the Army and Navy report dated April 25, 1943, this is radar's (radio detection and ranging) birth certificate. This would make radar an American invention. However, the report compiled by Taylor and Young (delivered on September 22, 1922, to the Bureau of Engineering, Department of the Navy) created very little interest at the time.

In Great Britain, Sir Edward Appleton and Barnett were carrying out a series of measurements to determine the height of the different atmospheric layers and, in particular, the height of the ionosphere (the ionized layer surrounding the earth wich reflects back certain kinds of radio waves). First, they used continuous waves, then pulsed waves, to distinguish between the different layers. At the same time, research was also going on in the United States; between 1925 and 1930 the phenomenon of reflection noted by Taylor and Young was used to measure the height of the Kennelly-Heaviside layer. At the Carnegie Institute, Gregory Breit and Merle A. Tuve were using the pulse method. L. A. Gebhardt constructed the pulse modulator; it was subsequently perfected by M. H. Schrenk, who fitted it with a multivibrator to shorten the pulses. At the same time, Young developed a more sensitive receiver. During this period Taylor and Young also measured the time it took for signals to travel around the earth by reflection from the Kennelly-Heaviside ionospheric layer. On June 14, 1930, one of Taylor's assistants, L. A. Hyland, observed that an aircraft crossing a line located between a transmitter and a receiver produced a disturbance of the radio waves; this gave a clear indication of the aircraft's presence. On November 5, a report on the radio location of moving objects was submitted to the Naval Bureau of Engineering. But the Navy was still not convinced. The transmitter and receiver were located too far away from each other; how could they be installed on the same ship? During this same period, work began in Japan, Germany, and France while the English, of course, continued their research. We have already discussed the work done by Yagi and Okabe on the magnetron. The first Japanese radar sets were used in October 1942 during the Battle of Santa-Cruz; these radar sets were the ones later captured in the Philippines and in Singapore.

From 1939 to 1941, the Japanese were carrying out research into the development of frequency modulated radars, but these never got beyond a 5-kilometer operating range. The only Japanese radar equipment capable of effective use throughout this period was the radio barrage system which could detect the presence of an aircraft in the general vicinity without exact position information. The Japanese Navy was looking into ways of developing high-frequency and ultra high-frequency radars. By the beginning of 1942, the Japanese were not far behind the English and the Americans in magnetron technique, but they had no idea how to use it. Even more seriously, there was almost no cooperation between Navy and Army; each branch had its own radar equipment, and there was total confusion as to whether an object detected was friendly or an enemy target. It was only in the autumn of 1944 that the Army and the Navy decided to pool their efforts, but by then it was far too late.

In 1929 a small research team known as the N.V.A. (Nachrichtenmittel-Versuchs-Anstalt) was set up in Kiel, Germany, under the leadership of Dr. Rudolf Kühnhold to study the problems of underwater sonar detection techniques on behalf of the Navy. Right up until 1933, no one had thought of using sonar echoes as a means of measuring the distance of objects on land as well as under water. In France, Henri Gutton and Pierret began to experiment with short waves; and Maurice Ponte, who had been working with Gutton, Sylvain Berline, and Hugon at the C.S.F. Laboratory since 1930, began his work on the magnetron. In 1931, Mesny and David, technical consultants for the French Military Signals Department, noted that a disturbance was created in communications whenever an aircraft passed through the zone between the transmission and the receiving stations. At the beginning of 1934, the first equipment using returning radio wave echoes to locate a moving obstacle was produced. In his memoirs, Emile Girardeau emphasizes that *the date for the creation of French radar is authenticated by the test reports for May to July 1934, as well as by the filing of patent no 3711922: a new system for locating obstacles and its applications in the name of the General Wireless Telegraph Company.* (Girardeau, 1967: 212.) The vessel used during the tests was the freighter *Oregon* and test results proved the possibility of detecting other vessels and the coastline itself at ranges of over 20 kilometers. Upon completion of tests performed in November and December 1934, radar equipment was installed on the *Normandie* in 1935.

French radars used ultrahigh frequencies (16 cm), while English radars between 1935 and 1940 used only high frequencies which were less accurate and called for much bulkier equipment, given that the higher the frequency, the better the angular resolution for an antenna of a given size.

But scientists were still unsure how to produce the very high power levels needed to achieve sufficient range. Finally, the magnetron, introduced by Maurice Ponte to English scientists in 1940, was hailed as "an essential contribution to radar." According to this view of the facts, then, radar was essentially a French invention.

On the English side of the Channel, Sir Robert Watson-Watt was still working along the lines first explored by Sir Edward Appleton. His main assistant was his wife, Margaret Robertson. Watson-Watt had been scientific consultant in telecommunications for the Air Ministry under Sir Robert Renwick. He made his first experiments on what the English called "radio-location" on a country road near the BBC station in Daventry.

Watson-Watt commented that women showed greater initiative and application even though, generally speaking, they were less well-trained in the mechanical side of things. As we can see from the many photographs taken at M.I.T., Westinghouse, and Western Electric during this period, many of the teams working on radar were in fact composed of women. In February 1935, Sir Robert Watson-Watt proposed to the Secret Scientific Committee presided over by Sir Henry Tizard that radio-location be given a trial run. On December 5, 1935, the Air Ministry signed an undertaking to build the first five radar stations to protect England's east coast; construction was finished in 1938. Between 1938 and 1940, work began on a number of other stations that were to play a decisive role in the Battle of Britain in 1940.

Just after the war, Watson-Watt gave a lecture on radar at the Palais de la Découverte in which he made no mention of the French contribution.

1

3

5

7

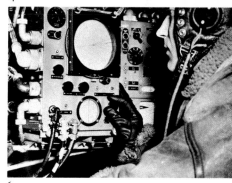

2

4

6

1. During World War II in England, operations rooms were largely staffed by WAAF personnel.
2. The "radar girls" shown here checking out electrical equipment were working in America during this period.
3. Malvern College in England was used for Telecommunications Research during World War II.

4. Here instructor Nora van der Green explains a problem to a drafting student at the MIT Radiation Lab.
5. Large reconnaissance photograph of the German radio direction finding station at Bruneval in Northern France. The direction finding apparatus may be seen in a shallow pit between the cliffs and an isolated house.
6. Indicator unit and controls.
7. Final assembly and wiring line (Hawthorne Company).

**The Second World War:
Radar**

1

2

1. Antenna of the first
complete radar installation –
"Topsy" – in a building at
Naval Research Lab in
Anacostia D.C. in the late
1930's. Its so-called
"directable" antenna was
mounted in such a way as to
allow it to be swivelled
around for 360° coverage.
2. The German "Jagdschloss"
radar.
3. A giant Würzburg with a
7 meter diameter reflector.

3

Progress in Great Britain, he said, *was direct and
fairly fast. Aircraft could be located at distances of
30 miles by June, 45 miles by August 1935, and 100
miles by March 1936. Large-scale tactical exercises
were undertaken and resulted in the decision to
change Great Britain's system of aerial defence and
base it on radar. These new developments were duly
passed on to the French in 1939. Tuned cavity
magnetrons and the corresponding progress made in
UHF receivers in Great Britain before the summer of
1940 marked the beginning of cooperation between that country and the United States in the field
of radar. The British magnetron thus became the
focal point for research developments in both countries. Demand for higher transmission power and
receiver sensitivity began to grow. Under the leadership of M.L. Oliphant, a group working at the University of Birmingham provided the essential component of the breakthrough. J.T. Randall applied the
tuned-cavity technique to the relatively inefficient
prewar magnetron, transforming it into a totally new
and extremely powerful device. R.W. Sutton, who
was attached to the Admiralty Signals Department,
produced a receiving tube that was no less innovative, while the Clarendon Laboratory team made an
essential contribution to the effective use of a system
of common antennas for transmission and reception
of pulsed radar signals. The use of UHF radar for
ground use or for installation on boats or aircraft
was now well within reach.*
(Watson-Watt, 1946).

We should not ignore the contribution made by
Dutch technicians from the Philips company who
had offered their services to the Mullard company at
the beginning of the war. But France, according to
Watson-Watt, contributed nothing. In his estima-
tion, radar was an English invention. In September
1940, Sir Henry Tizard led a British task force to the
United States, while American Commodore Jen-
nings B. Dow spent most of the following year in
England preparing to reorganize the Radar Branch
of the United States Radio Division. Over the last
few years the Americans had allowed themselves to
be convinced by the enthusiasm of Leo C. Young,
who had been working with R. M. Page on a 60
MHz pulse unit since 1934. A year later R. C. Guth-
rie joined the Young-Page team to help perfect the
modulated pulse transmitter and receiver units. Congress agreed to allocate a budget of $100,000, and in
June 1936, an official demonstration took place at
the Navy's research laboratory for the benefit of
senior Navy personnel. Rear-Admiral Bowen was
immediately convinced of the importance of this
research, and he was backed up in his resolve by
Admiral A. J. Hepburn after tests had been carried
out in tactical maneuvers in the Pacific. In 1937 further tests were carried out on the cruiser *Leahvy.*

What were the Germans doing on the eve of
1940, the year that was to see the most important
progress in the field of radar? They were working on
a badly-designed UHF radio wave program. Unlike
the Allies, they never succeeded in producing high
power oscillations at ultrahigh frequencies, they preferred to concentrate their efforts instead on the
development of infrared detector systems. However,
they had got off to a good start. An English radar
expert, Sir Robert Cockburn, admitted that in 1940
British radars were in many ways inferior to the
"Freya" and the "Würzburg" and that there was no
bombing control or navigational aid system comparable to the "Knickebein." Sir Cockburn also pointed to German superiority in the field of remote controlled weapons.

But for various reasons the Germans were some two years behind the Allies in radar research; their backwardness was mainly due to the German High Command's lack of interest in this field of research. In the summer of 1933 Rudolf Kühnhold's team began their research work on behalf of the Navy's Transmission Research Division in Kiel. In the fall of the same year Kühnhold learned that Philips had produced a transmitting tube that could operate at a frequency of 60 MHz and produce 70 watts of power. He then set about building a device that could use these tubes; the device, still in its infancy, was then transferred to the German Navy's wartime research laboratory in Pelzerhaken for testing. These tests were carried out in collaboration with Dr. Schultes, a phycisist with G.E.M.A. (Gesellschaft für Elektro-Akustische und Mechanische Apparate), a company set up in 1934 to concentrate on radar production. Kühnhold then conducted a demonstration of refracted wave reception using the ship Grille, which was anchored 12 km out in the bay of Lübeck. As luck would have it, a Junkers W34 seaplane flew past the research center at a distance of some 700 meters at exactly the same moment, and Kühnhold's device recorded its passage. The Navy was convinced and allocated 70,000 Reichsmarks for further experimentation. However, even at this point, the government and the Army paid very little attention to this new device, which they considered to have little practical potential.

It was notable that the German Navy became interested in radar first, just as in France and the United States. On September 26, 1935, Admiral Räder (Commander-in-chief of the Navy), Admiral Carls (Chief of Naval Forces), and Admiral Witzell (head of the German Marine Corps), arrived in Pelzerhaken to watch a demonstration. By 1936, the Germans were working on the interception of aircraft; and Wolfgang Martini, director of radio liaison, learned that the device could be used to detect aircraft at a range of 80 kilometers. He was already considering using "Freya" not only for defense purposes but also for guiding aircraft from the ground. In 1938, the Navy took possession of its first "Freya" alert devices. They had an operating radius of 120 km, functioned on a wavelength of 2.40 m, and produced 20 kw of power. When Martini was appointed general he made the most extraordinary efforts to have a large number of "Freyas" allocated to him, but with no success. By the outbreak of war, the Wehrmacht had only eight of the units: two on the island of Helgoland, two in Wangerooge, one in Borkum, and one in Nordeney, with two more in reserve. Their purpose was to protect the coast between Holland and Denmark. At that time certainly, the Germans were thinking much more in terms of a war of aggression − their cherished "Blitzkrieg" − than a defensive war.

The German Telefunken company had begun radar research in 1936. In just two years, the research team, led by Wilhelm Runge, developed the "Würzburg" which operated on a wavelength of 52 centimeters and which could locate aircraft flying at a range of 40 kilometers. However, the Würzburg would only become operational for antiaircraft artillery units during the summer of 1940. Some 400 Würzburgs were built during the war; after September 1941 the company produced the giant Würzburgs which had twice the operating range of the earlier models. At the end of 1941, the Lichenstein units for fighters came into production. But the Germans were falling behind on the development of UHF radar techniques.

During the same period, Telefunken developed a small airborne radar unit that was eventually installed in night fighters. At the end of 1937 the first shipborne radar, a "Seetakt," operating on a wavelength of 50 cm, was fitted on the Graf Spee. This very bulky equipment, like the Freya, gave fairly mediocre technical results because of its low output power. Thus, although the Germans developed fire-control radar before the English, they soon fell behind on quality. Early in the 1930s, the Lorenz company had built a system suitable for bad weather conditions, night flying, or blind approaches and landing. In 1933, a German radio-transmission specialist, Hans Pendl, transformed this system into an automatic air-to-ground bombing system. Within five years he had invented an automatic bomb-release system in which a radio signal transmitted from the target area triggered the device. The most important model was the "Wotan 1," codenamed "X." However, one disadvantage of the system was that it could not be used for bombing large areas, as the target proximity beam was often limited to a range of 2.2 km. The English, under Cockburn, began to produce a radar to counter Wotan 1. Telefunken then developed another blind approach bombing device, nicknamed "Broken Leg," that was more simply constructed, but less accurate than Wotan 1. Pendl then began to work on the blueprints for Wotan II (or "Y"), which would have a larger firing range.

The reason the Germans made little progress thereafter in the field of radar is undoubtedly the result of Hitler's strategy choices. Right up until 1944, they took almost no part in the race to find ever highter frequencies. They made no attempt to do any research into UHF waves or the use of the magnetron; they limited themselves to developing triode transmission tubes that could produce power at 50 cm wavelengths and to producing antiaircraft (and sometimes antishipping) Würzburg-type bombing-control radars and shipborne radars. It was not until after they had experimented with the H2S radars found on downed English aircraft that the German began to construct 10 cm radars (the "Rotterdam" and "Berlin" models), partly copied from the H2S. But by then, of course, it was far too late.

Churchill's Memoirs show that he considered the Homburg radar superior to English radar in certain respects. Churchill said also that it was English operational effectiveness rather than the innovatory nature of their equipment which most contributed to the victory of Great Britain.

While the Wehrmacht needed a great deal of coaxing to use radar, French military authorities showed hardly more enthusiasm. As Emile Girardeau points out:

As far as radar is concerned, we owe its creation and perfecting exclusively to private initiative between 1934 and 1940: the armed forces (land, sea and air) had no faith in it whatsoever. Over these six years they gave us no orders, nor did they initiate any research contracts; I had to take personal responsibility for committing large sums of money on behalf of my group of companies. (Girardeau, ibidem).

But the world finally woke up in 1939. In France, Admiral Darlan granted unlimited research credit and put the battleship Strasbourg at the disposal of the CSF team. In September 1939, General Herring, military governor of Paris, took the advice of Emile Girardeau and Maurice Ponte and decided to immediately set about building a radio-detection station on the outskirts of Paris in Sannois. The military establishment, as usual, procrastinated, and Girardeau decided to begin construction at his own

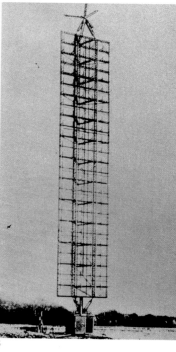

1. Dr. Hans Pendl.
2. General Wolfgang Martini.
3. The German "Wassermann" radar, an improvement over the "Freya".

The Second World War: Radar

1

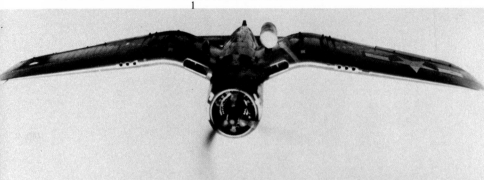

2

1. The MIT Radiation Lab revolutionized radar by using microwaves instead of long waves. Here, E.G. Bowen (seated), member of a British team which had carried out early magnetron research in 1940, is being shown an American-made copy by Radiation Lab Director L.A. du Bridge (left) and I.I. Rabi, Nobel Prize Winner (right).
2. Under the "elbow" of its inverted gull wing, this Navy Corsair carries the bomb-shaped antenna for its Western Electric AN/APS-4 radar. In the last year of the Pacific War, the Corsair joined carrier air groups as a fighter-bomber for the final assaults against the Japanese.
3. Normandy, July 1, 1944. In the shadow of the destroyed radar apparatus are the graves of British and German soldiers killed during the fierce fighting which took place before the positions were finally taken.
4. Radar-jamming streamers. The "shadows".

3

170   4

expense. The station went into service on June 8, 1940; three days later it was destroyed so that no information could fall into enemy hands! The United States, on the contrary, was already gearing up for war. In October 1939, the government awarded RCA a contract for the production of six airborne radar units built to the same designs as the Navy Laboratory's model. In August 1940, Rear Admiral Bowen, acting upon his belief that the United States would soon enter the war, persuaded Charles E. Wilson, Chairman of General Electric, to start up a radar-production plant. Within two weeks, Wilson had sent twenty of his Schenectady-based research workers to examine the Navy's model and had all his radio engineers transferred to the radar section. In October 1940, Bowen managed to convince Westinghouse executives A.W. Robertson and George Bucher to join the war effort. The Bell company followed suit. After Guadalcanal, shipboard fire-control radar played an increasingly important role in naval battles. Early, warning radar distinguished itself in the Battle of the Coral Sea by detecting the Japanese attack formations at a range of 70 miles. Radar was also to play an essential role in the Battle of the Atlantic by detecting enemy submarines.

It was a Westinghouse radar, the SCR 270, that warned of the approaching Japanese attack on Pearl Harbor. Unfortunately, no one believed the radar operator. This is how the historian David Woodbury describes the scene:

At seven in the morning of December 7, in an out-of-the-way hideout at Opana near Pearl Harbor, Corporal (now Lieutenant) Joseph L. Lockhard of the Signal Corps started a voluntary watch with a Westinghouse mobile SCR-270. Scanning the horizon for the next few minutes he located what seemed to be a large flight of planes in the north-west sector. He followed them for some time, plotting their rapid approach to the island. Alarmed, he notified his superior, a lieutenant who told him that the planes were nothing to worry about — they must be American bombers from the carrier "Saratoga" on patrol in the vicinity.

Lockhard was not satisfied. On a chart he plotted the course of the objects clearly shown by the pips on his radar's scope. The flight approached the island from the direction of Japan, became stationary over Waialua Bay, then split apart and swept in like a pair of pincers over Kaneohe and Pearl Harbor simultaneously.

That much of the story is history. But what did Lockhard do next? Returning to his station, after the blitz, at about ten o'clock, he stuck to his radar until he had plotted the retirement of the Jap attackers to the northwest for their rendez vous with their carrier force. That map of Lockhard's is now in Washington. It could have been used to pursue and destroy the Japanese carrier force. But it was not. The most it did was to clear the Signal Corps of blame later for failing to warn the island in time. Radar was still so secret that most Army officers had never heard of it. It was so new that those who did know of it had little confidence in it. (Woodbury, 1951: 148).

On October 11, 1940, the Radiation Laboratory was founded at M.I.T. Alfred Loomis, a physicist, ex-lawyer and ex-banker, whose most famous invention was Loran, a long-range navigational system, was named head of the Microwave Committee. If we take a look at the Microwave Committee flowchart published after the war, we note that the Director, Lee du Bridge, was also a physicist; in fact seven of the nine technical divisions were headed by physicists, the eighth by an astrophysicist. Only the ninth division had an engineer at its head. This was what inspired H. Guerlac to write: *The Radiation laboratory was an extreme example of the scientists' dream of a Republic of Scientists.* Legend has it that even the doorman was a physicist. This story arose because Charles Süsskind once found a mention of a well-known physicist who worked for the Committee, followed by the astonishing description: "in charge of visitors." The Radiation Laboratory's findings were later published in twenty-seven volumes which specialists consider the "Bible" of Radar.

In 1941, CSF recommenced tests with the Navy in Toulon on the Saint-Mandrier peninsular. By August 1942, their UHF radar could detect cruisers at 25 km and whalers at 6 km, with an accuracy of 25 m. In November everything had to be destroyed because of the German occupation. Nevertheless, the CSF laboratories secretly continued research into magnetrons and super high-frequency circuits right up until the end of the war. From 1940 to 1941 English-builtradar systems began to prove their effectiveness and two National Physical Laboratory physicists, E.G. Bowen and Wilkins, developed the first interception radars, extremely efficient in detecting high-flying aircraft, but somewhat less for low-flying aircraft. A second radar chain operating on lower wavelengths was set up in September 1940. In the meantime Wilkins developed the IFF (Information Friend or Foe) identification unit which could distinguish between friendly and enemy aircraft by the type of echo sent back to the screen. Bowen then developed the AI (Air Interception) radar for night fighters; this device allowed the pilot to locate enemy aircraft, and was used at the end of the Battle of Britain during the German night bombings. Before the end of the Battle of Britain, Denis Taylor had developed the GCI (Ground Control Interception) whose screen showed a graphic representation of the geographical detail surrounding the aircraft being tracked. Harold Lardner and W.S. Eastwood perfected the Elsie that was used for DCA ranging by bearings; a radar enabling coastal artillery to locate enemy shipping and also the CHL-system used in intercepting low-flying aircraft (the work of W. S. Butement and P. E. Pollard) were developed around the same time, as was Bowen and Hambury-Brown's surface submarine detection radar. Following Randall's successful research into UHF waves and the magnetron, Dee, Allen, Hodgkin, and W. E. Burcham together developed a UHF night fighter radar, while Dee also worked on the UHF ASV for guiding aircraft to enemy submarines. Dee, Skinner, and Lovell worked together on the H2S, which made it possible to detect moving objects in heavy cloud. The English were also working on long-distance navigation aids using pulse techniques, such as the GEE, developed by R. J. Dippy and G. C. E. Bellringer; this device, intended for aircraft and shipping, operated on short-wave frequencies with an operating range of 300 to 500 km depending on altitude. This was the device used in the bombing of Cologne and in the Allied Invasion. Loran was based on the same principle, but had much greater range up to 1,500 or 2,000 km at night. It was used in the bombing of Berlin. Another navigational aid, the "Oboe," was developed by F. C. Williams, A. H. Reeves, and F. E. Jones. It was a very accurate device used for bombing mines in the Ruhr valley and coastal gun emplacements in Normandy. The "Decca" system was a hyperbolic navigational system using several slave stations. It was used to position shipping during Operation Overload. Finally, Williams, Pringle, and Lines developed the "Rebecca-Eureka," a radio-

beacon which served as a homing station for radar-equipped aircraft. It was installed in enemy territory by Resistance workers or parachuted agents in order to facilitate the dropping of weapons or agents into occupied territory. To scramble German radar, the Allies used "windows," long coils of metallized paper that drifted around in the sky for a few seconds before falling, thus creating an artificial electromagnetic fog and sending out false echoes. In the field of counter jamming, the Germans were for a long time a step ahead of the Allies. World War II was the first "electronic" war.

The Russian research and development program lagged far behind the American, English, and French programs, even behind the German program, until Roosevelt launched his lend-lease exchange policy. This at least was what both Russian and non-Russian sources believed for a long time. However, Soviet texts subsequently began to refer to a prior research effort. In fact, A. S. Popov and P. N. Rybkins had begun work on the reflection of electromagnetic waves as early as 1895, but the real work began some thirty years later, when as a result of Vidchynski's 1923 research, a group of scientists formed around P. K. Oshchepkov, the Russian Watson-Watt, and began to work on developing ways of locating possible airborne targets. From the very beginning Toukatchevski, the People's Commisar for the Army and the Navy, encouraged their work. In 1934 the Soviet Academy of Science, led by A. I. Joffe, also turned their attention to the problem. The first Soviet radar research station was set up by Oshchepkov at the Leningrad Electrophysics Institute and in January 1934 the scientist inaugurated the first conference on the locating of objects by means of radar. The atmosphere was tense. Joffe was highly skeptical of the potential of high, very high or ultrahigh frequencies. Altogether, the Soviet project lacked any clear policy definition, a problem that was to bedevil it for a number of years. In fact, right up until 1943 the Russians were still debating the merits of pursuing radar research! In the first hours of the war, the Soviet Air Force lost 1,200 aircraft, 800 of them still on the ground. Had they used even rudimentary radar, much of this dreadful catastrophe could have been averted.

In 1934, the Russians built their first experimental model, the "Rapid," which was capable of detecting the presence of aircraft although not their exact position. The "Burya" (hurricane) operated on a wavelength of 18 cm (developing 6 to 7 watts of power) but was not yet at the operational stage. In 1937, the year of Oshchepkov's disgrace (he was forcibly removed from his work by a Stalinist purge and spent the next ten years in prison), Soviet physicists W. Muchin and O. I. Maljarev constructed a magnetron, the idea for which had already been explored by the famous radio engineer M. A. Brontsch-Brujevitch. This magnetron was further perfected by D. E. Malyarov and N. F. Alekseiev, whose 1940 article mentioned Japanese scientist Okabé and French researchers Gutton and Berline. But in spite of their efforts the unit did not function particularly well. The "RUS.1" radar constructed after 1937—38 was first used during the Russo-Finnish war (1939—40). This radar had a range of 80 to 100 km, but its lack of accuracy made for disappointing results. After January 1941 the "RUS-2" series came into production; it had a range of 130 km and used HF wavelengths. In 1942, the "Pegmatit" fixed radar played an important role in the defense of Leningrad, detecting aircraft approximately 30 to 40 minutes before bombing began. Russian inferiority in developing a successful radar system is explained by the chaos resulting from Stalin's internal policy, and the wholesale massacre of Red Army officers on the eve of the war. The Stalinist purges reached their height during this period and were hardly conducive to scientific progress. Scientists disappeared one after the other; Toukatchevski himself was murdered during the massacre of Red Army generals. Other factors were a certain absence of confidence in the technique of radar itself and the lack of a consistent policy on the part of the authorities. Brontsch-Brujevitch had to appeal to Jdanov for permission to begin research again after the vacillation that characterized the period after Toukatchevski's death.

One of the most important postwar achievements was undoubtedly the development of Doppler-effect radar. This technique was a well-kept secret of the English and the Americans long before it became known in France, where it was used for the first time in 1957.

It is based on the effect discovered at the beginning of the nineteenth century by an Austrian physicist, Christian Doppler, in the behavior of sound waves, a discovery that French scientist Hippolyte Fizeau verified for light waves in 1848. When the source of an oscillatory phenomenon (sound or light) and its receiver move away from each other, the frequency of the signal received (the number of oscillations per second) is different from the frequency of the transmitted signal. This difference is of course linked to the relative speed of the source and the receiver, but it is also dependent on the propagation speed of the signal through whatever medium it is traveling. This speed is in no way modified by the movement of the source, because it is fixed by the particular characteristics of the given medium. Signals are not accelerated in the way that the speed of a shell fired from a moving canon is increased by the speed of the canon. In the simple case of a moving source and a fixed receiver toward which the source is moving at a speed v, we can see that the travel time of a given oscillation between source and receiver will be shorter than the travel time of the preceding oscillation, because the source has moved nearer the receiver and the distance to be covered is thus smaller. The time separating two successive oscillations when transmitted is the time T of the phenomenon. Normally, if nothing is moving, this time remains unchanged at reception. But if the target moves nearer, a direct relationship exists between the distance traveled by the target and the time difference between the transmitted and the received signals. Thus, if an oscillation takes $t = \dfrac{x}{c}$ seconds to reach the receiver (x = distance and c = unchangeable propagation speed) the next oscillation will only take $t' = \dfrac{x - vT}{c}$ seconds, vT being the short distance traveled by the source at speed v during the period cycle time T. The time that elapses between the arrival of an oscillation at the receiver and the arrival of the following oscillation will correspond to the time T' as measured at the receiver. If we examine the "schedule" of these oscillations beginning with a givent time t_o, a simple calculation will show us that this time $T' = T - \dfrac{VT}{c}$ or $T(1 - \dfrac{v}{c})$, since:

$- t_o + \dfrac{x}{c}$ = arrival time of first oscillation — (1)

$- t_o + T + \dfrac{x - vT}{c}$ = arrival time of second oscillation — (2)

If we substract the time of the first from the time of the second oscillation, we find the following phenomenon : the resultant equation shows that the reduction in period cycle time or the increase of its reciprocal, the frequency, is higher when the source is moving faster (v higher) and as the source/propagation speed ratio increases. The shortening of the period is fixed by this ratio and does not increase with each new oscillation; this is what is meant by the highly significant Doppler effect. Leaving aside the famous example of the locomotive whistle that grows louder as the train approches and fades away as the train goes away into the distance, it is the basis of Doppler radar systems that measure the speed of tracked targets. An aircraft approaching a radar at speed v meets in one second a number of waves each cT centimeters long as well as waves normally received during that second, that is 1/T waves + c/cT or a total of $1/T : (1 + \dfrac{v}{c})$. But this is not all: the aircraft acts as a moving source by reflecting back the waves encountered and these arrive at the radar receiver showing a reduction in the period time. Thus, a double time reduction has to be done; this gives a measurement. The spacing between the spectral lines of light emanating from a star gives the measurement of radial speed in the same way (the projection of their speed over the line of sight).

Most of the basic radar techniques used in modern radar equipment for both military and civil use (airports, meteorology) were designed during or just after World War II. The development of very high powered klystrons, cross-field amplifiers, and the introduction of transistors and integrated circuits have obviously led to better signal control and processing and to increased power capacities. But in terms of frequency increases we have been marking time for the last thirty years. The first English radar chains went up to 15 MHz ; the first American radars built at the beginning of the war to 100 or 200 MHz. By the end of the war fire-control radar was operating in the 3 to 10 GHz range. Even today we are still using relatively low power at frequencies in the GHz range, partly because of atmospheric limitations, but also because of a lack of high power sources and receivers with a good signal-to-noise ratio. Component miniaturization has introduced yet another dilemma : the most important area of research involves finding a compromise between new forms of technology and high performance on the one hand, and simplification and economy of equipment on the other. The debate is far from being closed.

1

2

3

4

1. Radar control room.
2. The German SRE-MS
radar system (1978) − double
pattern scanner (9 m ×
4.5 m).
3. Military system in Zaire.
4. This first German
shipborne radar systems
greatly reduced false returns
from rain and waves
providing a good clutter-free
display.
5. Radar turret.

5

Sonar

by Harold Edgerton, professor at the Massachussetts Institute of Technology (M.I.T.)

Harold Edgerton is known for inventing the electronic flash and various other procedures used in photography. He has also done a great deal of research into the properties of sonar and explains the principles in this article.

Fig. 1. The depth-finder and the side-scan sonar systems. A side-scan "fish" towed behind a ship has two transducers with a shaped beam that is *narrow* along the ship's path and *wide* perpendicular to it. Targets at 200 meters and more, to both sides, can be recorded. At a ship's speed of 4 knots, about a square mile is surveyed per hour, for wrecks, rock outcrops, and geological features.

Fig. 2. This penetration sonar record was made with a 5 Khz system to disclose under-bottom features. Note that the vertical scale is different than the horizontal. This area, east of Aigion, Greece near the Selenious river, is where the ancient city of Helice may have been located before it was engulfed by a tidal wave and earthquake. The sub-bottom features it this record may be of geological origin.

Fig. 1

The word sonar is an acronym (SOund NAvigation Ranging) that was devised many years ago by those used underwater sound pulses, mainly to detect submarines in the sea. Visual methods of exploration in water are limited to very short distances by light absorption and diffusion. Fortunately, sound waves are not absorbed as much as light and, as a result, many useful tasks can be accomplished by sound pulses in ocean observation and instrumentation.

The maximum range for visual observation in the sea, under exceptionally clear conditions, might be as much as 30 meters. Sound, however, can be detected and used for many thousands of meters. Visual results are presented as a "picture," where the angles to the subject and other parts of the view are recorded on a two-dimensional plot. The photographic camera is used underwater exactly as it is used above water. Sonar results, however, are presented in travel-time from the transmitter to the target and then back to the receiver. The information comes back on a single channel, with travel-time and intensity as the quantities to be observed. The angle to the subject depends upon the direction in which the sonar transmitter, receiver, and beam angles are directed.

Pulses of sound created by a myriad of sources, from small "pings" to large dynamite "boom" explosions, are an important way to gain information from the sea and subbottom part of the earth. Echoes from sound pulses carry knowledge that cannot be obtained in other ways. As humans, we cannot extract information from echoes, as bats do by their inborn ability. Instead, we must resort to accoustical recorders, computers, analyzers, correlators, etc., all of which have been made practical by today's electronic devices.

An excellent historical account, especially dealing with underwater sound developments, is given in Frederick Hunt's *Electroacoustics*.

Sonar Depth Finder

The most widely used sonar instrument is the "depth finder," almost universally installed on ships. Sailors must be alert about water depth in which their ship is navigating. Those who are careless soon find that they are in serious trouble when their ship hits bottom or a submerged object that could cause expensive damage or sinking.

The sonar transmitter is commonly mounted on the bottom of the hull (Fig. 1). Usually, it is made of a piezoelectric crystal that changes shape when an electrical signal is impressed on the electrodes. Likewise, when a pressure or displacement signal strikes the crystal, a voltage is produced. The time between the applied pulse and the arrival of the reflected wave is a direct function of the distance to the target. Since the velocity of sound in water is essentially constant, about 1,500 meters per second, the depth finder can be calibrated to read water depth.

$$\text{Depth} = \frac{\text{Velocity of sound in water}}{2} \text{ x}$$

Elapsed time between the signal and echo.

Electronic circuits are required to drive the transducer and to amplify the echo signals. The simplest depth finders give only a reading of depth. The more complicated types have a recorder that plots the depth as a function of time. In this way, a ship approaching land will have a useful, continuous record of the water depth below the ship.

Hydrographic charts are made by accumulating data with ships equipped with accurate depth finders and precise navigation devices. From this information, equal-depth contour lines are prepared and charts drawn.

There is always a chance that the depth finder will miss a small peak target, such as a wreck or rock outcrop. Thus, important areas are carefully explored by a wire carried between two ships, at a required depth. Such measurements establish safe channels for large ships. The side-scan sonar to be described next is useful for finding irregularities on the bottom's surface.

Side-scan Sonar

A specialized type of sonar equipment, called side-scan, has been developed mainly for geological and biological exploration. It also has been remarkably useful for searching large sea areas for wrecks.

Figure 1 shows a side-scan unit, usually carried in a "fish" that trails behind and below the exploring ship. A beam fans out from each side of the fish, with a narrow beam width along the line of travel. The received echoes from the bottom and from targets are plotted on the ship's recorder, as a function of real time. In this way, correlation of the results is viewed optically, something like a map, but with some distortions due to geometry and ship speed. New models are being introduced that correct for the depth factor, geometry, and ship speed resulting in a "chart"-type picture that is easy to interpret. The bottom of the sea can be immensely complicated. It is safe to predict that hydrographic charts in the future will contain information that today is not observed by the depth-finding sonar.

Penetration Sonar

Many other specialized sonars produce information about the

subbottom, that is, the volume of sediment and rocks below the sea bottom. Such equipment is a very important tool to explore the unknown area of our world. Foremost in this is the seismic exploration for oil and gas deposits, both on land and in the sea. On the other end of the scale is the precise, high-resolution penetration sonar used by archaeologists to find small objects that are less than a meter below the surface. In addition, very high-frequency sonar of very short pulse length is currently being used medically to study the brain, internal organs, and the fetus.

The principle of penetration sonar is exactly the same as the depth finder, except that the sound's pulse length must be short enough to resolve the target that is being sought and a powerful peak power is usually required. One source, called a "boomer," uses a metal plate that is thrown outward and upward by a pulse of electrical current in an adjacent coil. Other types of penetrators use conventional crystal transducers, except the basic frequency must not be too high since the energy is attenuated, as the signal penetrates the sediments. Practical considerations limit the upper frequency to about 12,000 cycles per second (or hertz) for a penetration of 4 meters in soft mud. It has been more difficult to penetrate sand and sediment with gas contents.

Deep penetration sonar sources for geology and seismic exploration consists of explosions, air guns, sparkers, etc. Electronic correlation devices have been found to be important for the interpretation of the information.

Figure 2 shows a penetration record made on a submerged sediment delta in Greece. Note that the 5 Khz (5,000 cycles per second) sonar shows interesting and probably important targets below the bottom. This general area is of interest to archaeologists and historians since a city, named Helice, in the Bay of Corinth east of Aigion, Greece, according to ancient authors, was lost in an earthquake and tidal wave. It is hoped that further research will be accomplished to determine if humanmade structures or geological formations are the reason for the subbottom echoes that are disclosed by sonar.

Fig. 2

Radar

by Merrill Skolnik, Superintendant Radar Division, Naval Research Laboratory of the U.S. Navy.

Radar, which was born from the combined labor of French, English, and American researchers during World War II, has become one of the most important components of electronic warfare. What are the principles upon which it is based?

Radar is an electromagnetic device for the detection and location of reflecting objects such as aircraft, ships, satellites, and the natural environment. It operates by transmitting a known waveform, usually a series of narrow pulses, and observing the nature of the echo signal reflected by the target back to the radar. In addition to determining the presence of targets within its coverage, the basic measurements made by a radar are *range* (distance), and *angular location*. The Doppler frequency shift of the echo from a moving target is sometimes extracted as a measure of the *relative velocity*. The Doppler shift is also important in CW (continuous wave), MTI (moving target indication), and PD (pulse Doppler) radars for separating desired moving targets (such as aircraft) from large undesired fixed echoes (such as ground clutter).

The Doppler frequency shift is also sometimes employed to achieve the equivalent of good angular resolution. In addition to the usual measurements of range, angular location, and relative velocity, radar can obtain information about a target's size, shape, symmetry, surface roughness, and surface dielectric constant.

The origins of radar go back to the pioneering experiments of Heinrich Hertz who in 1886 demonstrated the similarity of radio waves and light. Hertz showed that radio waves could be reflected (scattered) from objects on which they impinge. This is the fundamental mechanism of radar. Based on these experiments, C. Hülsmeyer in Germany patented and demonstrated in 1904 a "Hertzian-wave" echo-location device for the detection of ships. Although the components used in his device were primitive by modern standards, the basic concept and philosophy were like that of classical radar. Radar did not evolve from the early work of Hülsmeyer since there was no serious need for it until much later, when the development of military aircraft progressed in the 1930s to where it represented a significant new threat. The maturing of the aircraft as a military weapon then led to the independent and almost simultaneous reinvention of radar in many countries as a means for the detection and tracking of aircraft. The demands of World War II accelerated the practical deployment of radar. Its development and improvement have been continuous ever since.

Radar has found many important applications. It is widely used for air traffic control, both for the detection of aircraft in flight and the control of aircraft moving on the airport grounds. Radar technology also forms the basis for aircraft navigation and landing aids, as well as for cooperative beacon systems (such as secondary radar, IFF, ATCRBS, and DABS). On board aircraft, radar is used for weather avoidance, terrain avoidance, and navigation. It is widely found on ships of the world for piloting and collision avoidance. Large, ground-based radars are employed for the surveillance and tracking of satellites and ballistic missiles. On board spacecraft, radar has been used for landing, rendezvous, and docking. It has been proposed for the survey from space of agriculture and forestry resources, and the monitoring of sea and ice conditions.

It is used for the ground-based observation of the weather and for the investigation of extraterrestrial phenomena such as the ionosphere, meteors, and aurora. The study of birds and insects has also benefited from the application of radar. Radar is used in law enforcement for the detection of intruders and for the speed measurement of vehicles. The military, whose support has made possible most of radar's major advances, employ radar extensively for surveillance, navigation, and the control and guidance of weapons.

The "typical" ground-based radar for the long-range detection of aircraft might operate at a wavelength of about 23 cm and radiate a peak power of about a megawatt (with an average power of several kilowatts) from an antenna about 10 meters in width rotating at 5 or 6 rpm. The early microwave radars used a magnetron oscillator as the transmitter. It is still widely used; but when high average-power or controlled-modulation waveforms are required, the transmitter is often a power amplifier like the klystron, traveling-wave tube, or crossed-field amplifier. The solid-state transmitter, usually based on the transistor, is also used for its potential in increasing reliability and maintainability. Low-noise receivers, such as the transistor or the parametric amplifier, have also found their way into modern radar practice. The impressive advances in solid state digital circuitry made in recent years have resulted in significant new capabilities in radar signal processing and data processing. Sophisticated Doppler processing techniques for MTI radar have been reduced to practice because of the availability of low cost, small-size digital processing technology. Another significant example of the benefits of

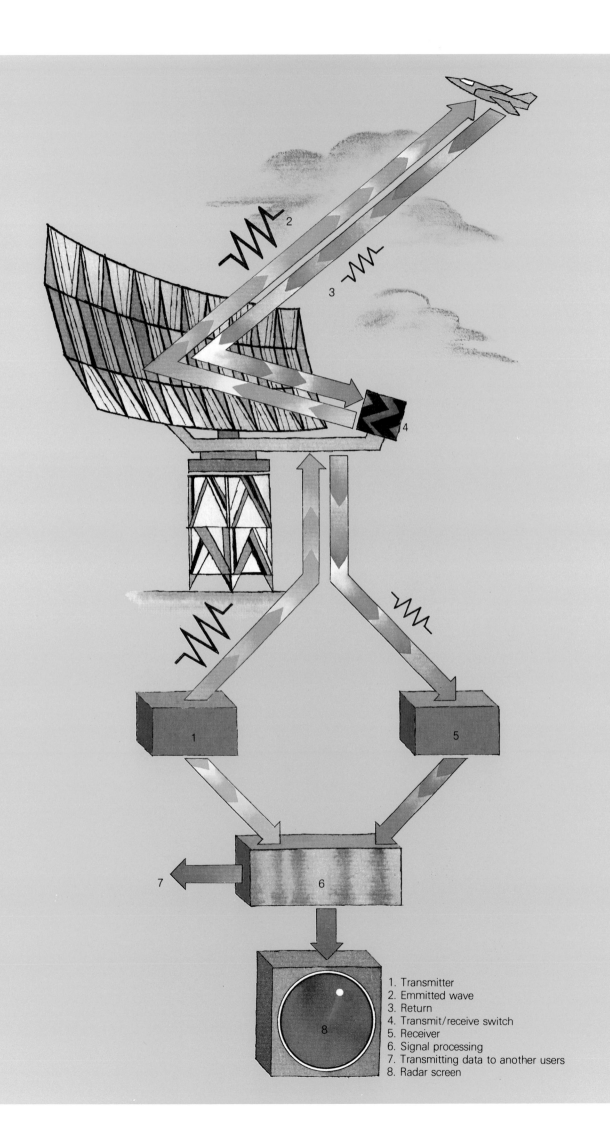

1. Transmitter
2. Emmitted wave
3. Return
4. Transmit/receive switch
5. Receiver
6. Signal processing
7. Transmitting data to another users
8. Radar screen

advances in digital circuitry is in the automatic detection and tracking (ADT) of aircraft targets. A single operator observing the output of a standard PPI radar display might be able to track but a handful of aircraft. However, the small size and low cost of digital computers make it possible to automatically and accurately detect and track several hundreds of aircraft simultaneously so as to present to the operator fully processed tracks rather than raw radar data. The parabolic reflector antenna has been and will continue to be the antenna most commonly employed with operational radar systems. However, the phased array antenna with agile eam steering controlled by electronic phase shifters has been of interest for radar applications because of the ease and rapidity with which its beam can be pointed anywhere within its coverage. The large cost and complexity of electronically steered phased array antennas have limited their application. Limited phased arrays whose beam is scanned over limited angular regions are more practical and have seen wide use as 3D air-surveillance radars, aircraft landing radars, and hostile-weapons location radars.

The range resolution of a radar can be of the order of a fraction of a meter, if desired; but the beamwidths that are practical with a microwave antenna limit the resolution in the angle coordinate (or cross-range dimension) to many orders of magnitude greater than this. It has been possible to synthesize the effect of a large antenna and thus overcome the cross-range limitation by employing the radar on a moving vehicle (such as an aircraft) and coherently storing the received echoes in an electronic or photographic memory for a time duration equivalent to the length of a large antenna. This technique for achieving cross-range resolution comparable to the resolution that can be obtained in the range dimension is called *synthetic aperture radar* (SAR). The output of such a radar is a map or image of the target scene. The use of a stationary radar to image a moving or rotating target by using resolution in the Doppler domain is called *inverse* SAR. It has been employed, for example, to image the surface of the planet Venus under its cloud cover.

Radar is generally found within what is known as the microwave region of the electromagnetic spectrum. It is possible, however, to apply the radar principle at HF frequencies (from several to perhaps 30 megahertz) to obtain the advantage of long "over-the-horizon" ranges by refraction of the radar waves in the ionosphere. Aircraft can be detected by one-hop ionospheric propagation out to ranges of about 2000 nmi. Radar has

also been considered for use at frequencies higher than the microwave region, at millimeter wavelengths. Laser radars are found in the IR and the optical region of the spectrum, where they offer the advantage of precision range and Doppler frequency measurement.

19
Particle accelerators and nuclear electronics

The explosion of the first thermonuclear bomb created an urgent need for research into phenomena bringing into play very high energy levels within very short time periods, and a complementary need for equipment to study the nature of nuclear reactions in the laboratory. The proton synchrotron was introduced at the end of the 1950s to obtain energy levels in excess of 1 GeV (10^9eV).* The proton synchrotron marked an extraordinary turning point in the field of high-energy physics. Another extraordinary breakthrough was the discovery and development of collision rings. The principle behind particle accelerators is very simple : they are used to project particles (electrons, protons, etc.) at enormous speed onto an immobile target, the nucleus, in order to break it up into fragments and thereby study its component parts. But this destruction, or breaking up, of the nucleus-target does not take place in a random or unorganized way as happens, for example, when a cannonball hits a rock. Rather, it is a passage from one ordered structure to another more simple structure, from one order to another. The development of nuclear power has made a reality of the old alchemists' dream of transmuting matter. High-energy physicists face two different challenges: accelerating ions of increasing orders of heaviness in order to obtain uranium beams, for example, and accelerating ever lighter particles, such as protons, at increasing energy levels.

Modern accelerators are the direct descendants of Crookes's and Röntgen's tubes. The kinetic energy acquired by the electron in the acceleration space between cathode and anode is applied in the same way with the difference that this kinetic energy is a few kiloelectron volts in the case of tubes and a few GeV in the case of accelerators. The first accelerators could be built only after techniques for producing structures capable of withstanding such high voltages were mastered. In fact a standard x-ray tube risks exploding if subjected to voltages superior to 400 kV. To eliminate this drawback, Robert Van de Graaff designed an electrostatic generator for accelerating protons at Princeton in 1931. This generator, which directly transferred the charge by means of a long insulated belt, enabled very precise measurements to be obtained because the voltage supplied (and therefore the flow of particles) was constant. Van de Graaff's machine is still used today for supplying other kinds of accelerators working at much higher energies or for disintegrating heavy nuclei such as nitrogen, carbon, etc., as the voltage stabilizing system has proved very practical.

In 1928, a Norwegian physicist, R. Wideroe, suggested that in order to avoid the phenomenon of accelerator tube disintegration, an electrical force variable in time could be used, by dividing the tube into sections with successive switching of the applied voltages. In 1932, John Cockcroft and E.T.S. Walton used the first (if we ignore Rutherford's pioneering although rather primitive experiments along these lines) linear accelerator at Cavendish to bombard lithium with protons accelerated to 0.8 megavolts and to obtain the first artificial transmutations of matter. In France this device was known as the Villard assembler, in Germany as the Greinacher multiplier; yet another example of the great difficulty inherent in determining the true parenthood of such inventions.

It was only after the war that particle accelerators became really effective instruments, mainly because of the work done by William Hansen at Stanford. An electronic beam was directed into one end of Hansen's rectilinear vacuum metallic tube, while at the same time a traveling wave was sent to accelerate the particles and travel at the same speed as them, thanks to the low mass of the electron in its rest state, (0.511 MeV).

This gave the possibility of propagating the electromagnetic wave along a circular section of copper tubing if the wavelength was less than half the tube's internal circumference. During this period, around 1930, Ernest Orlando Lawrence designed another accelerator, the cyclotron, that was circular rather than linear. Low-energy atomic particles were injected inside a cylindrical high-vacuum sleeve, surrounded by an electromagnet in the form of two semicircles. The particles revolved, accelerated by the electrodes and subjected to very high frequency voltages. They were maintained in their circular orbits by means of a magnetic field induced inside the metal sleeve sections. But with each acceleration a slight distortion occurred increasing the diameter of the

Linear accelerator control room at the CEA nuclear physics department near Saclay. The accelerator is controlled by a microcomputer.

* The electron-volt (eV) is the unit used to describe the energy acquired by an electron subjected to a voltage accelerator of 1 volt. 1 Megaelectron-volt represents 1 million (10^6) eV and is written 1 MeV. 1 Gigaelectron-volt represents 1,000 MeV (10^9eV) and is written 1 GeV. 1 teraelectron-volt represents 1,000 GeV (10^{12}) eV and is written 1 TeV.

Particle Accelerators and Nuclear Electronics

* The gauss is a unit of magnetism.

1. Early cyclotron designed by Lawrence and Livingstone.
2. Cockroft and Walton's apparatus for artificial disintegration of elements.

1

2

particle's trajectory; so that it eventually reached the walls of the cyclotron and was directed onto the target. In all of these developments, researchers had to deal with the problem of the tendency of the particle to move in a straight line and lose itself on one of the accelerator walls, thereby provoking dangerous secondary reactions for the observer. This rigidity has been calculated and is equal to:

$$\frac{A}{Z} \frac{\text{(mass of particle)}}{\text{(charge of the particle)}} \beta \gamma$$

where ß represents the ratio: $\frac{v \text{ (speed of particle)}}{c \text{ (speed of light)}}$

and γ represents the "Lorenz factor": $\frac{1}{\sqrt{1 - ß^2}}$

Although the electron is fairly easy to focus in a linear accelerator, thanks to the presence of magnetic focusing coils, the proton is much more stubborn, and only the cyclotron can control this with its magnetic fields. However, this then poses a problem of synchronization since, at a certain speed, the particle becomes subject to the laws of relativity (the particle becomes relative when its kinetic energy equals the energy of its mass).

As the electron mass is only 0.511 MeV, this means that the electron becomes relative when its kinetic energy attains this magnitude. It therefore follows that the electron cannot be very greatly accelerated, in contrast to the proton, whose mass is 931 MeV and which therefore becomes relative when its kinetic energy approaches 1 GeV. The invention of the frequency-modulated cyclotron or synchrocyclotron, on the heels of the discovery of phase stability by McMillan in the United States and Veksler in Russia, compensated for this drawback. The synchrocyclotron made it possible to control the trajectory of the relative particle by means of magnets placed in a circle or crown. One of the first devices of this type, the cosmotron built at Berkeley, could handle energy levels up to 6 billion eV or 6 GeV. The electron is generally used in electronic interactions to determine the internal structure of charged matter, whereas the proton is used to generate nuclear interactions. CERN's SPS (superproton synchrotron), completed in 1977, has an energy capability of 400 GeV. It takes up 7 kilometers of a circular tunnel, and is set 40 meters underground. It uses more than 1,000 electrical magnets to help guide the protons. The protons are accelerated by a special unit producing a radio wave that increases the energy of the protons by 2.5 million eV on each successive circuit of the tunnel. The protons travel 160,000 revolutions per 8.4 second cycle in the vacuum. Initial plans for the SPS were drawn up by Carlo Rubbia and followed up by Crowley-Milling. It was finally constructed after the fundamental mastery of electron cooling techniques achieved at Novosibirsk (Gersh Budker) and at CERN in Geneva (Simon van der Meer); these techniques paved the way for the grouping of intense beams of stable particles and research by means of antiproton beams.

A 50 GeV (BPS) proton sychrotron is now being built in Peking. Another project is underway in Tsukube, Japan, that requires two particle sources (a 12 GeV proton synchrotron and a 2.5 GeV linear electron accelerator) to produce collisions between electrons between 6 and 25 GeV and protons from 90 to 300 GeV in a ring 3 km in circumference. However, before tackling the problem of collisions rings, we should point out that the fixed-target machine currently producing the highest energy levels is undoubtedly the 3 TeV proton synchrotron in Serpoukhov, which uses superconductor magnets.

There is a trend to integrate machines that are difficult to modify into a new system. This has been done with the CERN 10 GeV proton synchrotron (now used as an injector for the SPS and ISR collision rings) and the DORIS (used as an injector for PETRA). In addition, some machines now in existence will be used for research work for which they were not originally designed. This is the case of the proton synchrotron which can be transformed into proton or antiproton storage rings for collision applications.

Up until now we have talked only of fixed-target machines or accelerators. But there is another type of machine where the target itself moves. In this machine, we gain in movement energy what is lost in collision probabilities, whereas in an accelerator, collisions are more numerous but energy remains low unless it is greatly accelerated (to an energy above 1 TeV or 10^{12} electron volts). We have seen that in fixed-target machines, the accelerated particles are projected onto the target once they have reached the required energy level. They can then be used either to observe interactions between the projectile and target-nucleus or, as is more common, to produce beams of secondary particles whose very variable nature and energy can be produced up to a certain threshold. The secondary beam is directed against a second target which is surrounded by detectors in order to observe the desired reactions. In the case of moving-target machines – intersection and storage collision ring type – the particles meet at a predetermined point. When the particles are of opposing charge, they travel in opposite directions inside a collision ring and meet at certain points of their travel. If the particles are of identical charge, two rings with intersection zones are used. These are known as intersection and storage rings (ISR), and were used to substantiate the existence of quarks through bombardment of protons and neutrons with electrons and very high energy neutrons.

Other kinds of machines besides CERN's ISR have also been developed, among them the CESR 8 GeV electron positon developed at Cornell, Desy's 19 GeV PETRA, the 18 GeV PEP at Stanford and a number of different projects undertaken in Europe. We should also mention Stanford's LEP; the work being done at Novosibirsk on prototype beam collision devices for nuclear accelerators; the two rival electron-proton collision machines (Desy's HERA and the Fermi Laboratory's CHEER) and the prototype proton-proton storage rings developed at Brookhaven (ISABELLE – 400 GeV) which introduced the problem of deflective superconductor magnets. Superconductors pose the major technical difficulty in the domain of high-energy physics. One of the leaders in this area of research is the Fermi Laboratory described by former director R. Wilson as the "world's largest consumer of superconductors." Superconductors may be used to produce fields of 40,000 or even 50,000 gauss.* Fermi's Tevatron, Brookhaven's ISABELLE, and the research work now being undertaken at Serpoukhov is furthering mastery of this technique in the field of magnets; while research teams at Karlsruhe, Cornell, and Standford are one more looking into high-frequency superconductor cavities. Various detection methods are in use to log particle trajectories and observe collisions.

One of the first methods, developed in 1949, was the scintillation detector. Charged particles pass through a luminous environment, such as polystyrene, containing a solid solution of an organic fluorescent substance in which atoms are excited. When the atoms return to their natural state they emit light, and scintillation effect can be observed in the matter. The photomultiplier converts the light signal into an electrical signal. The photons thus created in the scintillator are transmitted to the photomultiplier photocathode and liberate electrons which are multiplied by an electron multiplier. One can then measure the number of particles emitted and the passage time. Spark detectors use charged condensors that create a succession of flashes along the particle's trajectory in order to photograph its path. At the beginning of the 1960s Frank Krienen developed the White Spark Chamber with ferrite and wire core memories, followed by George Charpak's multiwire chambers which operate in accordance with a binary yes/no code (the particule has/has not passed). The drift chamber appeared at the beginning of the 1970s, followed by the picture chamber and the bubble chamber in which the liquid bubbles heated almost to boiling point are formed along the particle trajectory.

Over the last fifty years the energy levels of accelerators and all other "big machines" have increased by a factor of 10^7. Radioastronomers and astrophysicists have rediscovered in the cosmos the laws that high-energy physicists have discovered in the matter "tortured" in their fabulous machines. Accelerators have become one of the most noble instruments of scientific research and a spectacular illustration of the progress man has made in mastering techniques and giving full rein to his ingenuity. As superconductor techniques are mastered it is probable that high-energy physicists, who have already discovered quarks and charmed particles, will continue to uncover yet other hitherto unknown secrets of the universe.

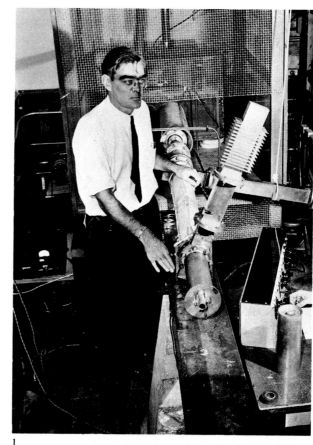

1

2

1. William Hansen at Stanford with the first linear accelerator.
2. The size of a neutron generator gives us an idea of the size of nuclear electronic devices.
3. CERN's synchro-cyclotron.

3

PART FOUR

The backstage role of science

Our future is too important to be left to experts and self-proclaimed specialists... today man has almost unlimited power to do whatever he wants with the building blocks of Nature's gigantic construction game. And he uses this power with a child's insouciance and lack of consciousness. Perhaps one day we will compare the different forms of madness man has unleashed in his tinkering with the seemingly limitless possibilities of science and technique with the religious epidemics that buffeted Europe at the end of the Middle Ages.

Georg Picht

Once upon a time...
James Clerk Maxwell

James Clerk Maxwell, the author of electromagnetic theory, opened up the path to theorical research later explored by J. J. Thomson and the avenue of applied research followed by Heinrich Hertz. Maxwell, son of a well-known Scots family, became the first director of the famous Cavendish Laboratory. He was educated at Edinburgh University and Trinity College, Cambridge and for five years taught physics and astronomy at King's College London. In his Memoirs *Recollections and Reflections,* Thomson quotes a phrase attributed to Maxwell: *I never try to dissuade a man from trying an experiment. If he does not find what he is looking for he may find something else.* And in fact, Thomson himself discovered the electron while pursuing his research into cathode rays and the conduction of electricity in gaseous atmospheres. Maxwell's work was based on the intuition of another English physicist, Michael Faraday, who suspected that electromagnetic waves existed, although he was not able to substantiate his hunch. But in order for us to fully understand the enormous breadth and importance of Maxwell's work, we must go back even further, with a brief stopover in the thirteenth century to glance at the research done by a monk named Pierre de Maricourt into the property of magnets. We then make a leap forward into the seventeenth century when an English doctor named Gilbert carried out similar research.

In 1780, Charles Coulomb established the first quantitative formulation of the laws of electrical and magnetic attraction. In 1819 a Danish scientist, Hans-Christian Oersted, accidentally discovered the effect of an electrical current on a magnetized needle, which provided proof of the long-suspected link between electricity and magnetism. Oersted was demonstrating to his pupils the heating effect of the current on a metal wire. Each time that he switched on the current, the needle of a compass lying nearby on the experiment bench deviated, thus appearing to show that the electrical current created a magnetic force. The French scientist Arago was passing through Geneva and heard of Oersted's experiments through the Swiss scientist De la Rive, who was most enthusiastic about them. Back in Paris, Arago gave an account of Oersted's experiments to the Academy of Science on September 4, 1820; and reproduced them

himself on the eleventh of that month. Ampère came to watch his experiments and was inspired to do similar research. About the experiments which Ampère conducted on September 18, 1820, he wrote:

I have reduced the phenomena observed by Mr. Oersted to two general facts. I have shown that the current in the battery acts on the magnetized needle just like the current in the connecting wire*. I have described the experiments through which I succeeded in observing the attraction or repulsion of the whole of a magnetized needle by the connecting wire. I have also gone into some detail on the way in which I see magnets i.e. as owing their properties solely to electrical currents in the plane perpendicular to their axis and to similar currents which are found on the earth. In this way I reduced all magnetic phenomena to purely electrical effects.

On the experiments conducted on September 25, 1820, he wrote:
I developed this theory further and observed new facts concerning the attraction and repulsion of two electrical currents without any magnet being interposed, a fact which I had observed with conductors bent in the form of a coil. (Massain, 1939: 137). Within this fifteen-day period Ampère had laid the foundations of a new science, *electrodynamics,* and set down a code of laws governing the interaction of magnets.

Ampere had identified magnets with circular electrical currents.

Eleven years later, in 1831, Faraday confirmed this theory and set down the laws governing electromagnetic induction. Faraday observed that when a magnet was placed in a circuit and moved, it not only induced a current into that circuit, but also that a circuit through which current was flowing had the same effect as the magnet on a neighboring circuit when one of the two circuits was moved in relation to the other. To his successors Faraday left the idea of a field made up of electrical, magnetic, and mechanical forces.

In 1867 Maxwell proposed his "displacement current" theory which represented his basic physical considerations on the movement of electricity in

James Clerk Maxwell (1831-1879) Scottish physicist whose research opened up a new era in the history of physics.

* That is the wire that joins the two terminals of the battery together.

capacitor (open) circuits. This displacement current only flowed very briefly between two isolated plates representing the capacitor and was caused by a movement of electrical charges. He noticed that it prolonged the conduction current supplied by an exterior battery which supplied the initial energy. This displacement current created a magnetic field; this induced an electrical field and the disturbance was then propagated at the speed of light. Maxwell demonstrated mathematically the theoretical possibility of the existence of electromagnetic waves and their propagation speed. He established that the electromagnetic field could be made to oscillate and move through the air in the form of waves. He also demonstrated that if these waves encountered a conductor in their path, they induced an electrical current whose period of oscillation was equal to that of the wave field.

Everything was now in place for the logical development of applying this theory to send "information," but Maxwell had no idea of how to transmit or receive carrier waves. It was not until twenty years later that Heinrich Hertz would take this second step.

1. Drawing taken from the English documentary "The Electron's Tale".
2. The Cavendish Laboratory during Thomson's reign.
3. The experiments which led Christian Oersted (1777-1851) to the discovery of the magnetic field created by electrical currents.
4. Michael Faraday (1791-1851), English physicist and chemist who discovered how to liquefy almost every kind of gas, also discovered electromagnetic induction and enunciated the theory of electrolysis.
5. Lines of force and polarized paper sent by M. Faraday to J. von Liebig around 1850.

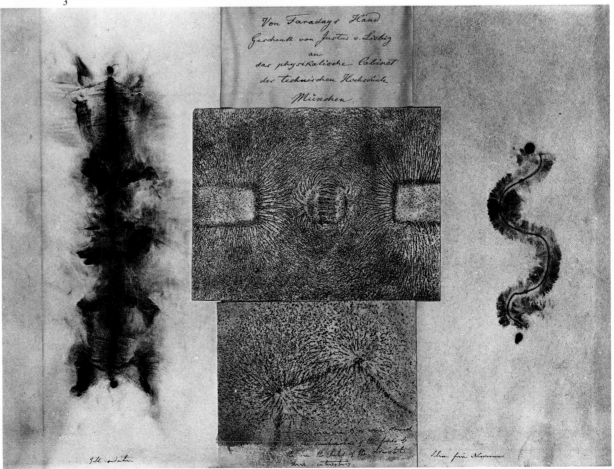

Maxwell's equations

by Pierre L. Aryl, Engineer and Author of Scientific Books for the Layman.

One of the fundamental theories underlying all developments in the field of electricity and electronics is Maxwell's electromagnetic theory. In this article Pierre L. Aryl reviews its genesis and implications.

(1) Maintained constant in two rectilinear conductors of infinite length positioned one meter from each other, such a current of one ampère produces a force of two millionths of a Newton (unit of force) between these conductors.

(2) Let us note that the "work" of the field $\vec{H}$ is the same regardless of its distance from the conductor. Over a circular contour of radius r the "work" is equal to C = $2\pi r\vec{H}$. Thus C = $2\pi \vec{H}$ for r = 1.
Since the field is inversely proportional to the distance, its size at any distance r' becomes $\vec{H}/r'$ and the "work" C' = $2\pi r'$. $\vec{H}/r'$ which is indeed equal to C' = $2\pi \vec{H}$ = C.

The Maxwell electric field theory is contained in his "wonderful" equations, both invention and synthesis. On their publication in 1864, they appeared as revolutionary as the theory of relativity forty years later. The electric field and the magnetic field are propagated through space in the form of waves, according to the new theory. Thus, the effect due to an electric charge does not appear instantaneously in the entire surrounding space at the moment when the charge that produces it appears in one point. Similarly, the effect of a magnetic field caused by the passage of an electric current only makes itself felt gradually by degrees. Actually, this propagation is so rapid – its velocity, as Maxwell demonstrated, is that of light – that the stated electric and magnetic effects seem instantaneous.

The Ampère Innovation: Unification of Electricity and Magnetism

Without undue simplification, the genealogy of the Maxwell equations can be traced back to the discovery by the Danish scientist Oersted. Due to the Volta cell, the physics laboratories of Europe had a constant source of electricity at their disposal which, in a few years, would reveal phenomena invisible to the transitory sources of electrostatic machines. When Oersted established in 1819 that a wire crossed by a current acts on the needle of a compass he concluded – and all the scientists at the time with him – that the wire is converted into a magnet under the action of the current. All except one – André-Marie Ampère, who wrote in September 1820: *The current in the cell acts on the magnetized needle like that of a connecting wire ... spirals and galvanic helices (i.e., coils) will produce in all cases the same effects as magnets ... whose properties are uniquely due to* electric currents *in the planes perpendicular to their axis ... all magnetic phenomena (reduce to) purely electric effects.* A brilliant interpretation!

The magnetic effect of an electric current is exerted in a direction perpendicular to the latter, so that a concentric "field" is formed around the wire. The orientation of this field in space depends on the direction of the current. The "Bonhomme d'Ampère," an old acquaintance of schoolboys, or the more prosaic rule of the right hand, pointed this out: if you hold the wire in your right hand, placing the thumb in the direction of the current, the other fingers close in the direction of the magnetic field. The intensity of the field is obviously not constant; it is inversely proportional to the distance with respect to the wire.

If two wires are brought together, they will behave like magnets – which they are not – by attracting or repelling each other according to the direction of the current passing through them. Furthermore, it is this electrodynamic force that is used to define – in our current "International System" of measurement – the unit of intensity of the electric current, that is, precisely one ampère.[1]

The magnetic state produced around the conductor is actually a strong "power," a latent power that is manifested only in the presence of another conductor if the latter is also crossed by a current. To describe the state of space governed by a magnetic field, Faraday thought of drawing "lines of force" – a concept that was to prove extremely inventive.

The magnetic state of space is called magnetic *induction*. For one given field, the induction will be stronger as the permeability of the medium increases.

Let us go back to the magnetic field around the conductor: Ampère showed that let' says the "work" that the field $\vec{H}$ does traveling along a closed path enclosing an electric circuit crossed by a current I is equal to I."[2] The mathematical expression for it is simple:

$$\oint \vec{H}\ \vec{dl} = I$$

The symbol $\oint$ is that of a "line integral," where $\vec{dl}$ denotes (Fig. 2) an element of the closed path small enough to be considered rectilinear and also small enough so that the field $\vec{H}$ can be regarded as maintaining a constant value during its displacement along this element. Let us multiply the component of $\vec{H}$ parallel to dl by $\vec{dl}$. Let us do the same for all the small elements $\vec{dl}$ constituting the total path around the electric circuit. Let us add up all these elementary products; the sum we obtain is precisely that which is represented by the line integral.

Let us still go a bit further, introducing a vector quantity that is absolutely essential in the Maxwell equations: the curl.

The path along which the field vector travels marks the boundaries of a surface: when this surface – designated ds – becomes infinitely small, but however small it becomes, the line integral continues to exist, we can define a "curl."

This new mathematical entity represents the limit of the ratio of the vector $\oint \vec{H}\ \vec{dl}$ to the surface $\vec{ds}$ when the latter tends to zero.

$$\text{curl } \vec{H} = \lim_{ds \to 0} \frac{\oint \vec{H}\ \vec{dl}}{\vec{ds}}$$

This curl vector is oriented in the direction of the forward movement of a

Fig. 1

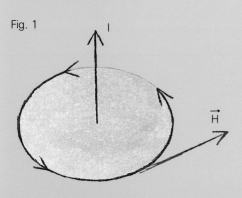

Fig. 2

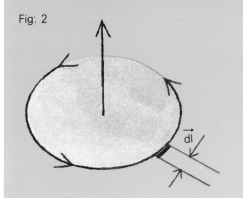

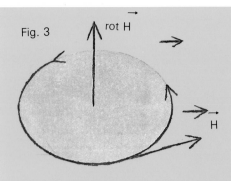

Fig. 3

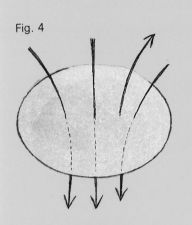

Fig. 4

corkscrew turning in the direction of the $\vec{H}$ vector along the path.

By comparing the curl $\vec{H}$ in Figure 3 to the current I in Figure 1, we see that these two quantities are positioned and oriented in an identical manner. What can be their relationship?

The Ampère theorem gives the equality between the work of the field $\vec{H}$ along a path enclosing an electrical circuit and the current of the conductor I:

$$\oint \vec{H}\,\vec{dl} = I$$

In general, this theorem obviously holds whether the current is unique or composed of several various currents in different directions whose algebraic sum is I.

. Let us introduce the concept of current *density* $\vec{J}$ at each point of the surface marked off by the path, more appropriately termed the contour enclosing the circuit. The total current I is equal to the product of the current density by the surface, which is indeed the total flux passing through the surface, expressed by this new integral

$$I = \oint \vec{J}\,ds.$$

In the Ampère relationship, we can therefore replace the current I with this integral of electric flux, which gives the line integral of the magnetic field as follows:

$$\oint \vec{H}\,\vec{dl} = \oint \vec{J}\,ds.$$

Now, we have another very convenient relationship (Stokes theorem) which indicates that the total "work" of the vector is equal to the flux of the curl of this vector across the surface bounded by the contour, that is, in our case,

$$\oint \vec{H}\,\vec{dl} = \oint \text{curl } \vec{H}\,ds$$

Given two quantities equal to a third and all equal to each other, we then have

$$\oint \text{curl } \vec{H}\,ds = \oint \vec{J}\,ds$$

from which it follows that

$$\text{curl } \vec{H} = \vec{J}$$

This is the differential expression of the Ampère theorem.

The Displacement Current: The Key to Electromagnetic Waves

Let us proceed immediately to set up the first Maxwell equation which produces the latter. Ampère studied electric circuits in which the charges, as we now know, are carried by electrons, which inturn are moved there by an "electric field" $\vec{E}$.

The vacuum devoid of such media should therefore not be the seat of a current. In the meantime, when we

charge a condenser in which the dielectric — insulating material separating its two plates — is not a conductor, the charging current is prolonged in this dielectric by a "displacement" current. Moreover, this current appears only if the "conventional" conduction current in the conductors varies with time. What is the nature of this displacement current? The displacement $\vec{D}$ of a medium — its "stress" to take an image — under the action of an electric field $\vec{E}$ becomes more intense as the specific inductive capacity (permittivity) of the medium increases. This electric phenomenon is very similar to the magnetic phenomenon: the displacement $\vec{D}$ is the homolog of the induction $\vec{B}$ and it is similarly connected to the field by a physical constant. We thus have the following symmetric relationships: in an electric medium $\vec{D} = \varepsilon\,\vec{E}$ with ε = permittivity, in a magnetic medium $\vec{B} = \mu\,\vec{H}$ with μ = permeability.

There nevertheless exists a fundamental difference between the two media: if the first medium has mobile charges, the second does not.

The time distortion of the displacement $\vec{D}$ of a medium under the action of a *variable* electric field constitutes the displacement current (or, more precisely, the density of this current per unit surface). It is written:

$$\frac{\delta\,\vec{D}}{\delta\,t}$$

Maxwell admitted — and this shows at what point his work goes beyond an elegant mathematical ordering — that *the displacement current produces a magnetic current around it in the same way as a conduction current around a conductor.*

Provided that we are dealing with *variable* conditions, the two types of current can coexist in the same medium and add their effects. The Ampère equation curl $\vec{H} = \vec{J}$ should therefore be written:

$$\vec{H} = \vec{J} + \frac{\delta\,\vec{D}}{\delta\,t}$$

in order to become *the Maxwell-Ampère equation.*

The Faraday Lines of Force

Let us now go back more than thirty years, keeping in mind this notion of variable conditions.

Having discovered the magnetic effect produced on passage of an electric current and proved the identity or, at least, the direct interconnection of the two electric and magnetic phenomena, scientists set out to determine whether there indeed exists an electric effect due to magnetic induction. Ampère came very close to the experimental

Fig. 5

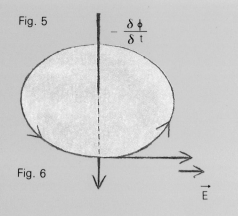

Fig. 6

$$-\frac{\delta\phi}{\delta t}$$

$$\vec{E}$$

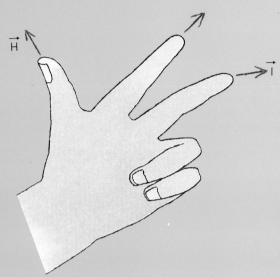

$$\vec{H}$$

$$\vec{I}$$

demonstration. Shortly afterwards, Colladon failed similarly by being overly careful. In fact, after establishing that no current appeared at the terminals of a wire wound around a magnet, this scientist wanted to see what would happen if he inserted a bar magnet into a coil. But, good experimenter that he was, not wanting to disturb his galvanometric measurements by the proximity of the bar magnet, he placed this device in an adjoining room under the supervision of a laboratory assistant. The latter — supposedly daydreaming — was not aware of the deviations that actually accompany the movement of the magnet.

Michael Faraday, an inveterate and shrewd experimenter, went over all these studies again, hoping to obtain electricity "through ordinary magnetism."

He finally arrived "at some positive results."

Faraday described at length his experiments with two separate coils of copper wires, one attached to a galvanometer, the other to a voltaic cell. When he drew the first coil near the second, the needle of the galvanometer turned in a given direction, and when he moved the coil away the needle pointed in the opposite direction. He also observed that as soon as the coils stopped moving either in one direction or the other, the needle of the instrument resumed its normal position. He wrote in 1831:

All my results tend to invert the sense of the proposition stated by M. Ampère that a current of electricity tends to put the electricity of conductor near which it passes in motion in the same direction," for they indicate an opposite direction for the produced current; and they show that the effect is momentary, and that it is also produced by magnetic induction, and that certain other extraordinary effects follow thereupon. The momentary existence of the phenomena of induction now described is sufficient to furnish abundant reasons for the uncertainty or failure of the experiments hitherto made to obtain electricity from magnets, or to effect chemical decomposition or arrangement by their means. (The Royal Society, 1831: 146).

These sentences describe nothing else but electromagnetic induction. This well-known phenomenon opens the way to the production of electric energy by non-chemical means. Conversely, electric energy could be converted into mechanical energy. This electricity of "motion" illustrates the principle of the conservation of energy, not very well perceived but already foreseen at the time. The inanity of hoping to obtain electric energy from a passive system (such as a coil wound around a magnet)

without any input of primary energy was definitively demonstrated.

The physical interpretation of Faraday's observations is immediate.

When an electric conductor encloses a variable magnetic flux φ or — which amounts to the same — a constant flux in which it moves, an electromotive induction force appears and is proportional to the variation of the flux. Moreover, for a variation of this flux in a certain direction — by increasing, for example — the current, in turn, forms in a certain direction; the current reverses if the flux changes in the opposite direction.

The current appears in a direction according to the law of the conservation of energy. This induction current actually behaves like any conduction current and creates a magnetic field around it, oriented in the known direction — the one indicated by the fingers of the right hand. This direction is the *opposite* of the direction taken by the variation in magnetic flux, which gives rise to the current.

This time it is the *left* hand that enables us to easily find the direction of the current due to induction by forming a rectangular Cartesian system with the first three fingers. The thumb is positioned in the direction of the magnetic field and the index finger is positioned in the direction of the displacement of the conductor; the middle finger thus indicates the direction of the current.

Let us therefore repeat, that the induction current tends to oppose the variation in the magnetic field which it produces. To overcome this opposition, it is necessary to supply mechanical energy to the electric conductor displaced in the constant magnetic flux (or, what amounts to the same thing, supply electric energy to the variable current which generates a variable magnetic flux surrounded by a fixed conductor.)

This opposition reaction is indicated in the equation by a *negative sign* of the term giving the time variation of the flux, and we write:

$$\text{electromotive force} = -\frac{\delta\Phi}{\delta t} = e$$

However, the flux Φ is the product of the induction $\vec{B}$ passing through the surface which marks the boundaries of the conductor, times this surface $\vec{ds}$ itself. And we can write:

$$e = -\frac{\partial\Phi}{t} = -\int\frac{\partial\vec{B}}{\partial t}\cdot\vec{ds}$$

Next, the electromotive force e is nothing other than the work of the electric field $\vec{E}$ along the contour surrounding the surface,

$$e = \oint E \cdot dl$$

Let us again introduce the curl of this electric field $\vec{E}$. The flux of that curl through the surface is equal, as we know, to the line integral of the field $\vec{E}$ on the contour surrounding the surface, and we will have:

$$e = \oint \vec{E} \; \vec{dl} = \int \text{curl} \, \vec{E} \cdot \vec{ds}$$

Equations (5) and (6) both give the value of e. Setting their right hand equal to each other, we find:

$$e = - \int \frac{\partial \vec{B}}{\partial t} \cdot \vec{ds} = \int \text{curl} \, \vec{E} \cdot \vec{ds}$$

from which we derive

$$- \frac{\partial \vec{B}}{\partial t} = \text{curl} \, \vec{E}.$$

This is the differential expression of the Maxwell-Faraday equation.

The Seat of Energy Does Not Lie Solely in the Point Source of Radiation

The two fundamental equations

$$\text{curl} \, \vec{H} = \vec{J} + \frac{\partial \vec{D}}{\partial t}$$

$$\text{curl} \, \vec{E} = - \frac{\partial \vec{B}}{\partial t}$$

are the ideal mathematical description of electromagnetic interaction.

Let us go back to this description:

A variable electric current — conduction current and displacement current — induces a variable magnetic field $\vec{H}$. The resulting variable magnetic induction $\vec{B}$ in turn induces a variable electric field, which produces a variable electric current which again induces a variable magnetic field, and so forth.

The process continues in the form of a self-sustained oscillation insofar as sufficient energy is supplied to the system and the dimensional elements of the circuit are adjusted to assure resonance. In a nonconducting medium, the process is not interrupted: the displacement current provides for the production of the variable magnetic field. If the magnetic field and electric field are propagated at an identical speed, the attachment formed between them will be capable of propelling itself through space. This is exactly what takes place.

However, they undergo a tranverse motion with respect of the displacement of radiation, which they determine (Fig. 7). Furthermore, they are positioned at right angles with respect to their maxima and minima coincide in time: they are in phase. Together they form an electromagnetic wave.

If we rotate the electric field over the magnetic field, the angular motion would always correspond to a corkscrew rotation, thereby assuring it forward movement in the direction of wave propagation.

It would not suffice to create electric and magnetic fields at right angles so that they are propagated in wave form. Still, they must have good velocity so that they are mutually generated: Maxwell calculated that this good velocity, suitable to assure regeneration of the magnetic field by the electric field, should be equal to the ratio of the units of electric and magnetic charges expressed respectively in the electromagnetic system and the electrostatic system. In a vacuum, this ratio is $3 \cdot 10^8$ m/s. When Maxwell concluded that there should actually be such electromagnetic waves which travel at similar velocity, it had already been known for a long time — due to measurements taken of all other types of phenomena — that this velocity of 300,000 km/s was precisely that of light. The mathematician therefore had an idea that light itself was electromagnetic radiation.

Maxwell wrote:

I was aware that there was supposed to be difference between Faraday's way of conceiving phenomena and that of the mathematicians, so that neither he nor they were satisfied with each other's language. I had also the conviction that this discrepancy did not arise from either party being wrong....

As I proceeded with the study of Faraday, I perceived that his method of conceiving the phenomena was also a mathematical one, though not exhibited in the conventional form of mathematical symbols. I also found that these methods were capable of being expressed in the ordinary mathematical forms, and thus compared with those of the professed mathematicians.

Faraday searched for the center of the phenomena in the actions really present inside the medium, whereas they [mathematicians] were satisfied to place them in a remote center of activity acting on electric fluids.

This medium, center of phenomena, or ether, should not be eternal. But the fantastic imagination of Maxwell firmly based on his wonderful equations, threw all of physics into confusion.

It was only after Maxwell's death that Hertz managed, in 1887, to demonstrate the quasiluminous behavior of electromagnetic waves generated by oscillating circuits. By reflecting these

Fig. 7

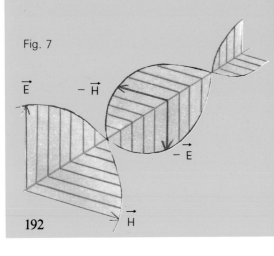

$\vec{E}$ $- \vec{H}$

$- \vec{E}$

$- \vec{E}$

$\vec{H}$

waves on metal plates, by producing standing waves − fundamental proof, − by passing them through prisms with a known refractive index for light waves, by arranging polarization networks, etc., he meticulously verified that the laws of optics apply perfectly to its invisible electric waves.

Thus, he proved at one time that free electric waves do exist and that light was only one of the manifestations.

The active transmission part of the devices used by Hertz consisted in the discharge of an electric arc. Such an arc constitutes an oscillating discharge process of sufficiently high frequency to determine a displacement current of sufficient frequency to generate a relatively intense induction capable of "launching" the process (since curl $\vec{H}$ is proportional to $\frac{\partial \vec{D}}{\partial t}$, a term indicating the speed of variation, that is, the frequency, of the state of displacement).

"In the sense of our theory," wrote Hertz, "we must correctly represent the phenomenon [of electric waves] by saying that, fundamentally, the waves which are being developed do not owe their formation solely to the processes at the origin, but arise out of the conditions of the whole surrounding space which latter, according to our theory, is the true seat of the energy." (Erskine-Murray, 1913 : 20).

This is a fascinating remark for us whose intuition often does not comprehend that the energy emitted by a tiny space probe reaches us so faithfully.

Magnetic properties of intermetallic compounds composed mainly of rare earths. These samples were prepared by melting in a levitation oven under an inert helium atmosphere. These is no crucible and therefore no contamination.

21

Wherein Heinrich Hertz
makes sparks fly...

In 1887 a young man of thirty verified Maxwell's theories through his own experiments. The personality of this young man, Heinrich Hertz, is little known to the general public. He was a most appealing, enthusiastic, hardworking, and romantic young man with an encyclopedic knowledge.

In 1878 Hertz met Hermann von Helmholtz in Berlin, and studied under him before becoming his assistant. Von Helmholtz was among the first scientists to establish the laws of energy conservation at the end of the nineteenth century. From 1869 to 1871, Helmholtz studied electrical oscillations; from then on he began to interest himself in the propagation speed of electromagnetic waves, whose existence Hertz proved from a problem Helmholtz had set him.

In 1883 Hertz became Professor at the Kiel Faculty and began to study Maxwell's theories. Helmholtz had been interested in them for some time, and it was he who alerted Hertz to the competition organized by the Berlin Academy of Science to prove experimentally the relationship between electromagnetic force fields and the polarization of dielectrics. Hertz returned to Karlsruhe in 1885 and began his experiments, at the same time continuing to teach physics. On November 13, 1886, he demonstrated the mutual induction of two open circuits and on December 2 he succeeded in making two circuits oscillate in resonance. On December 5 he wrote an initial summary of his experiments and sent it to Helmholtz.

Hertz noted that a metal wire in the form of an open loop produced a small spark at both ends when it was brought into contact with a solenoid through which the energy stored in a Leyden jar had been discharged. Each time the Leyden jar discharged, a potential difference was induced between the two ends of the open loop and a spark was produced. Then he observed that the spark could be generated in an empty space even if the jar was discharged at the other end of the laboratory and showed how the "waves" were propagated by using sheetmetal plates as reflectors, prisms as refractors, and parabolic mirrors as focalizers. By regulating the dimensions of the receiving loop, he saw that it could be tuned to any given wavelength.

On December 8, 1887, the same year in which Michelson proved that the "ether" did not exist, although Hertz remained unconvinced of this up until his death, Hertz wrote to Helmholtz that he had been able to observe the existence of stationary waves and to measure their length. Then he started to look into the propagation speed of electrical waves.

In November 1888 he discovered short waves which he was able to measure exactly using special small resonators. He summarized these discoveries in two letters addressed to Helmholtz.

In 1890 Hertz traveled to England where he met Crookes, Hughes (who had developed the microphone and the automatic telegraph), Lodge, W. Thomson, and Lord Rayleigh, although he did not meet J. J. Thomson.

Before his death, Heinrich Hertz had the pleasure of seeing his assistant, Philipp Lenard, achieve success in his research. He communicated his pleasure to Helmholtz:

In these last weeks, my assistant Doctor Lenard has made quite a strange discovery. He has succeeded in developing a method for hermetically sealing Geissler's tubes using extremely thin aluminum plates, so thin in fact that some of the phosphorescence-producing cathode rays generated inside the tube pass through them. He noticed that the rays could propagate through air and certain gases and that the latter had a dispersion effect on the ray which opens up a whole new area of research. (Hertz: 1928)

This "strange discovery" was in fact the photoelectric effect that Hertz had already predicted.

In his thesis written in 1897, Jean Perrin describing Lenard's experiments in the following manner:

Hertz had shown that thin metal sheets let cathode rays pass through them. Lenard had the idea of using such a sheet to close off a tiny window made in the glass wall of an empty tube and succeeded in finding a sheet strong enough to withstand atmospheric pressure and thin enough to let the rays through. In this way he was able to observe the effects of the rays as they penetrated into different

Heinrich Hertz's experiment with the spark arrestor.

195

ration process, and thus discover new properties.

The cathode rays were widely diffused when they arrived at the wall of the generating tube; but if the metal window was made small enough, a tiny concentrated beam of rays was emitted. If a second tube was placed next to the first tube, similarly closed off with a thin metal window, and the beam directed into this tube, the beam was widely diffused if it contained not very rarefied gases. If the tube contained very rarefied gases or air whose specific mass was less than $\dfrac{100}{1,000,000}$ of its normal specific mass, the beam remained concentrated and straight. Lenard obtained straight beams that extended to lengths of 1.5 meters, retaining their initial intensity. He believed that he had repeated for these rays the experiments which had earlier decided whether sound and light had their foundation in matter or in the ether, and since the last traces of matter appeared to be more than a hindrance than a help, he believed that his results proved the materialistic theory to be wrong.

Lenard also measured the deviations produced by a magnet on the rays upon leaving the production tube and observed that these deviations were independent of the nature and pressure of the gas in which the rays were moving, at least up to the limit where diffusion became so great that it prevented measurement. This very important property seemed to be irreconcilable with the bombardment hypothesis. The admiration with which these very important experiments were greeted prevented the faults in Lenard's reasoning from being discovered, and many physicists believed from his findings that the emission theory should be discredited. But cathode rays were very different from the light rays known up until then. In addition to their magnetic deviation which was still unexplainable, there was the singular law of absorption which Lenard had discovered in announcing that "the mass of an obstacle is the only thing which governs the disturbance that this obstacle causes to beam propagation." (Perrin: 1950)

Maurice Ponte comments that "the paradoxical irony of this whole story is that the electron was born in contradiction to all the theories of classical physics: when the small electrical charge moved, it emitted rays and had a braking effect, but because it was a negative electrical charge it was, according to the old laws on radiation, condemned to explode on the spot. It is amusing to note that all of modern physics is founded on a nonviable particle. At the time of Thomson's discovery therefore, the existence of the electron had been hypothesized but no one had yet managed to prove it. The classical theory of free electrons was proposed in 1902 by the German physicist P. Drude and then again in 1928 in quantum form by the German physicist and mathematician A. Sommerfeld. But it still left unexplained the difference between conductors and insulators. The electrons in the internal layers were considered to be fixed in the neighborhood of a particular atom, whereas the free electrons (detached valence electrons) moved independently of each in a space about the surface of the metal.

On January 1894, Hertz passed away, at the age of thirty-seven. He left behind him a very controversial question. Did Hertz have any idea of the technological implications of his experiments in the field of what was then known as Wireless Telegraphy?* In 1901 Karl Ferdinand Braun wrote an article entitled "Drahtlose Telegraphie durch Wasser and Luft" ("Wireless Telegraphy through Air and Water") in

1. Portrait of Heinrich Hertz (1857-1894).
2. Hertz's bipolar transmitter and receiver.
3. A letter in Hertz's handwriting.

* This question was developed further in an article published by Professor Charles Süsskind in the magazine *Isis*, 1965, vol. 58 n, 3, n° 185.

which he pointed out that: *As early as 1889, Hertz was asked by a Münich engineer, Huber, whether his waves could not be employed for wireless telegraphy. Hertz answered in the negative. If the question had been put to him two years later, perhaps his answer would have been in the affirmative.*

Huber in fact made no allusion to telegraphy. He was employed in a power station which supplied The Hague with electric light and was thinking about electricity and the telephone. What he actually wrote was:

I should be very interested to know whether it would not be possible, using your theory, to transmit the magnetic lines of force (the invisible ones of course) over a distance? I am thinking mainly about transformers and the telephone. One could, for example, excite the polepiece of an electromagnet at the focal point of a convex mirror. The magnetic lines (the magnetic field) are then received by the mirror of the second station and produce induction (secondary) currents by means of an induction coil.

Hertz's reply was dated December 1889:

Magnetic lines of force can be propagated as rays just as well as electric lines as long as their oscillations are sufficiently fast, for in that case they altogether coincide with the electric lines, and the rays and waves that occur in my experiments could be equally well called magnetic as electric. But the oscillations of a transformer or a telephone are much too slow. Take a thousand oscillations per second, which is surely a high figure, yet the corresponding wavelength of the ether would be 300 kilometers and the focal lengths of the mirrors employed would have to be of the same order of magnitude. If you could thus build convex mirrors as large as a continent, you might very well be able to set up the proposed experiments, but in practice nothing can be done, you would not perceive the slightest effect with ordinary convex mirrors. At least, that is what I suppose.

Thus, Hertz's reply was the only one possible at the time he wrote it. Scientists were then still unable to produce ultrashort wave oscillations at a useful power and modulation level. It was not until 1896 that Marconi described the first radio-telegraph system. What would have happened had Hertz lived long enough to become aware of the beginnings of wireless telegraphy? No one can say with any certainty, but right up until the time of his death he was still very interested in research, rather than applied physics. Be that as it may, Hertz was indisputably responsible for opening up a new field which at the beginning of the twentieth century every young researcher and engineer worthy of the name rushed to explore: the field of radio.

Wherein Heinrich Hertz Makes Sparks Fly...

1

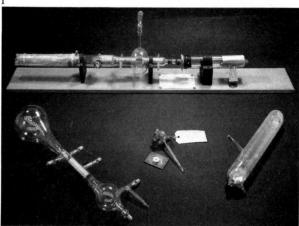

2

3

4

5

1. Hertz's original equipment (1886-1888). Left, polarization grid for electric waves. Right, resin prism for demonstrating the refraction of electromagnetic waves.
2. At back: P. Lenard's discharge tube with which he undertook the first quantitative measurements of cathode rays (from 1893 to 1898). Front right: Hertz's tube for studying cathode rays passing through thin metallic sheets. Middle: Lenard's first "window" tubes. Left: J.J. Thomson's tube (1897).
3. Other equipment used by Hertz between 1886 and 1888. *1* and *2*. Circular resonators. *3*. Spark apparatus under a glass bell jar for ultraviolet light tests. *4*. Device to measure electrical vibrations. *5*. Quadrangular resonator.
4. Hermann von Helmholtz (1821-1894) German Physicist and physiologist.
5. Philip Lenard (1862-1947) Nobel Prize for Physics in 1905.

197

Hertz's experiments

An extract from Henrich Hertz's Diary and Correspondence compiled by Dr. Johanna Hertz (Akademische Verlags-gesellschaft m.b.h. Leipzig, 1928 p. 164–165–166, 182–183, 201–202.

These four letters summarize the originality of the experiments carried out by Hertz. In the letter dated December 5, 1886, he summarizes his experiments on the mutual induction of two open circuits and on the setting of theses circuits in resonance. In those dated December 8, 1887, and March 19, 1888, he states the existence of standing waves and measures the wavelengths. The letter dated November 10, 1888, reports the existence of short waves.

December 5, 1886:

(Indeed) I managed to show very clearly the induction effect of an open rectilinear electric current on another open rectilinear current, and and I can hope that this will enable me to resolve, in the end, the various questions relating to this phenomenon.

I produce this rectilinear induction current as follows: a thick, straight copper wire 3 mm long is connected at both ends to two balls 30 cm in diameter or to two conductors of similar capacity. In its middle this wire is cut off by a 3/4 cm spark gap between small brass balls. Crackling sparks from a large inductor are next passed between these balls and – something hardly predictable at the start, of course – this induces electric oscillations in the rectilinear electric circuit which exert a relatively strong inductive effect on the environment. In a simple square circuit, 75 cm to one side, composed of one thick copper wire and having only one short spark gap, I still obtained sparks at a distance of 2 m next to the inductive circuit; by bringing one side of the square nearer to within 30 cm from the inductive circuit, the length of the sparks can be increased to 6 mm. Of course, different circuits behave differently, and here I believe to have successfully proved the phenomenon of resonance; the two circuits can be regulated, by modifying each one, to obtain a maximum effect, although the oscillation period is only of the order of one hundred millionth of a second. Just now, it is not surprising that the induction effect is also felt in a second rectilinear conductor, similar to the first and mounted parallel to the latter; furthermore, the ends of all the rectilinear metal wires or rods produce fine small sparks close by. Already during my first experiments I could again observe this effect in a parallel conductor, the distance between the two being 1.50 m, and I believe that, with correct regulation, this distance could be increased considerably. I also believe to have successfully produced stationary oscillations with two oscillations nodes in a simple system of wires.

In all these experiments, the danger of errors and incorrect interpretations is easily understood, but I have already found confirmation of these hypotheses in a sufficiently large number of experiments to be firmly convinced that these phenomena are interpreted correctly. Dear Counsellor, I do not want to impose on your time by stating the details; I hope that you will not consider it a lack of modesty that I am relating these experiments to you even before they have been completed. Expressing my profound respect, I remain, dear Counsellor, your devoted servant.

December 8, 1887:

In the meantime, I have successfully carried out several other experiments. For example, with the aid of the oscillations used in my preceding studies, I managed to produce standing waves with several nodes in taut rectilinear wires. Confining myself to four or five nodes, I could make the latter appear almost as clearly as the nodes of a vibrating cord. The largest of my devices gives – in rectilinear wires – waves whose length reaches nearly 3 m exactly. Their period, calculated from the potential and the capacity, was 1.5 hundred millionth of a second and, as a result, their propagation velocity would be about 200,000 km/s. But perhaps it would be better – since nearly all the theories prove it – to calculate the propagation velocity in the wires by proceeding as for the speed of light and, according to this, set the period of the device at 1 hundred millionth of a second.

I also succeeded in creating interference between the effects that are propagated through the wire and those which are propagated in the air. I thus hope to prove a finite propagation velocity of these effects. Up to now, however, the experiments are proceeding just as though the effects in the air are propagated much more rapidly than in the wires, hence also more rapidly than light.

Naturally, I still have to repeat these experiments and I also hope to perfect them in order to be completely certain of my position.

March 18, 1888:

I would now like to tell you the progress as follows: in the air, electrodynamic waves are reflected by solid conducting walls, the reflected waves interfere with the incident waves – the incidence is vertical – and give rise to standing waves in the air. In the first wave length, calculated from the wall, these phenomena are very pronounced and varied, and I believe that we cannot reveal the undulatory nature of sound in free space as clearly as the undulatory nature of this electrodynamic propagation. The first node can be determined very accurately and we can thus measure the wavelength in an extremely direct manner. As regards the device for which I indirectly determined a wavelength of 4.5 m, I now found a slightly higher value for it: approximately 4.8 m. The reflections that were not taken into consideration will probably create some disorder during these experiments.

November 30, 1888:

When you asked me in Berlin if I had carried out other experiments on [illegible] waves, I had nothing important to tell you at the time. Now, however, I have made some progress which firmly establishes the relationship between light and electricity and which I am therefore anxious to tell you about. First of all, a fortunate accident showed me that not only can we produce waves several meters in length, but that it is also possible to work with much shorter waves, which offers an immense practical advantage. I was able to confirm, favorably in part, may earlier results with waves in the air measuring 33 cm in length. I then repeated my experiments with these short waves, which involved sending this force over a distance by means of a concave mirror and producing a ray — all with great success. I placed my primary and secondary conductors at a focal distance from a parabolically curved metal sheet 2 m high and 2 m wide, and then found a ray issuing from the mirror having a well-defined width of about 1 1/2 m; it was discernible in a second concave mirror up to a distance of 16 m, and probably more....

Yesterday I also succeeded in proving the irregular reflection of the ray more clearly than I had hoped. If I juxtaposed the concave mirrors, there would be no effect in A and B; however, if I placed a flat metal wall in front of the concave mirrors, sparks would immediately appear in B and these would still be visible when the wall was moved 10 m away from the mirrors. I was also able to achieve a 45° reflection by using two adjoining chambers, as shown in the diagram. The fact that the wooden doors separating the latter are closed does not by any means hinder the formation of secondary sparks. On the other hand, the sparks cease if the flat reflecting wall is turned so that it leaves its correct position only by approximately 5°, from which it appears that the reflection is regular and does not scatter.

Continuation of Journal:
December 1, 1888:
Carried out experiments on reflection and the polarization of radiation. Expansion of these experiments.
December 2:
I am studying, without much success, the phenomena of refraction.
December 6:
Impatiently waiting for the prism.
December 7:
The prism arrived and was mounted in the electrotechnical laboratory. Experiments successful.
December 8:
Refraction experiments carried out with Dr. Schleiermacher.
December 9:
Completed a few auxiliary experiments and wrote a memo to the Academy. Took a walk.

A caricature which appeared in the "Berliner Wespen" (Berlin Wasp): the new invention of the telephone allows parliamentary deputies to follow debates from the comfort of home. "Call Bismarck — he'll say present."

Berliner Wespen.

Parlaments-Telephonie.

Durch das Telephon werden allen Klagen über Abwesenheit und Beurlaubung der Minister ein Ende gemacht. So wie der Ruf nach Bismarck ꝛc. ertönt, antwortet der Betreffende von Varzin ꝛc. aus: „Hier!" um sofort in die Discussion einzutreten.

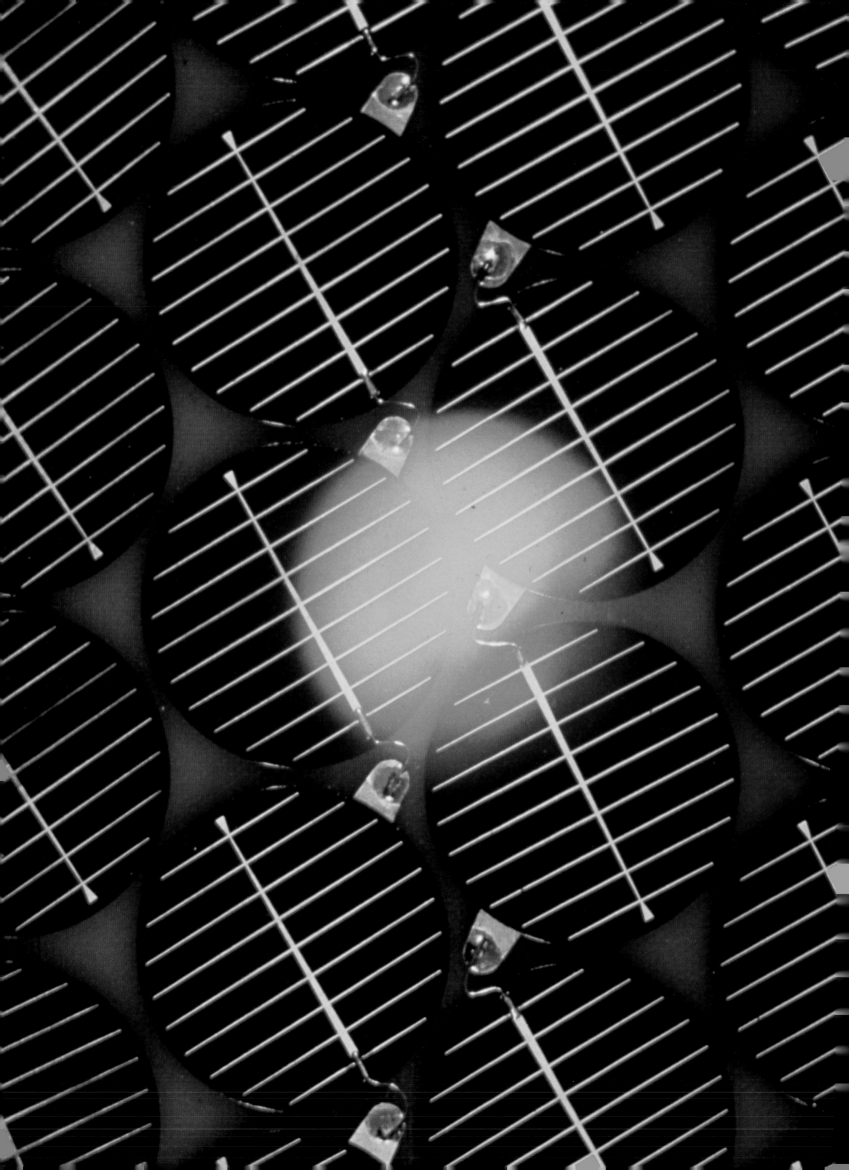

The photoelectric effect

The photoelectric effect may be described very simply. Particles of light or photons that bombard a body will liberate electrons from it. The energy of these electrons may either be amplified and modulated for signal purposes (photoemission or photoconduction) or be used to generate a source of electrical energy by means of direct photovoltaic conversion. In the second case we create an internal field and an internal potential difference which is expressed in volts. The photoelectric effect foreseen by Hertz and studied by Lenard was theoretically demonstrated by Einstein; it is one of the basic effects of electronic techniques. Therefore it is important to look into the origins of the photoelectric effect and to distinguish between its three basic forms.

Photoemission: very important in the electrical analysis of televised pictures, takes on renewed importance with the advent of stop-layer cells (used particularly in solar batteries). In this usage light ejects electrons from the bodies which it strikes.

Photoconduction: mainly used in semiconductor technology in which the conductibility of a body can increase when struck by light.

The photovoltaic effect: in which luminous energy is transformed into electrical energy within a series of conductors exposed to light.

The photoelectric effect was discovered almost by accident: On February 4, 1873, in a letter to the vice-chairman of the Society of Telegraph Engineers, cable engineer Willoughby Smith described his recent discovery, together with another operator named May, of a rather strange effect shown on a control galvanometer at the Irish transatlantic cable station in Valentia. The two men showed that a small variation of current could be observed across a selenium resistance when the resistance was lit up by the sun's rays. Willoughby Smith proceeded to duplicate the experiment using selenium plates placed in hermetically sealed glass tubes. Selenium resistances were used as high level resistors in control circuits at the coastal station during underwater cable laying operations and Willoughby Smith found that the resistance of these plates was reduced from 15 to 100 percent when they were exposed to the sun's rays. If the plates were plunged into water to cool them the readings did not change and Willoughby

Smith concluded that this must be due to light. This was the accidental discovery of the photoelectric effect, which had earlier been foreseen by Antoine Cesar Becquerel in 1835 in France and which would have quite extraordinary consequences in the field of electronics, because it concerned the possibility of liberating electrons from the nucleus of the atom. Two years later in Boston, G.R. Carey designed a system for transmitting images between a screen made up of a number of photoelectric cells linked by wires to a reception screen made up of tiny electric bulbs. This was one of the first preliminary steps along the avenue of research that would eventually lead to television.

In the course of his experiments, Hertz had observed that his spark arrestor worked well when he moved away from it but not when it was covered by his shadow. He deduced from this that sparks were more easily generated in the spark arrestor when it was illuminated by ultraviolet light. He then carried out a number of different experiments to verify that it was indeed the ultraviolet component of the main spark arrestor striking the pin of the micrometric spark arrestor that was the cause of the increase in spark length. He passed the light generated by the main spark equipment through a quartz spectrograph and found that only in the ultraviolet range of the spectrum did the spark length increase. He spent eight days writing a note on the phenomenon but never went into it further.

Hertz's pupil Lenard established, as we have already seen, that negative charges are expelled from metal by light, but also that the escape velocities are a function of frequency and not of light intensity.

In 1880 Wilhelm Hallwachs observed that when a metal plate was subjected to ultraviolet radiation it acquired a charge. He simplified Hertz's experiment and showed that under the influence of ultraviolet radiation, a body emitted negative electricity that followed the lines of force of electrostatic fields as it traveled along its path. Righi (an Italian wireless telegraphy expert) further modified Hallwach's experiment and designed a device which produced a current under the influence of ultraviolet rays. He demonstrated that by placing two similar cells in series, the deviation of the electrometer would dou-

A panel of solar cells.

The photoelectric effect

ble. He had in fact created a photoelectric battery.

In 1900 Max Planck announced that "irradiated heat is not a continuous and indefinitely divisible flow. This flow can be defined as a discontinuous mass composed of elements which resemble each other." In 1905 Einstein went even further and announced that the particles of light (later christened *photons*) accompanying the light waves donated their energy to the electrons by liberating them from bodies struck by light rays and that this energy was a function of frequency. Thus he introduced the notion of "light quanta" to explain the phenomenon of photoelectricity. Planck himself would never have dared admit so revolutionary an interpretation of his ideas, which put into question the very nature of light waves. In his remarkable book on Einstein, Ronald Clark writes:

For while light consisting of discrete packets of energy, as indivisible as the atom was still thought to be, conformed — if it conformed to anything — to the corpuscular theory of Newton's day, the idea also utilized frequency, a vital feature of the wave theory. Einstein had to face the embarrassing contradiction that Planck had tried to avoid; for some purposes, light must be regarded as a stream of particles, as Newton had regarded it: for others, it must be considered in terms of wave motion. (Clark, 1971: 69).

The resistance with which Einstein's theory on light was greeted only began to subside some ten years later. Einstein's idea was at the origin of wave mechanics, and Louis de Broglie was not reluctant to admit it: *I am convinced that the dualism of waves and corpuscles discovered by Einstein in his theory of light quanta was a completely general thesis and could be extended to all of physical nature, and it seemed to be undeniable that the propagation of a wave must of necessity be associated with the movement of any corpuscle, whether this be a photon, an electron, a proton or any other type.* During research into the photoelectric effect, two very precise laws were established. These laws were explained by Zworykin and Wilson in their book on the photoelectric cell:

The first law can be expressed as follows: the number of electrons emitted per time unit by a given photoelectric surface is directly proportional to the intensity of the incident light, for a given composition of the surface. The second law was even more surprising: the maximum energy of electrons emitted by a photoelectric surface is independent of the intensity of the light-excitor source but increases in linear progression with the frequency of the light. In other words, although the electrons emitted under the action of light possessed various different speeds, the maximum speed obtained was determined solely by the maximum frequency of the light excitation source. No matter how intense the source might be, this maximum speed did not change.

Even before the theory had been completely elucidated, various practical applications began to appear. One of the first applications of the photoelectric cell was the process known as telephotography developed respectively by German researcher Arthur Korn (1895) and French researcher Edouard Belin (1906). In Germany, Julius Elster and Hans Geitel observed that elements that were chemically electropositive (aluminium, magnesium, and zinc) produced photoemission effects. They showed that alkaline metals that were even more positive (sodium, potassium, and caesium) emitted electrons when subjected to visible light. They soon developed the first photoelectric cell, it

was not, however, particularly sensitive. The surfaces reacted to air and water vapor and formed layers of oxide and hydroxide. Elster and Geitel then amalgamated these different metals with mercury, which slowed down the reaction. Amalgamated potassium and sodium surfaces were then shown to be photoemissive not only when exposed to ultraviolet rays but also when exposed to visible light. They obtained their first photoelectric cell by enclosing alkaline surfaces in a vacuum-sealed glass case. The two men continued their work and discovered that some sodium and potassium crystals were more sensitive to light than pure metals. In 1912 they constructed bulbs from hydrogenated alkaline metal, the forerunner of modern photoelectric tubes. The photoconductive selenium cells descended from Willoughby Smith's cells were replaced by photoemissive cells or "phototubes" and by new types of photoelectric cells which appeared between 1920 and 1930. After 1940, photoconductive cells took on renewed importance *particularly as ultra-sensitive infrared beam detectors*. The development of photovoltaic cells, (particularly the solar battery) was a byproduct of research into semiconductor junctions (1944/1945) and silicon, and was only researched after 1954. These cells are discussed in the chapter on transistors. Unlike photoelectric cells they were designed not for producing signals but as a source of renewable energy.

1. Photoelectric cell with electron multiplication.
2. The "Satellite Power" system which can produce enough electricity for a large town from energy supplied by the sun. This satellite will take the form of a large platform with a set of solar

cells capable of converting solar energy into electricity. NASA and the Department of Energy in the United States are studying the technical and economic implications of this concept which is still at the design stage.

2

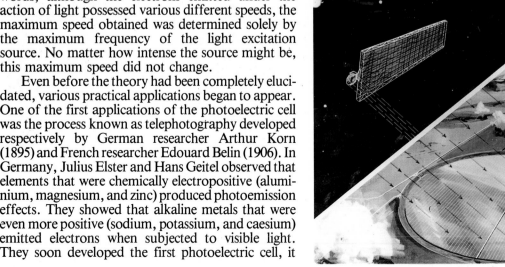

1

1. Physicist Julius Elster (1854-1920).
2. Physicist Hans Geitel (1855-1923).
3. Final checking of the solar generator panel for the European ECS information satellite (1981).
4. The Eole satellite equipped with a solar generator (1969).
5. This optical reflector serves as a target for ground-based laser systems. It is used to obtain exact measurements of the distance between the earth and the moon, gravitational influences, the distribution of masses on the moon, and continental drift.

The photoelectric cell

by Georges-Albert Boutry, founder of the Laboratories for Electronics and Applied Physics (L.E.P.)

The photoelectric cell is one of the basic devices used in electronics. What are the principles of photoemission and photoconduction?

The Fleming diode is a thermionic instrument: when one of its electrodes is raised to a sufficient temperature, it becomes capable of delivering a current. The diode known as the photoelectric cell is an instrument capable of delivering a current when one of its electrodes is *illuminated.*

A light flux is sent to one of the electrodes, i.e., the surface of this electrode is bombarded with photons (particles of light), which are able to convey all or part of the energy they are carrying to the electrons which may exist in this electrode. These electrons can then be liberated (if they were originally bound to the crystal lattice of the electrode) and accelerated. If the speed they take is of suitable direction and magnitude, they will leave the electrode and appear in the vacuum where they will be captured by the electric field, which is regulated by applying a potential difference in the appropriate direction between the two electrodes.

In a cell of this type, the sensitive electrode will deliver more current as the photon bombardment becomes more intense. The intensity of the current also varies with the type of metal used for the sensitive electrode: potassium and cesium, in thin films and complexes spread on a metal electrode, are the most sensitive combinations known.

The liberation of photoelectrons by photon bombardment can be observed not only in a vacuum, but also in all electronic semiconductors. This is how it takes place in silicon and germanium, well-known semiconductors used in transistor manufacture. Let us go back to the diode and bring the two electrodes constituting the latter closer together until they come into contact with each other. The experiment again shows that the contact will be photoelectric and that, by applying potential differences, the diode so designed will deliver a current. This principle can be used for the manufacture of diodes in which the two electrodes in contact are replaced by a nonhomogeneous semiconducting strip. In these diodes the semiconductor is the same everywhere, but the impurity added to it does not have a uniform concentration or character throughout the strip. This is how, in the case of silicon, the impurity responsible for conduction is composed, for example, of arsenic atoms on one side and indium atoms on the other. The experiment shows that a strip composed in this manner delivers a current if it is illuminated. The diodes complying with the above brief description are ordinarily called a semiconductor cell.

All that we have said up to now about photoemissive diodes and photodiodes assumes that the electrodes of these diodes are subjected to an electric field produced by a potential difference of external origin. If this potential difference is brought to zero, the current is not reduced to zero but merely decreases: the effect in question is very small in the case of photoemissive cells and hardly attracted the attention of physicists in the past. In semiconductor diodes, however, a potential difference inherent in this structure already exists between the part "doped" with arsenic atoms and the part "doped" with indium atoms. Let us therefore imagine that the photons arrive, enter the region where a field exists, and there convey − while disappearing − the kinetic energy they had to the electrons. The photon conveys its energy to an electron-hole pair which diffuses into this network where an electron discontinuity was produced, at the end of which there is an electric field (the junction). This field will separate the electron and the hole, thereby permitting their circulation in an external circuit and thus producing an electric current.

The most important question has been ignored in the grossly simplified account just given: it is not just any light that will produce the photoemission or photoconduction effect. In particular, the silicon cells are sensitive to light whose wavelength lies between approximately $0.4 \, \mu$m and $1.1 \, \mu$m: this entire set of emissions is contained in the visible and infrared spectoral region.

The question of efficiency is raised: the current supplied by the cell, discharging without any external electromotive force, is maintained only due to the energy of the photons which produces it in the first place. This current can be used in the circuit and, owing to its passage, will expend a certain amount of energy in order to be maintained: this energy, borrowed from the photons, cannot exceed the amount they bring in. We will define cell efficiency as the quotient of the energy supplied by the cell over the energy it receives from the light that bombards it at the same time. We now understand what a solar cell is: it is a semiconductor cell sensitive to the widest possible range of solar radiation. In this connection, the heterogeneous gallium arsenide and aluminum cell forms the converter of light energy into electric energy that we know how to build, which corresponds to a relatively low efficiency. It has reached 24 percent with difficulty under laboratory conditions, which means that, due to the numerous secondary phenomena that we cannot dwell on here, only 24 out of the 100 photons absorbed by the silicon will supply their energy to the current generated by the cell. This figure − at the time this report was written − constitutes the maximum that can be done on the laboratory research level (only a 13 percent efficiency is achieved on the industrial level). Research is being continued to find semiconductors with increasingly more suitable properties, and the task is proving difficult.

A solar cell with its connections.

23

J.J. Thomson
and the discovery of the electron

The true story of the electron begins in Cambridge, England, at the Cavendish Laboratory in 1897. The laboratory was then dominated by the personality of a distinguished-looking man of forty, famous for his absent-mindedness.

On April 30th, 1897, this man demonstrated his discovery of a body smaller than the atom, a negative charge which he called a corpuscle.

His name was John Joseph Thomson and the *corpuscle* he had discovered was the electron. His demonstration aroused enormous controversy. Later Thomson wrote:

At first there were very few who believed in the existence of these bodies smaller than atoms. I was even told long afterwaras by a distinguished physicist who had been present at my lecture at the Royal Institute that he thought I had been "pulling their legs." I was not surprised at this, as I had myself come to this explanation of my experiments with great reluctance and it was only after I was convinced that the experiment left no escape from it that I published my belief in the existence of bodies smaller than atoms. (Thomson, 1936: 23).

For a very long time Thomson refused to use any other term than "corpuscle" to describe this negative particle of electricity. But if we take a look at his memoirs, *Recollections and Reflexions,* we can read: *I at first called these particles "corpuscles" but they are now called by the more appropriate name "electron".*When he made this discovery, Thomson was already well known and respected in scientific circles. In 1893 the editor of Maxwell's work asked Thomson to take charge of a reprinting. At the same time, Thomson was engaged in publishing his *Notes and Recent Research in Electricity and Magnetism* which was nicknamed "the third volume of Maxwell's work."

From the moment that he first started research work at Owen College in Manchester, the young Thomson took a lively interest in Maxwell's research. (Maxwell had published his *Treatise on Electricity and Magnetism* in 1873.) Thomson joined the Cavendish Laboratory and worked with Lord Rayleigh, who had taken over Maxwell's chair in

Physics in 1879. In 1884, Rayleigh retired and Thomson took over his position. He began to look at William Crookes's work on cathode rays and carried out experiments on conduction through gases. Around 1895 Ernest Rutherford began to collaborate with Thomson in the study of temporary conduction of gases due to ionization under the effects of x-rays, which had just been discovered in Germany by Röntgen.

On December 28, 1895, Röntgen communicated his discovery to the Würzburg Society of Physics and Medicine. Like Thomson during the same period, he had carried out experiments on the conduction of electricity through gases and had observed a fluorescent screen of barium platincyanide produced on the side of a high-vacuum tube through which an electrical charge was passed. He noted that the rays that caused this fluorescence penetrated various different substances and could pass through them, even if they were opaque to ordinary light, and that they had an effect on photographic plates. Because he was uncertain of the nature of these rays he called them "x" rays. x-rays are in fact a form of electromagnetic radiation produced by the impact of electrons on a target. They are shorter than the waves in the electromagnetic spectrum. x-rays and the exploration of matter which they allowed, helped bring the "classical" period of physics − characterized by a belief in the light-carrying "ether" − to a definitive close. Thomson's discovery two years later, based on a quantitative reassessment of Jean Perrin's experiments concerning a particle smaller than the atom, sounded the deathknell of the mechanistic and continuous approach to matter. Thomson immediately applied Röntgen's discovery (which acted as a kind of powder trail leading to further research, and which was to be used almost immediately in medical circles) to his work on conduction through gases. Along with Rutherford and an Irish student, McLelland, who had just joined the little team, he worked on the applications of radio waves, which were the first experimental verifications of Maxwell's theory. Thomson's position with respect to Hertz was very critical. He was amazed that the German physicist had completely failed to discover the nature of electrical charges in the course of his experiments.

Sir J.J. Thomson talking with F.B. Jewett in 1923; the latter was to become Bell Laboratories' first President.

* Röntgen, the man who discovered X-rays, continued to oppose the theory of electrons right up until 1910.

J.J. Thomson and the discovery of the electron

1. J.J. Thomson studying cathode rays.
2. J.J. Thomson (arms crossed in the middle of the first row) and his students or visitors. Right: P. Langevin. Standing at back: Thomson and E. Rutherford.
3. Ernest Rutherford and J.J. Thomson taking part in a cricket match.
4. Heindreik Anton Lorentz (1863-1923) winner of the Nobel Prize for Physics in 1902, who invented the equations which allowed Einstein to discover his theory of relativity.

Hertz came to the conclusion that charged particles did not exist. He adopted the theory of the majority of German physicists who believed that flexible electrical currents, when fluctuating in the ether, were subject to magnetic forces and obeyed the laws of Ampère concerning the forces which act on electrical currents. However Hertz had never attempted to discover the nature of negative electricity. It is obvious that Thomson judged Hertz in the light of his own research. He does seem to have ignored, or not to have been aware of, some of the statements made by another German physicist whom he attacked. This was Helmholtz, Hertz's former teacher, who wrote in 1881: *If we accept the hypothesis that elementary subjects are composed of atoms, we cannot avoid concluding that electricity also, positive as well as negative, is divided into definite elementary portions which behave like atoms of electricity* (quoted by Thomson's son, Sir George Paget Thomson). In fact, Hertz had adopted to the letter Maxwell's position on the subject: that the existence of what he called "a molecule of electricity" was unthinkable. Thomson, like Perrin, was fortunate in being at a watershed created on the one hand by Maxwell's electromagnetic theory and on the other by William Crookes's research into cathode rays. He began by looking into Perrin's experiment on cathode rays and the measurement of their charges:

It seems to me that to the experiment in this form it might be objected that though the experiment shows that negatively electrified bodies are projected normally from the cathode and are deflected by a magnet, it does not show that when the cathode rays are deflected by a magnet the path of the electrified particles coincides with the path of the cathode rays. The supporters of the theory that these rays are waves in the ether might say, and indeed have said, that while they did not deny that electrified particles might be shot off from the cathode, these particles were, in their opinion, merely accidental accompaniments of the rays, and were no more to do with the rays than the bullet has with the flash of a rifle. (Thomson: 1897).

Thomson was making an allusion to the theories of the German physicist and, more particularly to the theory advanced by Wiederman, observing, as Perrin had done in his 1897 thesis, that *if matter is transported along cathode rays, this matter has as little to do with the phenomenon as the cannonball fired by the canon has to do with the sound which accompanies its departure.*

For contemporary physicists this was indeed the crux of the matter, and Thomson was well aware of this. The scientific world was split between the partisans of the wave theory and those who favored the ether (that imaginary but very practical substance dear to classical physicists) and the partisans of the particle theory. The second group had to prove that particles of electricity were not a secondary phenomenon, but that the cathode actually emitted particles charged with negative electricity. Despite what Thomson later wrote, this was the intention of Jean Perrin, although the French scientist did not immediately think of establishing the relationship of the charge to the mass of the "corpuscule." Thomson wanted to verify that the cathode rays transported a charge, that this charge accompanied the rays when they were deflected by all electrical field of known intensity, and that these deviations indicated the presence of a charge. He then wanted to measure the energy of the rays and by combining these measurements with those of the magnetic deviation, to determine the speed and the relationship of the charge to the mass. He also wanted to determine the same quantities by combining electrical deviation and magnetic deviation. Thomson then decided to modify Perrin's experiment:

Two coaxial cylinders, with slits cut in them, the outer cylinder being connected with earth, the inner with the electrometer, are placed in a discharged tube, but in such a position that the cathode rays do not fall upon them unless deflected by a magnet; by means of a magnet, however, we can deflect the cathode rays until they fall on the slit in the cylinder. If under these circumstances the cylinder gets a negative charge when the cathode rays fall on the slit, and remains uncharged unless they do so, we may conclude, I think, that the stream of negatively electrified particles is an invariable accompaniment of the cathode rays. (Thomson, 1897).

It was at this point that Thomson began to believe that Hertz had been mistaken in stating that deviation was impossible in an electrical field. He began to look at the German physicist's experiments, and observed that Hertz's error was due to the ionization of the gas, whereby the positive ions were attracted by the negative plate in the negative ions by the positive plate. This tended in both cases to neutralize the charge of the plates. Thomson decided to work at higher levels of vacuum. He finished by establishing not only that his "corpuscles" were the smallest bodies known, but also that this affair of the cathode rays was not just a simple anomaly in the field of conduction through gases. For Thomson, the corpuscules were a universal component of all matter, the key to a whole system.

We do not intend to mention all the basic points made in the report that Thomson presented to the members of the Royal Institute. We shall content ourselves with noting some of his calculations.

Thomson established that the force f acting on the hypothetical "electron" corpuscle of charge e and moving at speed v (the current is then ev) equals $f = Bev$, B being the induction of the magnetic field to which the corpuscle is subjected. Its trajectory is then curved, f being perpendicular to the speed, and the corpuscle described an arc of a circle radius R. If the mass m is subjected to a centrifugal force

$$f = \frac{m\,v^2}{R}$$

that is equal and opposite to the force determined by the magnetic field. This gives:

$$Bev = \frac{mv^2}{R}$$

from which the relationship of mass to

charge: $$\frac{m}{e} = \frac{Br}{v} \quad (1)$$

We can see that in Thomson's notation, the symbol 1 designated the product of the measurable and known magnitudes, induction B and radius R. This single relationship does not allow us to deduce $\frac{m}{e}$ since v is unknown.

But the electron is accelerated by an electrical field E and its kinetic energy $1/2\ mv^2$ is equal to a potential energy Ee. This gives us a second relationship including m and e:

$$\frac{mv^2}{2}\ Ee$$

from which

$$\frac{m}{e} = \frac{2E}{v^2} \quad (2)$$

We can eliminate the value v between these two relationships (1) and (2) and in function of the magnitudes which are known and measured we find:

$$\frac{m}{e} = \frac{B^2\,R^2}{2E}$$

J. J. Thomson, a good thermodynamics man, looked for (and found) the second relationship by means of a thermocouple. He established that the specific charge of the "corpuscle" was about 2,000 times higher than that of the ions belonging to the lightest electrolyte, hydrogen; that is, that their mass was around 2,000 times smaller than the mass of hydrogen atoms (1/1836 times smaller to be exact). Thomson thus deduced the relative value of the electron. Later, Millikan was to establish the real electrical charge and in consequence the value of the electron mass (its absolute value). This development flashed like a bolt of lightning across the peaceful horizons of classical physics, already slightly disturbed by the discovery of x-rays. It opened up the minds of scientists to the quantum theory, Einstein's theory on photoelectricity, and Niels Bohr's research on the atom. Jean Perrin wrote: *The problem of the atomic structure is now of immediate concern since it has been proved that the atom is no longer the base element of matter.* J. J. Thomson supposed that the atom, neutral in its entirety, was made up of a homogeneous electrically positive sphere inside which electrons occupied positions of attraction and repulsion tending to balance each other out. Jean Perrin left this theory behind as early as 1901 when he proposed an atomic structure similar to that of the solar system, and very close to the structure eventually laid down by Rutherford and later by Bohr.

Jean Perrin

An extract from Jean Perrin's 1895 report to the Académie des Sciences and J. J. Thomson's paper on "Cathode Rays" delivered at the Weekly Evening Meeting, Friday, April 30, 1897, to the members of the Royal Institution of Great Britain.

These are the two original papers delivered by Jean Perrin and J. J. Thomson, now considered as marking the true debut of the electron in contemporary physics.

Presented below are excerpts from these two texts, beginning with that of Jean Perrin.

In order to verify whether cathode rays are negatively charged, I turned to the laws of influence, which allow us to establish the introduction of electric charges inside a closed conduction shell and to measure them. I then forced cathode rays into a Faraday cylinder.

To this end, I used the vacuum tube shown in Figure 1.

ABCD is a metal cylinder closed on all sides except for a small opening in the middle of the side BC. This will act as the Faraday cylinder. A metal wire, soldered at S to the wall of the tube, connects this cylinder to an electroscope.

EFGH is a second metal cylinder, permanently connected to the ground and pierced with only two small openings at β and γ. It protects the Faraday cylinder from any external effects.

Finally, an electrode N is located at approximately 0 ^{m}.10 in front of FG. The electrode N served as a cathode; the anode was formed by the protective cylinder EFGH: a cathode-ray beam thus penetrated the Faraday cylinder. Invariably, this cylinder was charged with negative electricity.

The vacuum tube was placed between the poles of an electromagnet. When the latter was excited, the cathode rays not refracted no longer penetrated the Faraday cylinders. Consequently, this cylinder was not charged: it was charged as soon as I stopped the excitation of the electromagnet.*

Briefly, the Faraday cylinder is negatively charged when the cathode rays enter it, and only when they enter it: the cathode rays are therefore charged with negative electricity.

It is possible to measure the amount of electricity delivered by the rays. I did not complete this study, but I will give you an idea of the size of the charges obtained by saying that for one of my tubes, at a pressure measured at 20 microns of mercury, and for only one interruption of the primary winding of the coil, the Faraday cylinder received enough electricity to raise a capacity of 600 C.G.S. units to 300 V.

Since the cathode rays are negatively charged, the principle of the conservation of electricity prompts us to search for the corresponding positive charges. I believe that I found them in the same region where the cathode rays are formed, and discovered that they go in the opposite direction, rushing headlong into the cathode.

In order to verify this hypothesis one need only use a hollow cathode with a small opening through which part of the positive electricity drawn can enter. This electricity will then be able to act on the Faraday cylinder inside the cathode.

The protective cylinder EFGH with its opening β satisfies these conditions;

consequently, I used it this time as the cathode, the anode being the electrode N.

The Faraday cylinder is thus invariably charged with positive electricity. The positive charges were of the order of magnitude of the negative charges obtained earlier.

Thus, at the same time as the negative electricity is radiated from the cathode, the positive electricity moves toward this cathode. I carried out a study to determine whether this positive flux formed a second system of rays absolutely symmetrical to the first.

To do this, I constructed a tube similar to the preceding one (Figure 2), the only difference being a diaphragm positioned between the Faraday cylinder and the opening β. This diaphragm had an opening β' so that the positive electricity entering through β could act on the Faraday cylinder only if it also passed through the diaphragm β'. Then I repeated the preceding experiments.

With N as the cathode, the emitted cathode rays readily pass through the two openings β and β' and caused considerable separation of the gold leaves in the electroscope. However, when the protective cylinder is the cathode, the positive flux — which according to the previous experiment enters through β —is not able to separate the gold leaves except at very low pressure. Replacing the electroscope with an electrometer, we see that the action of the positive flux is real but weak, and increases when the pressure decreass. In a series of experiments at a pressure of 20 μ, it raised a capacity of 2,000 C.G.S. units to 10 V: this same capacity was raised to 600 V at a pressure or 3 μ over the same time period.**

I was able to suppress this action completely by means of a magnet.

These results, as a whole, do not appear to be easily reconcilable with the theory that turns cathode rays into an ultraviolet light. However, they agree well with the theory that makes them a physical radiation and which, I believe, could be expressed as follows:

The electric field near the cathode is intense enough to break some of the remaining gas molecules into pieces — into ions. The negative ions move toward the region where the potential increases, acquire a considerable speed, and form cathode rays; their electric charge and, consequently, their mass (on the basis of one valence-gram per 100,000 coulombs) is easily measured. The positive ions move in the opposite direction; they form a diffuse brush sensitive to the magnet and no radiation, strictly speaking.

* All these experiments were equally successful with an induction coil or with a Wimshurst machine.

** The rupture of the tube temporarily prevented me from studying the phenomenon at lawer pressures.

Fig. 1

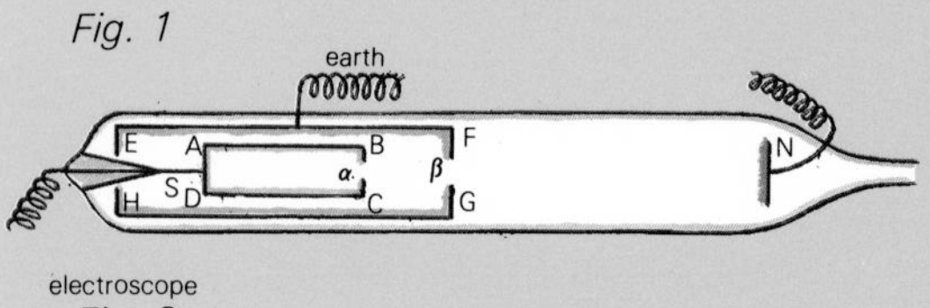

Fig. 2

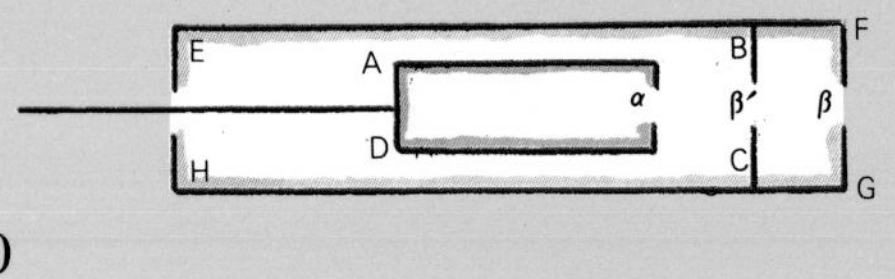

and J. J. Thomson's experiments

I will now try the experiment. You notice that when there is no magnetic force, though the rays do not fall on the cylinder, there is a slight deflection of the electrometer, showing that it has acquired a small negative charge. This is, I think, due to the plug getting negatively charged under the torrent of negatively electrified particles from the cathodes, and getting out cathode rays on its own account which have not come through the slit. I will now deflect the rays by a magnet, and you will see that at first there is a little or no change in the deflection of the electrometer, but that when the rays reach the cylinder there is at once a great increase in the deflection, showing that the rays are pouring a charge of negative electricity into the cylinder.

The deflection of the electrometer reaches a certain value and then stops and remains constant, though the rays continue to pour into the cylinder. This is due to the fact that the gas traversed by the cathode rays become a conductor of electricity and thus, though the inner cylinder is perfectly insulated when the rays are not passing, yet as soon as the rays pass through the bulb the air between the inner cylinder and the outer one, which is connected with the earth, becomes a conductor, and the electricity escapes from the inner cylinder to the earth. For this reason the charge within the inner cylinder does not go on continually increasing: the cylinder settles into a state of equilibrium in which the rates at which it gains negative electricity from the rays is equal to the rate at which it loses it by conduction through the air. If we charge up the cylinder positively it rapidly loses its positive charge and acquires a negative one, while if we charge it up negatively it will leak if its initial negative potential is greater than its equilibrium value.

I have lately made some experiments which are interesting from the bearing they have on the charges carried by the cathode rays, as well as on the production of cathode rays outside the tube. The experiments are the following kind. In the tube A and B are terminals. C is a long side tube into which a closed metallic cylinder fits lightly. This cylinder is made entirely of metal except the end furthest from the terminals, which is stopped by an ebonite plug, perforated by a small hole so as to make the pressure inside the cylinder equal to that in the discharge tube. Inside the cylinder there is a metal disc supported by a metal rod which passes through the ebonite plug, and is connected with an electrometer, the wires making the connection being surrounded by tubes connected with the earth so as to screen off electrostatic induction.

If the end of the cylinder is made of thin aluminum about 1/20th of a millimeter thick, and a discharge sent between the terminals, A being the cathode, then at pressures far higher than those at which the cathode rays come off, the disc inside the cylinder acquires a positive charge. And if it is charged up independently the charge leaks away, and it leaks more rapidly when the disc is charged negatively than when it is charged positively; there is, however, a leak in both cases, showing that conduction has taken place through the gas between the cylinder and the disc....

... Another effect which I believe is due to the negative electrification carried by the rays is the following. In a very highly exhausted tube provided with a metal plug, I have sometimes observed, after the coil has been turned off, bright patches on the glass; these are deflected by a magnet, and seem to be caused by the plug getting such a large negative charge that the negative electricity continues to stream from it after the coil is stopped.

An objection sometimes urged against the view that these cathode rays consist of charged particles, is that they are not deflected by an electrostatic force.

If, for example, we make, as Hertz did, the rays pass between plates connected with a battery, so that an electrostatic force acts between these plates, the cathode ray is able to traverse this space without being deflected one way or the other. We must remember, however, that the cathode rays, when they pass through a gas make it a conductor, so that the gas acting like a conductor screens off the electric force from the charged particle, and when the plates are immersed in the gas and a definite potential difference established between the plates, the conductivity of the gas close to the cathode rays is probably enormously greater than the average conductivity of the gas between the plates, and the potential gradient on the cathode rays is therefore very small compared with the average potential gradient.

We, can, however, produce electrostatic results if we put the conductors which are to deflect the rays in the dark space next cathode. I have here a tube in which, inside the dark space next the cathode, two conductors are inserted, the cathode rays start from the cathode and have to pass between these conductors. If, now, I connect one of these conductors to earth there is a decided deflection of the cathode rays, while if I connect the other electrode to earth there is a deflection in the opposite direction. I ascribe this deflection to the gas in the dark space either not being a conductor at all, or if

the gas in the main body of the tube.

Goldstein has shown that if a tube is furnished with two cathodes, when the rays from one cathode pass near the other they are repelled from it. This is just what would happen if the dark space round the electrode were an insulator, and so able to transmit electrostatic attractions or repulsions.

So far I have only considered the behaviour of the cathode rays inside the bulb but Lenard has been able to get these rays outside the tube. To do this he let the rays fall on a window in the tube, made on thin aluminum about 1/100th of a millimeter thick, and he found that from the window there proceed in all directions rays which were deflected by a magnet, and which produced phosphorescence when they fell upon certain substances, notably upon tissue paper soaked in a solution of pentadekaparalolylketon. As the window is small the phosphorescent patch produced by it is not bright, so that I will show instead the other property of the cathode rays, that of carrying with them a negative charge. I will place this cylinder in front of the hole, connect it with the electrometer, turn on the rays, and you will see the cylinder gets a negative charge.

From the experiments with the closed cylinder we have seen that when the negative rays come up to a surface even thick as a millimeter, the opposite side of that surface acts like a cathode, and gives off the cathodic rays; and from this point of view we can understand the very interesting result of Lenard that the magnetic deflection of the rays outside the tube is independent of the density and chemical composition of the gas outside the tube, though it varies much with the pressure of the gas inside the tube. The cathode rays could be stauted by an electric impulse which would depend entirely on what was going on inside the tube; since the impulse is the same the momentum acquired by the particles outside would be the same; and as the curvature of the path only depends on the momentum, the path of these particles outside the tube would only depend on the state of affairs inside the tube.

From Lenard's experiments on the absorption of the cathode rays outside the tube, it follows on the hypothesis that the cathode rays are charged particles moving with high velocities, that the size of the carriers must be small compared with the dimensions of ordinary atoms or molecules. The assumption of a state of matter more finely subdivided than the atom of an element is a somewhat startling one, but a hypothesis that would involve somewhat similar consequences – viz. That the so-called elements are

– has been put forward from time to time by various chemists.

Let us trace the consequence of supposing that the atoms of the elements are aggregations of very small particles, all similar to each other; we shall call these particles corpuscles, so that the atoms of the ordinary elements are made up of corpuscles and holes, the holes being predominant. Let us suppose that at the cathode some of these molecules of the gas split up into these corpuscles, and that these, charged with negative electricity and moving at a high velocity, form the cathode rays.

I have endeavored by the following method to get a measurement of the ratio of the mass of these corpuscles to the charges carried by them. A double cylinder with slits in it, such as that used in a former experiment, was placed in front of a cathode which was curved so as to focus to some extent the cathode rays on the slit; behind the slit in the inner cylinder, a thermal junction was placed which covered the opening so that all the rays which entered the slit struck against the junction, the junction got heated, and knowing the thermal capacity of the junction we could get the mechanical equivalent of the heat communicated to it. The deflection of the electrometer gave the charge which entered the cylinder.

Thus, if there are N particles entering the cylinder each with a charge e, and Q is the charge inside the cylinder,

$$Ne = Q$$

The kinectic energy of these

$$\tfrac{1}{2} Nmv^2 = W$$

where W is the mechanical equivalent of the heat given to the thermal junction. By measuring the curvature of the rays for a magnetic field, we get

$$\frac{m}{e} V = I$$

thus

$$\frac{m}{e} = \tfrac{1}{2} \cdot \frac{Q\,I^2}{W}$$

In an experiment made at a very low pressure, when the rays were kept on for about one second, the charge was sufficient to raise a capacity of 1.5 microfarads to a potential of 16 volts.

Thus

$$Q = 2.4 \times 10^{-6}$$

the temperature of the thermo junction, whose thermal capacity was 0.005 was raised 3.3°C. By the impact of the rays, thus

$$= 6.3 \times 10^{5}$$

the value of I was 280; thus

$$\frac{m}{e} = 1.6 \times 10^{-7}$$

This is very small compared with the value 10^{-4} for the ratio of the mass of an atom of hydrogen to the charge carried by it. If the result stood by itself we might think that it was probable that e was greater than the atomic charge of an atom rather than m was less than the mass of a hydrogen atom. Taken, however, in conjunction with Lenard's results for the absorption of the cathode rays, these numbers seem to favour the hypothesis that the carriers of the charges are smaller than the atoms of hydrogen.

Fig. 1. Deflection of cathode rays into a cylinder. Here C is the cathode, the rays go through a slot in the anode A. D and E are metal vessels, the outer cylinder D is earthed to guard the inner one E from stray effects. E is insulated except for a connection to an electrometer. When the cathode rays are deflected onto the slits in D and E a charge is measured (J. J. Thomson, *Conduction of Electricity through Gases,* 2nd Ed.).

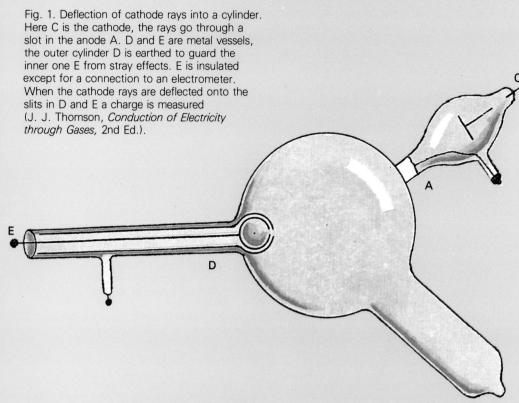

Fig. 2. m/e tube. The electric field between the plates is found from the voltage of the battery. The field of the magnet is measured in an auxiliary experiment. The velocity of the rays is found by balancing the two deflections (G.P. Thomson, *Inspiration of Science*).

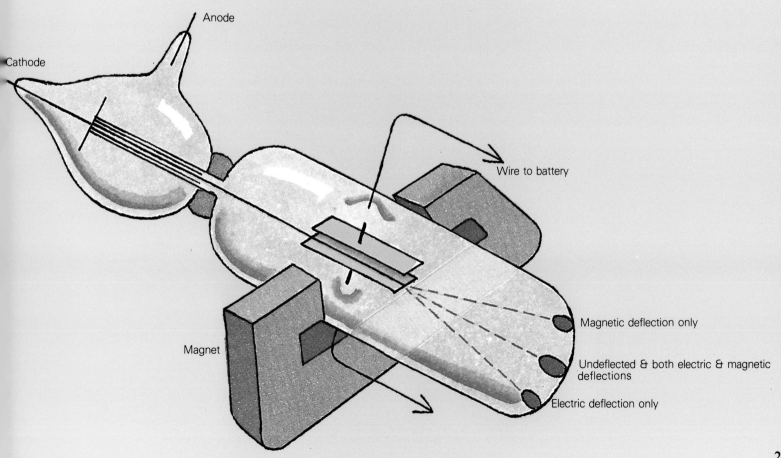

24

Jean Perrin's work in France

In 1897 J. J.Thomson was almost overtaken in his race toward the electron by a young French scientist named Jean Perrin. Perrin was fourteen years younger than Thomson and a man of completely different temperament.

While still a young man freshly employed at the Ecole Normale Supérieure, Perrin had charged headlong into the continuing battle of the cathode rays being fought by wave and particle partisans; and he repeated an experiment first attempted by Crookes. To eliminate the secondary effects that had hindered Crookes's research, he designed a device that was later taken up by Thomson, the "Faraday cylinder." His son, Francis Perrin, later described this experiment in detail:

Jean Perrin understood that only by conducting his experiments in conditions which totally eliminated any possible interference by a secondary phenomenon could he hope to identify and define an electrical charge with no risk of experimental error. The experiment consisted, as Faraday had shown, of introducing charge carriers through a small opening inside an almost closed electrical enclosure, so that an equal charge appeared on the outside surface of the enclosure; this charge was easily demonstrated if the enclosure was itself placed inside a second conductive enclosure, also closed, thus forming an effective electrical screen against outside influence. Perrin created these conditions by placing a Faraday cylinder inside a discharge tube and noted that a negative electrical charge did in fact appear on the outside surface each time (and only when) a beam of cathode rays penetrated it. He had thus found irrefutable proof that cathode rays carried a negative electrical charge, and eliminated the only serious objection to their hypothetical construction from material particles traveling at high speeds and negatively charged. While developing the consequences of this hypothesis, Jean Perrin predicted and verified the action of an electrical field on cathode rays, and understood that by comparing this action to that of a magnetic field, it was possible to calculate the relationship of the charge to the mass, or the specific charge of the particles making up the cathode ray. Unfortunately for Jean Perrin, it was this last calculation which he neglected to do and which J. J. Thomson did. Had he published the order of magnitude which could be deduced from his observation in the report that he gave to the Academy of Science in December 1895, Jean Perrin would have beaten his English colleague to the draw.

We should note that although Thomson made many allusions to Perrin's work, he made very little mention of the work of the German scientist Wiechert, even though Wiechert was working with Kaufmann on measuring the mass and electrical charge of "corpuscles." At the beginning of 1897 Wiechert had given a long speech to the Königsberg (today Kaliningrad) Scientific and Economic Committee, in which he suggested that cathode rays corresponded to particles of electricity which he described as "indestructible" and "electric" atoms, and that these electrical atoms were components of matter in the same way as chemical atoms were. He also attempted to determine their speed by comparing the passage time of cathode rays in the tube with the oscillation period of a Hertzian electric oscillator. But he could not generate high enough frequencies and only achieved this after the completion of Thomson's work. Nevertheless, he too had come fairly close to the discovery.

Perrin's work did not stop with his hypothesis of calculating the relationship of charge to mass and the specific charge of the particles making up the cathode ray. He went on to study the properties of x-rays, establishing that they made the gases through which they traveled conductive, and that they discharged electrified bodies from surfaces which they struck. In both cases, Perrin produced electron emission. He then definitively proved the existence of atoms, demonstrating through experimentation the truth of Avogadro's law* which established molecular magnitudes by means of observing an emulsion in which granules big enough to be visible to the microscope were suspended in a liquid. Perrin obtained the Nobel Prize for his work in 1926. If Thomson gave the Cavendish Laboratory an international reputation by attracting all the great researchers of the time, Jean Perrin, for his part, devoted the last ten years of his life to encouraging scientific research in France. He was behind the foundation of the CNRS (National Center for Scientific Research), the Palais de la Découverte (Discoveries Museum), and the Paris Astrophysics Laboratory. Above all, Perrin had, like many other scientists working at the beginning of the twentieth century, inherited all the enthusiasms and passions of the nineteenth century:

We are now just beginning to become aware that we are one of the shock waves which succeed each other in pursuit of an ideal which never ceases to grow in magnitude; perhaps this can console us for the pettiness and troubles of a too-brief existence, in which a few beams of light coming from Infinity may serve to give us some small serenity.

Jean Perrin (1870-1942), Nobel Prize for Physics in 1926.

"Avogadro's law," the basis for molecular theories, teaches us that equal volumes of all gases under identical conditions of pressure and temperature, contain an equal number of molecules. This number, the Avogadro number, has been established today as 6.02252×10^{23}.

Quanta

by Professor Pierre Auger, Membre of Institute.

We cannot talk about the genesis of contemporary physics and electronics without mentioning the revolutionary effect on the minds of physicists and biologists alike of the discovery of quanta.

Future historians of science will perhaps wonder, with a certain astonishment, how the physicists of the nineteenth century were able to carry out increasingly more precise spectroscopic studies while being content with admiring the superb series of spectral lines, attributed quite naturally to the electromagnetic vibrations in the atoms, and comparing them with the elementary "currents" which Ampère had devised to explain magnetism, without being — if I dare say — dumbfounded by the contradiction between the light emissions of specific wavelength and the Maxwell laws, which require a continuous expenditure of energy by a system of electric charges in motion. It is obvious that the physics of the continuous laws of electromagnetism were incapable of interpreting these spectral lines. But the greatest minds of the latter half of the past century remained deeply attached to the continuous laws of mechanics and electromagnetism.

Nevertheless, the discovery by Jean Perrin of the electric charge of the electron in cathode rays, the discontinuous carrier of electric current, came to replace the old "fluid" and introduced a fundamental discontinuity in a part of science where it had no place. Only chemistry, with the discontinuous laws of combinations, and biology, with the existence of indivisible living cells on all levels, would have been able to foresee the generalization of these basic discontinuities of nature which emerge when the increasingly finer analysis of phenomena is pushed far enough. In such a case, should the new discontinuity become evident and assert itself through chemistry or through spectral lines ? No surprisingly enough, the "quanta" should emerge through studies in the field of thermodynamics and general relationships between the temperature of solids and their radiation. And their "inventor" himself believed for many years that this was a simple trick of calculation, and that continuity should reappear by some means or other.

It was then that, in the fifteen years following the work of Planck, several experimental discoveries — let us rather say theoretical discoveries concerning experiments — gave quanta a physical reality. In 1905 Einstein explained the photoelectric effect, the removal of electrons from an illuminated metal surface, by the absorption of light composed of energy quanta (proportional to the frequency of light vibrations), each absorbed quantum ejecting an electron with the corresponding kinetic energy. Einstein regarded luminous radiation as being composed of "Lichtquanta," or fragments of wave trains containing energy which should be emitted or absorbed at one time.

Then, in 1913, Niels Bohr proposed his model of the atom based on the "planetary" model with a central nucleus and electron orbits, already proposed many years before by Rutherford, but giving it a "quantic" stability which did not permit the application of classical electromagnetism. This time the energy quanta were not connected directly with electromagnetic waves, but defined the orbits of the electrons: this was a materialization of the Planck oscillators. Electron orbits had already been used in 1911 by Langevin to interpret paramagnetism and diamagnetism, but without connecting their stability with quantum states. With the Bohr atom and his interpretation of the series of spectral lines — like the famous Balmer series — the door was opened to an immense chapter of physics and chemistry, atomic spectroscopy and, shortly, molecular spectroscopy. We could say that this revolution was "Pythagorean" because small whole numbers reigned supreme in the definition of electron orbits.

It is quite interesting to note that, during these first fifteen years following the announcement of the discontinuity in light wave emissions in December 1900, there emerged another theoretical system destined to have profound repercussions in physics — that of Einstein's relativity. But this time the repercussions would be in the field of the continuum because the exchanges between different objects in relative displacement at constant speed with respect to each other obey laws of a geometric nature in a four-dimensional universe: three dimensions of space, one of time. There could be no question of "quantizing" the kinetic energy of an object in uniform motion, even if an electron was involved, since this motion has no absolute value. Quantization therefore appeared to be connected with the periodicity of motion, light vibrations, electromagnetic oscillations, closed electron orbits, and, soon, molecular rotations and vibrations since these types of movement have an absolute meaning. The fifteen years that followed this initial period of development in the field of quantum theory saw the emergence of new concepts.

Perhaps the first to emerge was wave mechanics, arising from the works of Louis de Broglie published in 1923 in the proceedings of the Academy of Sciences. It is obviously the corpuscular nature of the "photon," the Einstein "Lichtquanta" closely associated with the electromagnetic waves that constitute radiation and carry its energy, that made the physicist consider the possibility of a somewhat immense relationship between a well-known corpuscle – the electron – and a new type of wave motion, this time not electromagnetic. The notion that the stable orbits of electrons could, in some manner, be compared to the states of standing waves in resonance in closed systems was unusually attractive. However, what one could not suspect at the beginning was the very rapid and intensive development of mechanics, worked out by various theoreticians such as Schrödinger and Dirac. It is also quite remarkable that another quantum mechanics, proposed by Werner Heisenberg and developed concurrently with the theory proposed by de Broglie, indeed leads to the same results – the interpretation of intra-atomic and, soon afterwards, molecular phenomena.

From a philosophical, or psychological, point of view, it is rather fascinating to see transformations as revolutionary as those of quanta, associated waves, and matrix calculus assert themselves within a few years despite obvious misgivings, similar to those which greeted the principles of relativity. The concept of antimatter emerged soon afterward, also in consequence of quanta, and it also came up against difficulties in being accepted by certain scientific circles that were convinced only by experimental proof of the positive electron!

Another result of the discovery of quanta was the new conception of probabilities. Up to the end of the nineteenth century, scientific thought was essentially determinist, Laplacian as it were. If phenomena were still unpredictable, it was because our means of measurement were inadequate due to either a lack of precision – qualitative insufficiency – or a lack of observation and measurement points – quantitative insufficiency. In the first case, for example, the prediction of a number coming up in roulette could have been made possible if the exact measurement of the speed of the roulette wheel, the speed imparted to the ball, the friction between the two, etc., had been carried out far enough. In the second case – that of meteorological forecasts, for example – they suffered, and still suffer from a shortage of observation points, on the oceans for example. Thus, the entire kinetic theory of gases was based on the large number of atoms whose positions and speeds were assumed to be "known" in theory.

At the present time, with quantum physics, probability is in a way removed from specific "events" and has become a simple physical quantity that can be calculated as such. A radioactive atom splits at a certain moment according to its average lifetime, but this moment is not defined by the precise, specific arrangements of the different constituent elements. This is what Heisenberg described as fundamental indetermination: there is simply no sense in speaking of the exact position of an electron in an atom; what is important is the probability of locating a certain site through experiment. Furthermore, the very concepts of the position and the speed of particles are becoming relative because the more precision is sought in one, the more is lost in the other since the product of the two approximations represents a universal quantum constant.

Extended to the case of the roulette, the exact measurements of speed and position which we are discussing in the determinist view of the world would be impossible together and if the probability of a number coming up is calculable, if it has meaning, then the prediction of this occurrence under these of those conditions has none, in all its rigor.

The quanta revolution is becoming familiar to young physicists, but also to biologists. For example, photosynthesis is currently the subject of analytic studies which reveal the action of a series of light quanta, and it has become possible to determine the number of "photons" that trigger the excitation of a retinal cell. And so, we are bringing quanta up to the human scale.

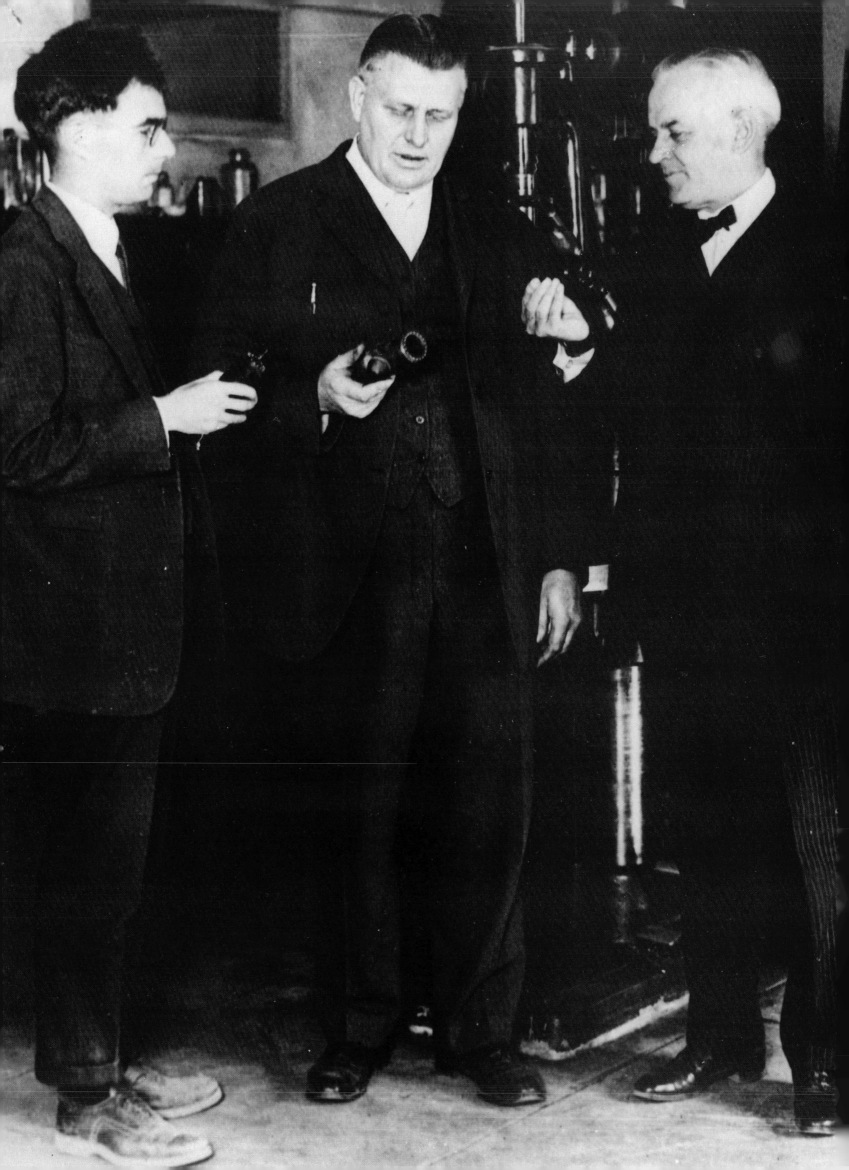

25

Millikan's drop of oil

It was Maxwell who had put forward the principle of the propagation of electromagnetic fields and Thomson who proved the existence of the electron. However, it was Andrew Millikan who made the theory viable in determining the respective values of the electron's charge and mass.

Strangely enough, Millikan became a physicist through the influence of his Greek professor at Oberlin College, a man passionately interested in science. At the University of Chicago, Millikan had the good fortune to work with Michelson, who received the Nobel Prize for Physics in 1907 for his spectroscopy research, the first American to be so honored. Millikan followed in his footsteps in 1923. After finishing his doctorate in physics, he spent a year in Europe where he studied under Max Planck and Henri Poincaré and in 1906 started teaching at the University of Chicago. He was aware that although for ten years he had devoted what free time he had from his official obligations to research, he had not yet published any decisive results. Should he remain a teacher and abandon research? He made his decision in favor of research in 1907. Scientific circles at the time were all interested in the challenge of determining the mass and charge of the electron. Thomson estimated, as we have seen, that the mass of the electron was less than one thousandth that of the lightest atom (the hydrogen atom). But nothing was known about the numerical value of the electron charge, about the nature of the electron itself. The skeptics, such as English writer G.K. Chesterton, continued to sneer: "This is the latest of the hypotheses which tomorrow will be seen as pure abracadabra."

The charge-mass relationship was considered to be a constant but this constant did not show that all electrons were similar particles. The argument was seized on by the adversaries of the particle theory who held that the charge observed was only the statistical average of different electrical energies. Millikan thus set himself the task of demonstrating that electrons were all identical and fundamental particles. He began by taking another look at the experiments done by a Cambridge physicist, C.T.R. Wilson. Wilson had produced an ionized cloud between two horizontal plates by means of x-rays. Then, without applying an electrical field, he used a chronometer to measure the time the cloud took to fall between the two plates. He then repeated the operation, this time complementing the gravity forces by applying an electrical field which accelerated the fall of the charged droplets. But his results showed little uniformity.

Millikan modified the method by calculating the fall time for drops of oil, bringing them to the top or braking their fall under the action of a magnetic field. As the mass and voltage of the droplet was known, the electron charge could be calculated. The reason that the experiment took so long to do is that a droplet − whether charged with one or twenty electrons − fell at the same speed, and the radius of the electron is about ten billion times smaller than the radius of a hair! This gives us some idea of the difficulties involved and the patience Millikan needed to successfully complete his experiment. In September 1909, at a meeting of the British Association in Winnipeg, he communicated his first results. He had already spent two years watching droplets of oil fall, and it was only in 1913 that he considered his experiments to be complete. He had succeeded in showing that the charge calculated was always a whole multiple of a single value. This charge would be recalculated with even more accuracy in 1928 by Erik Baecklin, using a method involving both x-rays and crystals. But the goal had been achieved: the electron existed, and Millikan had proved it. In his book *The Electron* he quoted Lord Kelvin:

When you can measure what you are speaking about and express it in numbers, you know something about it, and when you cannot measure it, when you cannot express it in numbers, your knowledge is of a meagre and unsatisfactory kind. It may be the beginning of knowledge, but you have scarcely in your thought advanced to the stage of a science.

After completing his demonstration, Millikan returned to his research on photoelectricity, and succeeded for the first time in verifying Einstein's formula in its tiniest details. Millikan's scientific stature is considerable, even if it has been strongly contested.

Form left to right: Ira S. Bowen (1927: gaseous emissions from molecules), Jules Pearson (who made Millikan's instruments) and R.A. Millikan.

The cunning Great Britain (represented by a monkey) covets the discoveries of the Siemens cat (1882 caricature). The civil and military powers were extremely interested in scientific research.

In 1917 he was appointed head of research at the National Research Council. As a Lieutenant Colonel in the Army Signals Corps, he directed work on meteorology, communication, and underwater detection. After the war the astrophysicist George Ellery Hale and the chemist Arthur A. Noyes asked him to become head of the newly-founded California Institute of Technology in Pasadena. One of Millikan's recruits was J. Robert Oppenheimer. After receiving the Nobel Prize, Millikan finally became recognized as a leading and influential figure in scientific circles.

Another of Millikan's important achievements was that he was one of the first scientists to understand the importance of industry to scientific progress, even encouraging his students to take up careers in industry. He himself accepted a position as consultant to Western Electric and declared: "No efforts toward social readjustment or toward the redistribution of wealth have one thousandth as large a chance of contributing to human well-being as have the efforts of the physicist, the chemist, and the biologist toward the better understanding and better control of nature."

It is perhaps strange that two men whose personalities and ideas were as different from each other as Jean Perrin and Andrew Millikan, both shared the belief that science was the future savior of humanity. Of course Millikan also foresaw the dangers inherent in increased government aid leading to the politicization of science. After the war he became more and more isolated in his opposition to the huge credits given to research organizations. These two problems are still very contemporary ones, although the early ideals of total progress and an egalitarian society have become blunted. But as scientists become increasingly involved in research undertaken for the benefit of technology and in particular to satisfy the requirements of the army, they do, as Millikan feared, risk losing their autonomy. Electronics is one of the best examples of this growing tendency.

Millikan's work on the electron was continued by James Franck and Gustav Hertz (not to be confused with his famous uncle, Heinrich Hertz) who received the Nobel Prize in Physics in 1952 for their work on the collision of an electron and an atom. They showed that for an atom to be ionized, the electron must have a certain minimum energy measured in electron volts, which varies according to the ionization potential of the different gases. These measurements showed quantitatively that a series of spectral lines thus obtained corresponds to a series of stationary states in the atom's internal energy, thereby verifying the theories of Bohr and Planck. Gustav Hertz's strange destiny is still a mystery. He was a professor in Berlin and joined the Philips company at the age of 33. He worked for Philips from 1920 to 1925, then became professor at Halle before returning to a teaching post in Berlin. When war broke out he was research director with Siemens. He disappeared from sight after the Russians entered Berlin, and no one knew whether he was dead or working for the Russians. All we know is that he died in East Germany sometime in the 1970s.

In 1924 Louis de Broglie proposed for his doctoral thesis a general theory relating to the liaison between corpuscles and associated waves, bringing back to life an old quarrel and offering a possible compromise solution. He affirmed that the movement of a corpuscle was always associated with the propagation of a wave and that the corpuscle's wavelength varied inversely to its speed. One of the cases envisaged by Louis de Broglie, in which, according to wave mechanics, the electron movement no longer follows the laws of classical dynamics, is that in which the wave associated with the corpuscle comes up against various obstacles in the course of its propagation. He explained it as follows:

To recognize the way in which things must happen, let us always be guided by analogy with light. Suppose that we project a radiation of a given wavelength onto a device likely to give rise to interference. Because we know that radiation contains photons, we can say that we are directing a group of photons onto the device in question. In the region where the interferences take place, the photons break up in such a way that they are found concentrated in the area where the intensity of the associated waves is at its greatest. If we now direct onto the interference device, instead of a radiation, a stream of electrons traveling at the same speed whose associated wave is the same wavelength as the light wave used earlier, the wave will interfere just as it did in the first experiment since it is in fact the wavelength which governs the interference. Thus it is perfectly sensible to think that the electrons will concentrate where the intensity of the associated wave is the highest, and that is in fact the principle foreseen by the theory of wave mechanics.

In 1937 Louis de Broglie's theories were confirmed by experiments carried out by an English scientists, Sir George Paget Thomson (son of J. J. Thomson) And also by two Americans, C. J. Davisson and L. Germer, who beat French scientist Maurice Ponte, handicapped by a breakdown at the ENS laboratories, to the Nobel Prize winning post. They sent a beam of monokinetic electrons onto a nickel crystal, thus obtaining a diffraction index similar to that obtained with x-rays, which in turn confirmed the wave motion theory of electrons. Even if today, with the discovery of quarks, antimatter and charmed particles, the electron does not seem to be quite so simple, and even if some scientists are now questioning the whole nature of the known universe, the discoveries of Thomson and Millikan, along with those of the other great physicists of the first half of the twentieth century, were still responsible for opening the way to a great technical revolution, without which the second great revolution of our era, molecular biology and genetics, could never have taken place.

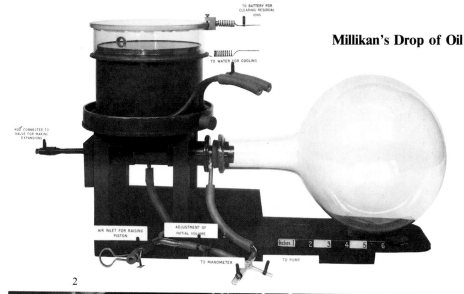

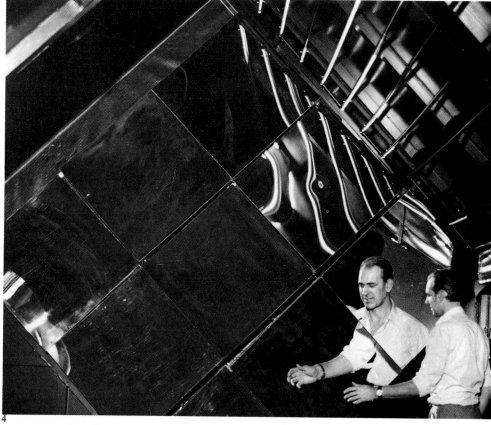

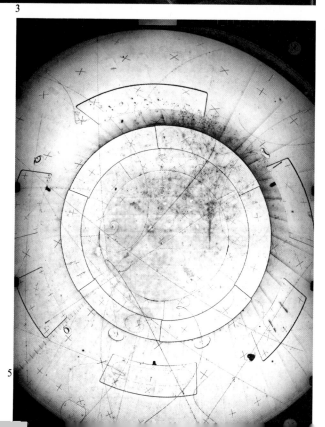

4. This picture, which was taken inside a Cherenkov particle counter gives an idea of the size of detectors used in relation to CERN's 400 GeV S.P.S.

5. Shower of secondary particles created by the collision of a high energy neutrino (A) against a proton of liquid hydrogen filling the huge BEPC European bubble chamber.

6. Popular humor surfaces in this caricature in a Berlin newspaper in November 1878: the bridesmaid "Lamp" says to the maiden "Gas flame": You can't have your cake and eat it.

1. Werner Heisenberg chatting with Marie Curie. Millikan is behind Dr. Curie.
2 and 3. An old detector (C.T.R. Wilson's expansion chamber – 1912) and a modern wire chamber type particle detector which detects the position of particles in space by the induction of a weak current into the wires coming into contact with charged particles. The current is collected at the points where the wires are anchored and transmitted via amplifiers towards a data processing computer.

221

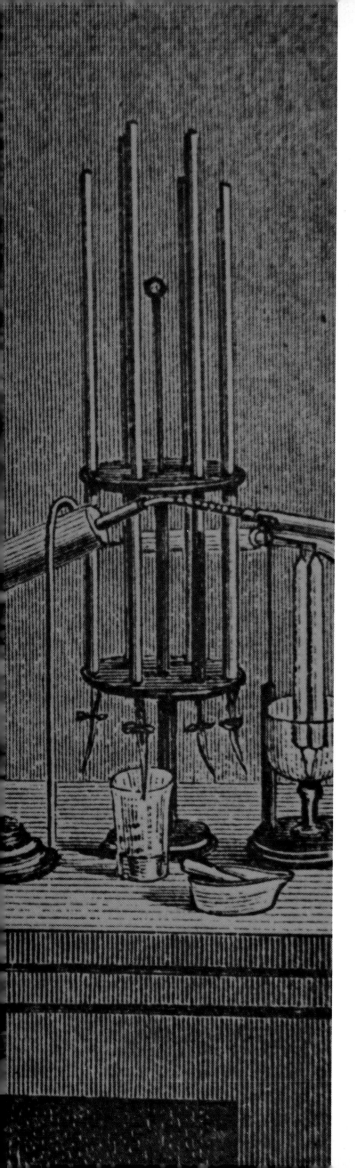

The knights of the electronic adventure

Can thought continue to abstain from thinking the thinker, when after having hidden from view for so long, he is now announcing his presence by shaking the foundations of all that is?
Martin Heidegger

26
From solitary pioneers to organized teams

In the beginning, discoveries were made by a handful of isolated men, all of whom, as individuals, possessed a broad, even encyclopedic knowledge of their world. We have only to read the diaries or the letters of such men as Hertz, Thomson, Perrin, Millikan, and Van der Pol, to become aware of the range of their knowledge; a range which inspired them with remarkable modesty *vis à vis* their personal achievements and which was at the origin of their unshakable belief that all scientific discoveries merely accentuated the mystery of creation. For the first important scientific steps in the field of electronics – the discoveries of the electron, the triode, and the transistor – we need only quote a handful of names: J. J. Thomson, Jean Perrin, Lee De Forest, William Shockley, John Bardeen, and Walter Brattain, although the last three were not isolated workers. Their research program had been put together by Kelly, Research Director at Bell Laboratories, who had been Millikan's assistant when the latter began his work on the electron.

In contrast, the invention of integrated circuits cannot be attributed to any one individual scientist. We may talk of Fairchild or Intel, with Robert Noyce, or of Texas Instruments with Jack Kilby, or discuss the question of who invented what first. But the individual has vanished: Goliath has succeeded David, Money has replaced Adventure, and the isolated inventor has given way to the team. The fate of Farmer Farnsworth in the domain of television when compared with the prodigious success of Zworykin, heavily backed by Westinghouse, General Electric, and RCA, serves as a typical example. There is sometimes a certain chauvinism or a desire for secrecy on the part of various countries who wish to claim the credit for such and such an invention. In fact, many inventions were made simultaneously in a number of different countries, and are quite definitely complementary. The chapter on radar contains some very good examples of this.

First of all, electronics is the story of men with a passion who joined forces with other dedicated men, such as the group formed at the Cavendish Laboratory around J. J. Thomson. J. J. inaugurated the afternoon tea ceremony over which Mrs Rose Elizabeth Thomson and two other ladies presided. This idyllic atmosphere should not obscure the fact that the Cavendish Laboratory was a training-group for scientists. For many years dominated by James Clerk Maxwell, it was in the first half of the twentieth century the cradle of a number of important breakthrough such as Rutherford's discovery of the structure of the atomic nucleus and the proton, the discovery of the neutron by James Chadwick, and the invention of the particle accelerator and its use in light atom transmutation by Cockroft and Walton. A favorite activity was composing songs, the most famous of which was sung to the tune of "Clementine". It was called "Ions Mine" and the chorus was:
Oh my darlings, oh my darlings
Oh my darlings ions mine
You are lost and gone for ever
When just once you recombine.

The history of electronics abounds in anecdotes, new developments and extraordinary personalities. Above all it is the story of missed opportunities and personal dramas, as in the case of one of television's pioneers, John Baird.

John Logie Baird was born in England in 1888. At the age of twenty he gave a demonstration of his great ingenuity by fitting the family house in Dunbartonshire with an electrical supply, the first in the town, and at the same time began to do some quite promising experiments on the telephone. Ironically enough, Baird had earlier undergone treatment for conjunctivitis and deficient view. Despite this handicap he decided to experiment with television, and tried to transmit images from one room to another. To earn his living, he turned his hand to a number of different trades: he sold shoes specially-designed to eliminate trench foot in soldiers; he tried to produce diamonds from coal; he sold secondhand cars, jam from Trinidad, honey and soap from Australia. At the same time he was also taking part in radio broadcasts; and in the spring of 1923 he met his future patron, Will Day, a cinema and radio shop proprietor who set him up in a laboratory in Soho at the end of 1924. Baird founded his television company in partnership with Will Day. Unfortunately for him, he had opted for electromechanical television, and, despite his great inventive talent, he was beaten to the post by Zworykin in the U.S. and the Schönberg

Electronic mapmaking. Behind any technical progress are the men whose job it is to invent, perfect, control and decide.

1. The only photo in the world which features most of the pioneers of the computer. Standing, from left to right: D.W. Davis (Ace), T.H. Flowers (Colossus), Grace Hopper (first programmer, Harvard Mark I), J.H. Wilkinson (Ace), T. Kilburn (Manchester Mark I), T.R. Thomson (Leo), M.V. Wilkes (EDSAC), Cecil Marks (Chairman of the British Computer Society), A.W.M. Coombs (Colossus). Seated, left to right: Elaine Hartree (wife of Professor Hartree who designed the first English differential analyzer), P.C. Williams (Williams storage tube, Manchester), E.A. Newman (Colossus), D. Wheeler (EDSAC) and K. Zuse (Z 1 to Z 4).

1

2

3

and Blumlein team in England. Baird missed his rendez-vous with electronics because he deliberately refused to marry this new technique to the concept of television.

This was not the case with the German researcher Konrad Zuse, although he too was left behind by the tide of history. Zuse, isolated from contemporary English and American research because of the war, worked alone in his parents' living room in Berlin. His first calculator, the Z1, was built in 1936 thanks to the encouragement and money offered by Dr. Kurt Pannke. In 1941 Zuse developed the Z2 and the Z3, the first calculators controlled by a program. He offered his invention to the German chiefs of staff who could see no future potential for it. The Germans, as we have seen in the field of radar, still believed in the blitzkrieg, and the two years it took to produce a calculator seemed to them an absurd waste of time. When the Americans entered Berlin, only the Z4 survived the bombing and it would soon become outdated by newer developments.

Electronics is also the story of missed rendez-vous and unexpected revelations. For example, the inventor of the cathode ray, Ferdinand Braun, hardly knew how to count. One of his pupils in Karlsruhe was a brilliant young zoologist, Jonathan Zenneck, who has left a rather extraordinary portrait of the brilliant inventor in his informal memoirs:

He was not always very rigorous in his calculations, so much so that the mathematicians used to amuse themselves by saying that the operation always succeeded when he had made two faults which cancelled each other out.... The story goes that in one of his courses he was supposed to calculate 25 multiplied by 2. He multiplied 30 by 2, arriving at 60, and concluded that the result must be approximately 50. Calculation bored him: the only thing that interested him was the problem and the solution. He one day wrote a phrase about Faraday that could just as easily describe himself: "He had no mathematical training, but it was perhaps for this reason that he grasped facts in the most naive and direct way.

In 1888 Ferdinand Braun met another famous pupil of his old teacher Helmholtz, but his meeting with Heinrich Hertz turned out to be a missed rendez-vous. In April of that year Hertz was passing through Tübingen where Braun was teaching. He was on his way to the Black Forest to relax for a while after the rigors of his work on Maxwell's equations. Braun invited him to lunch with one of his friends, Leo Graetz. Later Hertz wrote to his parents: *I had already discovered the results which are common knowledge today, and would willingly have discussed them, but I didn't have the opportunity as no one questioned me about it, although enough of my work had been divulged to excite the curiosity of a physicist.*

In fact, during this period many physicists shared Braun's uneasiness with Hertz's ideas and experiments. Only a small number of them had repeated his experiments to verify their accuracy. However it does seem strange that the implications of Hertz's research should have taken so long to arouse Braun's interest. It was some twelve years before he used Hertz's discoveries to develop a wireless telegraphy system, and it was not until September 1889 that he formally congratulated Hertz on his work.

Electronics is also the story of men who were superstars, repressed poets, mysterious, or quite obvious geniuses. One of the superstars is of course Marconi, the brilliant businessman with the appeal of a great romantic. He was a handsome man with a dreamy manner, slim, not very tall, with chestnut

2. The most famous physicists of their time continued to meet over a number of years at the various Conferences organized by the Solvay Institute. In this 1927 photograph from left to right: Back row: A. Piccard, E. Henriot, P. Ehrenfest, E Herzen, T. de Donder, E. Schrödinger, E. Vershaffelt, W. Pauli, W. Heisenberg, R.H. Fowler, L. Brillouin. Middle row: P. Debye, W. Knudsen, W.L. Bragg, H.A. Kramers, P.A.M. Dirac, A.H. Compton, L. de Broglie, M. Born, N. Bohr. Front row: F. Langmuir, M. Planck, M. Curie, H.A. Lorentz, A. Einstein, P. Langevin, C.E. Guye, C.T.R. Wilson, O.W. Richardson.

3. John Logie Baird (1888-1946).

hair and a regularly-shaped nose. In her childhood memoirs, Daisy Prescott writes of her cousin Guglielmo Marconi:

My first recollection of Guglielmo Marconi is of a small boy dressed in a blue sailor suit, standing in the middle of the drawing room. He was a remarkably pretty little boy of five years old with a fringe of golden brown hair lying low on his forehead, under which a pair of wide open and wonderfully intelligent deep blue eyes looked questioningly about him.

He always struck me as being just what an artist would love to choose for his model, or a sculptor for his chisel.

Guglielmo was always of a gay and bright disposition, very affectionate, and passionate at times. He was never tired of trying to invent something, even when he was quite a little boy, and he used to come to his Mother saying, "Come Mamma, and look at what I have made in the garden.

Second to his love for science, Guglielmo loves music, and if he had not given up his life to inventing he would very probably have dedicated it to that beautiful art. He frequently accompanied his mother on the piano when she sang, or played duets with her, as she is passionately fond of music and an expert musician. He had the privilege of being carefully instructed in music whilst at Bologna by Professor Rudolph Ferrari, an enthusiastic musical genius, who greatly admired his young pupil's quickness and talent, and was very sorry when the boy's love for electricity took him away to England for many long years.

Marconi was born in Bologna on April 25, 1874, the son of an Italian father and an Irish mother. His mother, right from the beginning, had blind and unquestioning confidence in her son's genius and helped him as best she could. She wrote to her cousins who arranged for him to meet the Chief Post Office Engineer for London, Sir William Preece, who gave the young man all the support within his power. The Marconi legend grew fast. After making his acquaintance, Emile Girardeau, founder of the French CSF company, wrote: *The real Marconi is even more cheerful, more human and more multitalented than his legend.* In his *Souvenirs of a long life,* Girardeau sketched the brilliant portrait of a perfectly-mannered gentleman who loved women, independence, and his laboratory. His love of dancing, dinners at the Savoy, and compromising women did not appeal to everyone and the Baron de la Chevrelière, Marconi's Paris agent, was scandalized to see "the famous Marconi, Nobel Prize winner, Italian senator and member of the London Royal Academy" carrying on in such a fashion. Marconi provided more scope for the gossips when he divorced the Honorable Beatrice O'Brien (by whom he had had three children) and had himself made a naturalized citizen of Fiume in order to marry a young lady related to a colonel in the Vatican guard.

In the often lyrical memoirs which he published in 1950, Lee De Forest wrote:

That original litigation with the Marconi Company worried me not a little. The Marconi stock jobbers used the newspaper publicity to the utmost possible extent to discourage investors from continuing to finance the growing American De Forest Company. The newspapers in those years flamed with grandiloquent claims and counterclaims, flamboyant advertisements, notices of countersuits for patent infringement, libel and slander.

Once again Marconi, anxious to hold onto his empire, had unleashed the dogs of war. And yet De

New Herald on his death in 1937. It must also be noted that the man who was known as the "Father of the Radio" was not reluctant to pay homage to himself:

Unwittingly then I had discovered an Invisible Empire of the Air, intangible, yet solid as granite, whose structure shall persist while man inhabits the planet; a global organism, imponderable yet most substantial, both mundane and empyreal.

Marconi prided himself on his gift for poetry, and not content with writing a single verse, as J. J. Thomson had done for the Ions ballad, he composed a pseudo-Lamartine Nocturne entitled "O star above the Western Hill" followed by a Personal Lament: *It is long years now since that day of tears and fond farewell, whereon I kissed her lips in pathos of adieu, in agony no lips can tell.* The history of electronics offers not a few such surprises!

Mysteries too, such as that surrounding the personality and work of the English mathematician Alan Turing. The circumstances surrounding his death in 1954 are far from clear and some of the work he did for the Foreign Office is still classified top secret.

A book published in 1959 by his mother Sarah allows us to catch a glimpse of the character of the adolescent who started experimenting in the family cellar from the age of twelve. In 1927, at the age of fifteen, he began to carry out research in advanced mathematics, totally ignoring elementary mathematics. One of his teachers complained that Turing was trying to build the roof before erecting the foundations. At Christmas that same year, he gave his mother a summary of Einstein's theory of relativity to help her understand the subject. His acquaintances, like Michael Woodger or Tom Kilburn describe him as having a secretive, unsociable personality with a large general knowledge but an inability to communicate with others. And yet Turing was to play an essential role in the shadow war waged by the secret services during World War II.

It was Turing who broke the ENIGMA code used by the Germans to code messages exchanged between Enigma machines. Turing designed an electromechanical calculator, the "Bomb," which was used to intercept Enigma messages from April 1940 onward. In his remarkable book, *Bodyguard of Lies,* Anthony Brown describes the lengths Churchill had to go to preserve Turing's code-breaking discovery. In 1936 Turing went to the United States where he met John Von Neumann two years later. Von Neumann invited Turing to become his assistant, but Turing refused and returned to England on the eve of the war. One thing is certain: the meeting of these two minds had a critical effect on the development of calculators. It has even been rumored that Turing and Von Neumann met again secretly during the war at the same time as work was underway on the Los Alamos bomb. Von Neumann himself continually stressed the importance of Turing's fundamental article "On Computable Numbers" (1936). Even during the war Turing's behavior was rather bizarre: he was once accosted by the police while he was taking a stroll wearing a gas mask and explained that the mask filtered out pollen and thus alleviated his hay fever. He used to correspond with his friends in code, taking the matter so seriously that no one but the person to whom he had given the key could decipher his missives. The mystery behind Turing's personality was never deciphered; if anything, it grew even deeper. After the war he fell into a very deep depression, living alone and eating only what he could grow in his garden. One morning he was found dead in his bed, a half-eaten apple sprin-

1. G. Marconi with his wife Cristina during a visit to the Emperor of Japan as part of a world tour in 1933-1934.
2. The Count von Arco (1869-1940).
3. Adolf Slaby (1849-1913).
4. Alan Turing (1912-1954).
5. Lee De Forest – he became the world's first disc jockey in 1907 when he broadcast his favorite melody, the William Tell Overture, from his Laboratory in Parker Boulevard, New York.

229

Forest did not hold it against him and even paid homage to the inventor in an article published in the kled with cyanide in his hand. Was it suicide or accident? His friends and family opted for the second hypothesis, arguing that the cyanide had no doubt been accidentally applied since it was one of the products he was using in his research into electrolysis. One of his friends had even seen him eating bayberries and that he had claimed people should accustom their bodies to poison. It is likely that the mystery will deepen even further when the British Foreign Office archives are actually made available to the public.

John Von Neumann, who was also involved in a great deal of the research and controversies of his time, is a much better known figure.

In the 1930s the town of Göttingen was a kind of meeting place for the scientific elite and it was there that Von Neumann, holder of a master's degree in physics and a doctorate in mathematics, met a young American student named Robert Oppenheimer. Von Neumann taught in Berlin and later in Hamburg, but one fine day, took the boat for the United States. In 1930 and 1931 he taught at Princeton, Americanized his first name to John, and married an American. In 1933, the year in which Hitler came to power in Germany, Von Neumann requested American citizenship. In 1928, when he was still in Göttingen, he had founded a new branch of logical science, the "game theory." Alan Turing in England was also extremely interested in this new science and in devising effective "game theory" solutions. War was threatening in Europe, and Von Neumann once again became interested in defense problems. In 1940 he became consultant for a number of different organizations, including Vannevar Bush's OSRD. Von Neumann's work on computers led him to become interested in neurology psychiatry, and the human brain. In June 1954, when Senator McCarthy was lighting the fuse for the anticommunist witch hunt, Von Neumann testified for Oppenheimer before the House of Representatives, where he spoke out against the dangerous confusion then reigning in the minds of Americans between scientific opinions and political opinions. He declared his support for Oppenheimer. On October 1954, President Eisenhower appointed Von Neuman to the Commission on Nuclear Energy as an appeasement gesture. But the following year he was afflicted with bone cancer and admitted to the Walter Reed hospital, where he died after six months of extreme discomfort. Von Neumann was no doubt one of the true geniuses of his age. An emigrant from old Europe, hostile to all forms of dictatorship, he had devoted his life to what he imagined to be one of the greatest advances ever made in the development of the human brain.

The history of electronics is sometimes also the history of the brusque and unexpected arrival on the scene of a whole generation of new men, whose ideas and approach completely upset the economic chessboard. Two examples of this phenomenon are the denizens of Silicon Valley and the Japanese marketing men. From 1955, with the foundation of the Shockley Semiconductor Company, that part of California around Santa Clara and the San Francisco Bay — the cradle of flower power, LSD trippers and Timothy Leary, Alan Watts disciples and adepts of Zen Buddhism, and the favorite country of English-language science-fiction stories — saw the emergence in ten short years of a universe as strange and mythical as Hollywood had seemed in its day. This brave new world was Silicon Valley, with its huge companies (Intel, Hewlett-Packard, Ampex, Varian, Fairchild Semiconductors), its great world premières (the microprocessor and the microcomputer), its international public, its own museum (the Foothill Electronics Museum), its special slang, and its superstars — the ten or twelve grand masters of business and technology (Robert Noyce, Jack Kilby, Lester Hogan, or Gene Amdahl). For, as a *Time* journalist once wrote: *progress in chip technology moves so fast that only engineers newly graduated from university could fully understand it.* It is certainly true that there exist very few men who can follow or even anticipate new inventions.

William Bradford Shockley, born in London on February 13, 1910, is a brilliant, touchy, and controversial figure whose often unusual technological or sociological opinions have earned him a certain amount of notoriety at Stanford and other places. People who know him speak about him with a mixture of reserve, amusement, and irritation. He left Bell in 1954 to found his own company, attracting some of the most talented and well-known researchers around, including Robert Noyce and Gordon Moore, co-founders of Intel, and Jean Hoerni, inventor of the planar transistor. However differences in opinion, psychological clashes, and the feeling among some of these researchers that they were moving into a field that would be quite unprofitable in the short term (Shockley was interested in researching the scientific properties of semiconductors and in producing switching diodes rather than transistors) soon led eight of them to leave Shockley in 1957 to found the Fairchild Semiconductor company. Shockley baptized them the "eight traitors." In 1958 the Shockley Semiconductor Laboratory itself turned to production as the Shockley Transistor Corporation, but it was too late. Shockley's business career came to an end in 1959 when the company was taken over by Clevite and later, in 1965, by ITT.

It would be difficult to imagine two people more different in style than William Shockley and his Nobel Prize co-winner, John Bardeen. Bardeen was born on May 23, 1908 in Wisconsin. He is a researcher with an extremely rigorous scientific spirit who cares little for making an impression and who does not seem to be exaggeratedly proud of his many successes in scientific research, despite the fact that they are solid enough to have twice won him the Nobel Prize. Bardeen left Bell in 1951 to create a semiconductor research department at the University of Illinois. One of his first students, Nick Holonyak, took up when Bardeen returned to Illinois in 1962, and others joined the staff to make the university one of the leading semiconductor research centers in the world. Holonyak has distinguished himself by his work on silicon-controlled rectifiers (SCR), on the first laser diode operating in the visible spectrum and also through his participation in the commercial development of light-emitting diodes as consultant to the Monsanto company. Bardeen has since devoted himself to work on the theoretical aspects of conduction of electricity in solids.

A sense of personal adventure may no longer be paramount in the Valley, but a new breed of risk-takers — the entrepreneurs — have nonetheless made their appearance in the last ten years. These men are following in the footsteps of W. Hewlett and D. Packard who in 1939 started a company with 480 dollars and their first oscilloscope, put together in their garage. These young companies are all cottage companies whose first products were for the most part put together in garages, bedrooms, or living rooms.

The legend of the Valley is exemplified in the story of Steve Jobs, a barefoot young man who in seven years was transformed into an undisputed hero of high technology. Jobs is a true child of the Valley. At the beginning of the 1970s, while still a student at Homestead High School in Los Altos, Jobs was fascinated by technology – but also influenced by the tail end of the Flower Power movement and its spin-offs: Timothy Leary, prophet of LSD, and Alan Watts with his Westernized Zen. Jobs, a vegetarian Buddhist, who took time off to travel around India before coming home to found his company, may have seemed an unlikely sort of technician. But with his friend Steve Wozniak, who was working at Hewlett-Packard – Jobs was then working for Atari – he designed a gadget, a small personal computer, which Wozniak then built. Jobs sold his Volkswagen truck, Wozniak his little calculator, and with the 1300 dollars thus earned, they founded "Apple" to market the home computer. "Apple I" was sold for the strange sum of $666,666, 666 being the name of the mythical Beast of the Apocalypse.

When Jobs decided to increase the size of his company, he called on Regis Mac Kenna, Public Relations man for Valley companies. Mac Kenna put him in touch with a potential backer called Don Valentine, himself one of a new breed of backers, the "venture-capitalists," who act as a kind of intermediary between investers and new companies in need of long-term assistance to find their feet. This breed of backer will accept risks that no ordinary banker would countenance. To begin with, Valentine was taken aback by the vegetarian Buddhist side of Jobs, but he soon caught on to the essential. The team needed a sales manager, so A. C. Markkula, ex-marketing manager of Intel, invested $250,000 in the business and became an associate.

However the Apple story is also instructive in that it shows the dangers of success: starting up and increasing sales turnover from $2.7 million in 1977 to $200 million in 1980 isn't enough – you must also know how to maintain the leading edge. In 1980 "Apple III" was a failure and it will be difficult for the company to recover completely. Apple may still lead in personal computers, but Tandy and Commodore are beginning to catch up, and even more disturbing, the two giants – IBM and Rank Xerox (nicknamed the "worm" because it wants to devour the "apple") – have not denied their intention of swallowing up Tom Thumb. The market is certainly an interesting one – Americans brought one million personal computers in 1981 and this figure is expected to triple in 1982.

Another interesting aspect of the history of electronics is the very different evolution in different countries for both political and geographical reasons. The two most obvious examples are no doubt the gigantic Soviet Union and miniscule Japan, in a sense the David and Goliath of our story. And just as in the Bible story, the huge Goliath eventually lost the battle to his tiny challenger.

During the Stalin era in the Soviet Union, Yablitchikov and Popov were showered with praise, while the work being done by Edison and Marconi was deliberately ignored. This chauvinism was the source of a joke that it was Ivan the Terrible who really invented x-rays – after all, did he not have the power to see right through his courtiers? Politics interfered with scientific research as, for example, when Stalin condemned cybernetics as an "imperialist development" and a "Jewish invention" (one of its famous exponents was Norbert Wiener). As a consequence, the Soviet Union began to lag behind in the development of computers. However, many Russians did distinguish themselves in the field, although they all did so overseas: Vladimir Zworykin

1

2

3

4

5

1 and 2. Steve Wozniak and Steve Jobs are multimillionnaires at the ages of 32 and 27. Today Jobs is in charge of Apple and Wozniak has gone back to University.
3. Sandy Kurtzig, 35, President of Ask Computer, one of the rare women "entrepreneurs" in Silicon Valley. Staff members have pinned their motto to the wall: "Equal Bytes for Women".
4 and 5. Don Valentine, the venture capitalist who believed in Apple and Tom Perkins, friend of Jim Treybig (Tandem). Without these venture capitalists who took riskes that no conventionnal banker would have accepted, microelectronics would not be what it is today.

1 and 2. Basov and Prokhorov.

dents at the Sverdlovska Academy of Medicine in Leningrad during the war.

Popovski describes Basov as a brilliant scientist, even in his youth, but totally lacking any moral scruples.

Since the tragic period of the war, he says, *Basov distinguished himself by a total absence of ethics. During a year and a half I was able to examine him at leisure and was able to note his profound ignorance of anything that could be called a conscience. He showed this when quite recently, after he had become a member of the Academy, winner of the Nobel Prize and the Lenin Prize, he put forward a most interesting idea for the Academy to study. The only trouble was, as another member suggested, that the same idea had in fact first been conceived by Sakharov in the 1950s. I myself can remember that stormy session in 1948 when Lyssenko officially condemned genetics. I interrogated Basov about his feelings on this subject in the Library of the Agricultural Academy. His answer was quite unambiguous: what did it matter if Stalin destroyed biology or physics, they still needed physicists to build the bomb. In the case of the invention of the laser, it seems clear that he was inspired by an article written by an American outlining the bases of this new science, but which did not supply all the answers. Basov set his team to work on solving the problems and published his complete findings. This is most probably how he came to invent the laser. One of Basov's most remarkable traits was his smile – he was nicknamed the smiling shark. An amusing detail is that Basov detested Prokhorov who returned the sentiment. They even had to divide the Institute into two sections where they could each work on their own issues. To sum up, Basov is, to my mind, a very talented scientist whose talent has been warped at the service of his career.*

It is true that research in the fields of radio (encouraged by Lenin, who supported the first Radio Research center at Nijni-Novgorod) and television (the first transmission took place in 1939), plus the extraordinary success of Russian satellites, are all proof of the ingenuity and talent of Soviet scientists. But the USSR has lagged behind in computer technology and particularly in the field of printed and integrated circuits. The first Eastern bloc calculator was developed around 1948–51 by S. A. Lebedev at the Ukraine Scientific Academy. The program was supported by Admiral Alex Berg, one of the first men to rehabilitate cybernetics after the death of Stalin in 1953. Berg was subsequently to become head of the Soviet Cybernetics Academy. But it was only in 1954 that Soviet-built computers first became really operational (the Ural series, BESM-1, Strela Lem 1).

Research continues at the Academy of Science and at the Institute of Mechanics and Instrumentation (a division of the Ministry for Radio and Industry) with I. S. Bruch, with Y. Y. Vassilievski in Armenia (where the Computer Institute had been founded in 1956–1957 by S. N. Mergelyan), and in Belorussia at the Physics and Mathematics Institute under the leadership of Ivan V. Lebedev. One of the major problems in the Soviet Union is the extremely mediocre quality of components. Miniaturization is only just being introduced. According to information gathered by Sarah White in the course of an enquiry on the situation of Soviet science and technology, we have been able to list five generations of computers: from 1955 to 1960 (electronic tube computers); from 1960 to 1965 (solidstate systems); from 1965 to 1970 (the beginnings of miniaturization); and

in the field of television and the electronic microscope; Isaac Schönberg, a disciple of Eisenstein before he emigrated to England, Boris Strelkoff, who participated in the development of television in France; David Sarnoff, Chairman of RCA, and J. V. Atanasoff, prominent in the field of calculators. In the Soviet Union itself we should mention Popov and Boris Rosing (the first teacher of Zworykin, Schönberg, and almost every other Russian scientist,) Basov, Prokhorov, the physicist Alex Joffe (who in November 1947 signed a polemic against Einstein, with whom he had once been friendly in Berlin days), and the pioneers of the Soviet space program. In his book *The USSR: Manipulation of Science* Mark Popovski, a Soviet journalist who was expelled from the Writer's union in 1977, and who now works for the Smithsonian Institute, gives the following figures:

In 1914, if we add together all the doctors, conservatory professors and teachers at the great seminaries, we come up with a total of about 11,600 scientific men in the country. Then there was the 1914 – 1918 war followed by famine and epidemics which decimated the scientific population. And yet at the beginning of the Second World War, according to official statistics, there were already 89,300 scientists, almost ten times as many as in 1914. The 1973 figure was ten times higher again, and the latest statistics show that the USSR can boast some 1,200,000 scientists!

The great need for scientists, particularly for military purposes, has led to an accelerated program of scientific education to the detriment of the quality of this education. Poposki also points out that in order to safeguard their future careers, students resign themselves most of the time to seeing their study directors assume the credit for all discoveries made:

The type of researcher who freely works on his own ideas has almost disappeared in Russia, one of the few exceptions being a small circle of theorists, most of whom are physicists. The young scientist who joins the ranks of these million workers understands very quickly that without power, he is nothing at all! For these reasons, it is fairly difficult to arrive at a clear view of the history of Soviet science and the history of the last fifty years of electronics.

Mark Popovski knew Basov, one of the inventors of the laser, quite well, as they were both stu-

from 1970 to 1975 (the beginnings of large-scale integration, specialized processors, and increased miniaturization). The introduction of optical electronics is supposed to have begun from 1975 on. This at least is the official research program, although it is unlikely that it was actually followed.

As for Japan, it is today in the forefront of technology and a veritable electronic superpower, but for many years its products were seen merely as "talented imitations." The reality is far more complex. It is true that at the beginning of their history, most of the large Japanese companies were associated with German, American, or English companies, Toshiba with General Electric, and NEC with Western Electric, for example. The "Japanese miracle" only really took hold after the Korean War 1950 – 53), but there were a number of early Japanese researchers and inventors pursuing research long before that time, men whose names are unknown in Europe and the United States. One of these men was Nagaoka, who in 1904 proposed an atomic structure model in which electrons revolved around a central nucleus; another was Seiichi Kaya, who studied the magnetic properties of certain materials. Yet another was Kinjiro Okabé, who developed one of the world's first magnetrons in collaboration with Hidetsugu Yagi (inventor of an antenna design in 1926 which is used even today as a TV reception aerial). We should also mention Kenjiro Takayanagi, one of the pioneers of television; Isaac Koga, who worked on quartz oscillators; and Leo Esaki, inventor of the tunnel diode and winner of the Nobel Prize.

To recognize the true importance of the Japanese in the field of electronics, we need only take a walk down Denkigai street, the electrical section of Tokyo's electronic kingdom, the Akihabara district, where almost 500 shops offer some 25,000 different articles for sale, from the most sophisticated hi-fi outfit to the tiniest miniaturized calculator. We could sum up the amazing success story of Japanese electronics by taking a look at the adventure of four men who were each at the origin of four of Japan's most successful electronics companies: Hisashige Ghi-emon Tanaka, one of the founders of Toshiba, a man who can stand as a symbol of Japan's passage from feudalism to the modern world; Namihei Odaira, founder of Hitachi and the first to suggest manufacturing purely Japanese equipment to meet Japanese needs; Konosuke Matsushita, who popularized the PHP (Peace and Happiness through Prosperity) slogan throughout Japan and fathered the new Japanese philosophy of creating well-being and prosperity through work; and Masaru Ibuka (Sony) who with Akio Morita promoted the image of a creative, aggressive, and highly competitive Japan.

In 1897, when he created the Shibaura Seisakusho company, one of the two companies which merged to create Toshiba, Ghi-emon was seventy-five years old and already famous for his mechanical inventions: dolls, robots, and clocks. Photographs taken at that time show him to be a distinguished old gentleman with the traditional pointed beard and kimono. He was to die well before the end of the Meiji period (which had begut seven years earlier, and which marked Japan's passage from feudalism to the moders era). He belonged, therefore, to the traditions of the ancient era. But his factory, although tiny, was built in the European style. The European influence had already made itself felt in Japan,* and the Japanese were both fascinated and repelled by this threat of "colonization."

But this was also an era in which the authorities actually promulgated an "anti-invention" law to prevent the people from trying to change a so-called "static" society, a time in which rivers were crossed either on horseback or the back of another man, depending on one's social rank, because it was forbidden to build bridges. This deliberate attempt to hold back the future was already beginning to show signs of cracking during Ghi-emon's youth, and in 1858 the Meiji Restoration inaugurated a period of change and development similar to that which took in Europe during the Age of Enlightenment. A new slogan was adopted straight away: "Absorb the new knowledge from the West." The beginnings of wireless research in Japan date back to 1896 during the Meiji period. The international prestige Japan won at the successful conclusion of the Sino-Japanese war (1894 – 1895) and the Russian-Japanese war (1904 – 1905), made her a pertner of the Western nations. An important first step had been taken, although for the duration of the Meiji dynasty Japan was held back by one very important constraint: the country depended completely on Europe and the United States for its industrialization. The first person to foresee the dangers inherent in this policy was the founder of Hitachi, Namihei Odaira, for whom this concern had always been a major preoccupation. He started work at the Kosaka mining company and trained at the Hiroshima hydroelectrical company. When the head of Kosaka recalled him to design electrical equipment for a new mining company to be set up near the fishing village of Hitachi, Odaira developed an electrical motor of completely Japanese design. Compared with contemporary German turbines and American machines, it was perhaps not a very wonderful device, but it was the first piece of modern electrical equipment constructed by a Japanese for Japanese needs. In November 1910 Odaira founded Hitachi Seisakusho, and the jeers abruptly turned to praise when the advent of World War I meant that the large Japanese companies were cut off from their usual foreign suppliers. In the 1920s Odaira was one of the first Japanese to promote the idea of research laboratories attached to industrial companies.

Thus, it took only two generations for Japan to open up to the outside world, to start industrialization, and to produce "made in Japan" products. For so sudden a change to be accepted and correctly oriented, it was obviously necessary to find a new ideal, a new moral approach, in other words, a clearly-defined goal. This was achieved in 1937 when Konosuke Matshushita, an industrialist whose name is a legend both in Japan and throughout the world,

* It was in 1853 that Commodore Perry's American fleet entered the Bay of Tokyo for the first time. In 1858 the Japanese signed a friendship and trading treaty with the United States, then subsequently with the English, the Russians, the Dutch and the French. In 1859 the ports of Kanagawa (today Yokohama), Nagasaki, and Hakodate were opened to trade.

As well as the magnetron, Dr. Yagi invented the famous Yagi antenna.

233

From Solitary Pioneers to Organized Teams

created his Seven Commandments: to serve the company, to be kind, to maintain harmony and cooperation, to fight for progress without losing courtesy or humanity, to assimilate new techniques, and to respect the boss. Matshushita was one of the new men created by the Meiji era. His name is a most felicitous one; it means "the fortunate man under an umbrella pine." He began his working life early; at the age of ten Matsushita was selling bicycles for an Osaka merchant, and at fifteen he joined the Electricity Company. In 1918 he founded his own company by scraping together 100 yen (then worth about $50). He suffered a preliminary setback; this only inspired him to work harder and finally the young man found the solution that would lead him to continuing success right throughout his life. In 1930 he baptized his radio receivers (the best in Japan) with the significant name "Winner." In 1946 he launched the PHP (Peace and Happiness through Prosperity) magazine, and founded the Institute bearing the same name. In February 1962 he was the first Japanese to make the front cover of *Time*. He summed up his philosophy in the following words: "I realized that work should not make one think only of profit. The goal of work is to make people happy by eliminating poverty and giving well-being to all." What Perrin or even Millikan saw as a generous ideal had become the working philosophy of a whole nation.

Masaru Ibuka and Akio Morita were the new men of post-war Japan, thirsty for the poetry and new inventions of a world from which they had long been cut off. Ibuka was a researcher, a man of vision and innovative talent. Morita was a physicist who had specialized in electronics and was therefore in a position to make Ibuka's dreams a reality. It was the fantastic alliance between these two very different and complementary personalities that helped Japan rise out of the ashes of defeat and climb to a level of international competition by means of two key activities: innovation and advanced technology. Their goal was to design highly sophisticated products that would answer society's needs, thus fulfilling the earlier goals of Odaira and Matsushita.

Abstract composition by photographer Erich Hartmann using microelectronic components.

Conclusion

A.F. If your house was on fire, what favorite object would you make sure to save?
J.C. I think I would save the fire.
Jean Cocteau

Electronics disturbs and confuses because it compels us to break with former habits of work, leisure, and daily occupation. This break with traditional activities has aroused a number of fears, some perfectly justified and others little more than myths. One of the most frequently expressed anxieties is that man will become subject to the law of the mighty computer, reduced to little more than a computer file of his past, his impulses and desires and personal life. But what we tend to forget is that right down through the ages it has been the habit of political authorities to keep files and to try to manipulate individuals — the only difference is that the means are now a little more sophisticated. The computer therefore poses the problem of a shift in political balance, particularly in the democratic countries, by playing off the old centers of power against new institutions. This fear is also related to the image men have of themselves in a civilization in which social exchanges are often based purely on appearance. From this point of view it would be men themselves who may tend to reduce themselves to a simple combination of the information given by the computer. This need for identification seems to stem from the same spirit that makes us identify ourselves to others by giving our name, our age, and our profession. People do this every day, and we know that it is a very approximate means of social communication. The computer is merely an extension of this same desire.

Another very frequently expressed fear concerns the changes in the employment situation which any technological upheaval necessarily generates. Generally speaking, this fear is not very soundly based, and the old adage "the machine kills employment" is not necessarily correct. First, according to a number of statistical studies, the number of unemployed people in the United States has not increased since automation of industry first began, and in Japan between 1948 and 1976 the number of job opportunities actually rose by 60 percent. Another important point is that some sectors have even been saved by the advent of electronics (watches and typewriters being two examples), and that this in turn safeguarded jobs. The problems are actually more on the level of necessary strategies for job transfer

and the retraining of the labor force for new markets (manufacture of data processing equipment, software, telematics, etc.). This urgent priority has been held back by ideological or even demagogical fears and taboos which must be recognized for what they are. The old saying "the machine kills employment" depicts the machine as a Machiavellian and perverse being; what is even worse it suggest that work, rather that well-being, is an end in itself. This doctrine may be legitimate in a planned economy; but we have for a long time been aware of its effects on the standard of living of workers and above all on the quality of their lives. Why, for example, in a country like the Soviet Union, have some techniques such as the laser, satellites, and electronic weapons systems been perfectly mastered, while others — such as data processing — have remained extremely weak? The first reply that comes to mind is that the Soviet Union, anxious to preserve its position as a world power in comparison with the United States, has concentrated its energies toward techniques of defense or aggression. But there may also be another reason. The machine liberates man from repetitive daily tasks, thereby giving him more possibilities for initiative and a perhaps disturbing freedom of spirit. We could go even further and say that the responsibility of programmers would be even greater and less easily controlled than the isolated initiatives of individuals working alone. In this sense, the machine can become synonymous with the broadest definition of liberty rather than a symbol of bureaucratic oppression. It is up to man himself to decide to what use he is going to put computers and it is a political choice of great importance.

We could also point out, to the credit of the West, the rise in living standards that accompanies mass production, the possibility of reinvesting extra funds; and the necessity for a competitive economy both on the domestic and the international level.

At the end of the eighteenth century, the concept of happiness advanced by Saint-Just was a completely new idea in Europe. Two centuries later happiness has gone from an idea to an ideology, from an ideology to a personal ideal and from there to a reality well within our reach. Electronics calls first and foremost for a new strategy of job transfer, but it also suggests a modification of behavior and desires, and this is perhaps its most revolutionary aspect. It allows man to reduce the time he spends working and to improve the conditions of his work. Above all, it allows him for the first time to dominate machines and systems — not with the force of his arms, as he has done with steam machines and electrical machines — by the force of his intelligence. It is also the first time that competition on a world scale tends to be based on the idea of quality of content and no longer that of the product, in fields such as telematics.

We cannot continue trying to assimilate this world, modified by electronics, with the mental structures of the past. We can no longer "enter the future by going backwards." As the sociologist Alfred Sauvy says: "The predominance of dogma and assumption and smug, closed attitudes is what is blocking society and preventing it from compensating for the large gaps between it, the techniques it has developed and its own social ideals." Paradoxically, man is afraid of abandoning the sphere of mechanical activity to the machine. He is in the grip of the same atavistic anxiety as the scribes of the Middle Ages with their hostility to printing. It seems as if he is afraid of his new liberty, of this possible obligation to engage in activities concerned with the mind, with creativity and imagination. Forced to question his own nature and role, he falls into a kind of metaphysical anguish.

This sense of aversion and refusal is demonstrated by the great caution with which the new services offered by electronics are greeted: television is too often used to retransmit plays or circus acts, or discussions which would be just as well broadcast by radio, because people have not succeeded in matching the new technique with a new content. Video tape recorders are often used merely to record television programs; electronic music often reproduces the sounds made by traditional instruments instead of finding its own new sounds. Perhaps this prudence and caution is dictated by the fact that new prophets have not yet replaced the old: MacLuhan has not replaced Gutenberg; nor Cage, Mozart; nor Peter Foldès, Rembrandt.

These fears are also conditioned by the abolition of frontiers and the telescoping of time. Radio allowed men to carry on a dialog with men at the opposite end of the earth, and to receive information from the farthest reaches of the universe. Television allows him to participate in the life and death of other human beings on this planet from the comfort of his armchair. The electron microscope and the radiotelescope allow him to explore the infinitely large and the infinitely small. The particle accelerator subjects nature to a kind of torture in order to uncover her secrets. Satellites push back the frontiers of the explored universe and the computer can multiply the speed of calculation and accelerate decision making.

In the era of satellites and telematics, national independence will be even more difficult to maintain and will tend to be replaced by the interdependence of all nations in the world. But this interdependence may trigger an accelerated culture shock and a telescoping of time. Will it be possible to preserve the "souls" of the different nations, to allow them to maintain their own specific cultural identity? One of the solutions proposed to end the cleavage between poor and rich nations, between the owners of energy and the masters of technology, will be technology transfer in exchange for raw materials. But this will mean that some races and peoples will have to make a leap of several decades, even centuries. There is the risk of provoking a violent rejection by cultures not prepared for the idea of the image as — we have seen from the followers of Islam or other peoples whose ancestral customs forbid pictorial representations of reality. If we fail to respect the national identity of countries to whom we want to "offer" these new technologies, we risk creating serious and violent crises. The interference of technology and access to information also pose the problem of monopolies held by one large country or a multinational company which can then dominate distribution networks and data banks. These problems may perhaps be solved by the setting up of a system of international laws, but also by an instinct for coherence and competition: this is what happened when the French group Matra-Hachette or the German group Bertelsman set themselves up in opposition to the *Los Angeles Times* which possesses its own forests, paper factories, newspapers, television channels, record houses, publishing companies, and video networks. But even such regroupings can pose the problem of freedom of creation and expression *vis à vis* the industrial or political powers. Creative people — writers, producers, and artists — must become aware of these problems and arm themselves with the means of protecting these freedoms. It is a great temptation, and one often succumbed to, to put science and politics in the same bag and say: "It's all impossible to understand." But this is the best way of delivering oneself over to the powers that be, which for their part, are never likely to grant a jot more than the people demand.

To avoid this danger of "homogenized" information, both workers and consumers must be vigilant and try to grasp the new mechanisms at work in the system in which they are obliged to live. We should not forget that electronics did not suddenly appear out of the blue. It must be seen within the context of the Western attempt to conquer and dominate nature and as part of the great journey towards knowledge and technical progress that inspired the French Encyclopedists in the eighteenth century.

Some ten years ago, Joseph Needham, in his book on Chinese Science, asked a fundamental question:

Why did modern science, the mathematization of hypotheses about Nature, with all its implication for advanced technology, take its meteoric rise only in the West at the time of Galileo? (Needham, 1969: 16).

He replied to this question by making an allusion to the legal practice of the Middle Ages which did not hesitate to hang a cock accused of laying an egg.

The Chinese were not so presumptuous as to suppose that they knew sufficiently well the laws laid down by God for non-human thins to obey, to enable them to indict an animal at law for transgressing them. On the contrary, the Chinese reaction would undoubtedly have been to treat these rare and frightening phenomena as chhien kao (reprimands from heaven), and it was the emperor or the provincial governor whose position would have been endangered, not the cock. (Ibid., 329).

This little anecdote is quite illustrative of the behavior of the Western world, the heir to the logical Euclidian tradition. It is true that another whole current of Greek throught is in fact opposed to that Euclidian vision, the foundation for our own technological era, reinforced by Cartesian dualism that considered the universe as inert matter quite distinct from man and capable of being enslaved by science: a system of thought which provided a framework for Newtonian mechanics. Quantum physics has renewed the debate by changing our angle of observation and integrating the observer himself into the universe which he observes. Neverthless the language of probabilities is still a mathematical language, as distinct from poetic or philosophical language which cannot be reduced to fractions. The danger lies in losing sight of this basic distinction: science cannot reflect on science, and scientists and technicians alone cannot resolve the global problems posed by technology.

The profound revolution that quantum physics has wrought in our way of *scientifically* regarding the world, is that we are no longer concerned with suggesting explanations of nature, but rather in establishing scales of probability. Even the particle itself is no longer an atom of matter; it does not exist, it is only manifested as a "possibility of existence." In March 1927 Werner Heisenberg enunciated the principle of uncertainty according to which one could not simultaneously determine both the speed and the position of a particle. This brought into question one of the basic principles of conventional science: determinism. On September 26 of the same year, Niels Bohr confirmed the principle of complementarity: a phenomenon such as light can appear, depending on the experimental equipment used to observe it, as a particle-type entity or a wave-type entity. These two complementary aspects of the same phenomenon, almost impossible to study in the same experimental situation, are necessary to the total comprehension of the quantum phenomenon.

The principle of complementarity led to the rejection of another basic cornerstone of classical science. This is *objectivity,* the possibility of describing physical reality independently of its observation. Faced with these abstractions, some physicists and researchers have referred to texts belonging to the

Oriental mystical tradition. Many of them, including Crookes and Lodge, Costa de Beauregard and Brian Josephson, have been attracted by parapsychology. This is where the ambiguity arises: we cannot really talk about "reality," rather we are dealing with possible or probable reality, although it is reality all the same. We are concealing a completely different approach that would tend to relate anything other than the "reducible" reality of physics, even quantum physics, to thought itself, which of its very nature escapes the bonds of science.

The scenarios drawn up by the Club of Rome, MIT, or various German futurologists, even though they do provide a kind of useful navigational aid for the future, fall down precisely because of the rigidity of this attempt at crystallization of thought. All these scenarios have quite often been thrown out of kilter by the advent of the irrational, the so-called human factor. These errors perhaps prove the impossibility of ever making medium or long-term forecasts. In any case they have only served to increase our confusion.

This confusion is further aggravated by the fact that the machine is tending to become less a "manipulated object" and is increasingly becoming part of the fabric of our daily lives. The same terminal can perform and integrate a number of functions, the idea of the "machine" is giving way to the idea of "services." The machine processes and transmits information of all kinds, thus it can no longer be considered as a neutral object totally divorced from its user, but rather as a vehicle for an "intelligent" program or even a "dialog" with man. But in fact all these computer-related terms — programming, intelligence, dialog, and memory can lead us into error because they may bring back to life the terrors aroused in another by such less sophisticated age by such fantastic monsters as Golem and Frankenstein and Goldorak and Doctor Strangelove. These fears may even be exaggerated by the fact that electronics, born of scientific theory the layman can understand only with the greatest of difficulty, has succeeded in making the machine almost invisible, the size of a "chip," and that it brings into use techniques (x-rays, infrared, and ultrasound) almost impossible to imagine without recourse to the kind of abstraction with which the average human being is not at all familiar. But the truth is that intellectual comfort has never been one of the rights of man. There is no question today of creating the "electronic man" or machine that can solve every problem put to it by means of its computing skills, such as the machine already conceived by Raymond Lulle in the Middle Ages for demonstrating the existence of God. We should perhaps ask ourselves whether the reason for man's present disorientation is that he believed for more than four centuries that science would resolve all the questions he put to it, but now finds himself even less well-armed than before to solve the ancestral riddle of the world and his own way of looking at it. Modern Everyman suffers from vertigo, is entrapped in a maze of confused or over-complex ideas, and is in the process of turning away from the basic options before him because of his inability to comprehend them.

Let us reconsider Brecht's famous verse:
Behind the drums
trot the calves
the calves that supply
the skin of the drums.
We have to avoid creating a scenario in which man trots behind the machines whose intelligence he has supplied without preserving his own ability to think. Today, because of the multiplication of techniques and the rapidity of inventions and innovations, decisions escape the experts, who by their very nature are incapable of following the intricate thread of Ariadne out of their labyrinth, of grasping all the links in the chain. They, more than the layman, are incapable of achieving an overall and homogeneous view of the situation. We must reflect on the destiny of our technical and scientific society in nontechnical, nonscientific terms. Despite the nightmares described in science fiction, machines will never dominate man, although it is possible that man, having spent all his ingenuity in creating them, has none left to imagine their creative use, and by refusing to see them for what they really are — extraordinary tools to simplify daily life, sometimes provided with artificial intelligence but lacking innate power to think — will accord them a divinatory and devision-making power that they were not designed to wield.

REFERENCES

Introduction
Werner Heisenberg. *La nature dans la physique contemporaine*. Gallimard, 1962, p. 18.

PART ONE

Chapter 1
Philippe Quéau. "Les ordinateurs à image". Le Monde-Dimanche, January 11, 1981, p. XIII and XIV.
Chapter 2
"L'Avenir du Futur", émission de J.-P. Guirardoni, diffusion February 23, 1981, TF1.
Chapter 5
Alec H. Reeves and E. Maurice Deloraine. "The 25th Anniversary of Pulse Code Modulation," *IEEE Spectrum*, May 1965, p. 58.

PART TWO

Chapter 7
Jean-Pierre Noblanc. "Les Semiconducteurs III et IV face au défi de la microélectronique." L'Echo des Recherches, May 1980, p. 45.
Ernest Braun and Stuart Mac Donald. *Revolution in miniature*. Cambridge University Press, 1978, p. 45.
Chapter 8
Ernest Braun and Stuart Mac Donald. "Revolution in miniature." Cambridge University Press, 1978, p. 49 - 50.
Chapter 9
Sophie Séroussi. "Le laser ou la lumière apprivoisée," Le Monde-Dimanche, November 2, 1980, p. XIV.
Charles Townes, interview, in Laser Focus. Advanced Technology Publication, inc. 1981

PART THREE

Chapter 12
William C. White. "The Early history of Electronics in the General Electric Company," unpublished, 1953—1955.
Chapter 13
Lee de Forest. *Father of Radio*. Wilcox and Follett, Chicago, 1950, p. 3.
William C. White. "The early history of Electronics in the General Electric Company," unpublished, 1953—1955.
Correspondence between Sir Appleton and B. Van der Pol (Philips documentation).
B. Tellegen. "Some Milestones in Electronics," Wireless World, February 1979.
Chapter 14
W.C. White "Some Events in the Early History of the Oscillating Magnetron." Journal of the Franklin Institute, vol. 254, n° 3, September 1952.
Emile Girardeau. *Souvenirs de longue vie*. Berger-Levrault, 1968, p. 218.
R. Warnecke and P. Guénard. *Les tubes électroniques à commande par modulation de vitesse*. Gauthier-Villars, 1951, p. 1 and 2, p. 20.
Chapter 15
L. Marton. *Early History of the Electron Microscope*. San Francisco Press, 1968, p. 2.
Chapter 18
Emile Girardeau. "Souvenirs de longue vie." Berger-Levrault, 1968, p. 212.
Sir R. Watson Watt. "Le Radar." Conférence au Palais de la Découverte, 1946.
David O. Woodbury *Battlefronts of Industry. Westinghouse World War II*. New York: John Wiley and Sons, 1951.

PART FOUR

Chapter 20
A.M. Ampère, Exposé à l'Académie des Sciences, cité par R. Massain, in "Physique et Physiciens." Editions Ecole et Collège, 1939, p. 137.
Erskine-Murrey : *A Handbook of Wireless Telegraphy*, p. 20. London: Crosby, Lockwood and Son. 1913.
James Clerk Maxwell. *A Treatise on Electricity and Magnetism*, Vol. I, Third Edition, Preface to 3d edition, page IX, London: Oxford University Press, Geoffrey Cumberlege 1st edition 1873 reprinted 1955.

- presentation of line integral
Maxwell's wonderful equations, Ch. VI, p. 109.
Electrons, waves and messages, *The Art and Science of Modern Electronics*, by John R. Pierce, Garden City NY: Hanover House, 1956;
- mathematical expressions adopted
Equations de Maxwell, Appendice, pp. 523-526.
Electrotechnique à l'usage des ingénieurs, vol. I
Electricité fondamentale, A. Fouillé, Paris: Dunod, 1969
- displacement of the electromagnetic fields, graphs
Physics, Section 31—11 The mechanism of electromagnetic radiation, pp. 562—563,
by Physical Science Study Committee
Boston: D.C. Heath and Cy.
Chapter 21
Heinrich Hertz. *Erinnerungen. Briefe. Tagebücher*. Leipzig: Akademische Verlagsgesellschaft. 1928, p. 249.
Jean Perrin. Œuvres scientifiques. CNRS, 1950.
Charles Süsskind, Hertz-Huber correspondence, Isis, 1965, vol. 56, 3, n° 185.
Chapter 22
Ronald Clark. *Einstein. The Life and Times*. The World Publishing Cy, 1971, p. 69 Stock.
Chapitre 23
J.-J. Thomson. *Recollections and Reflexions*. Bell and Sons, 1939, p. 23.
J. J. Thomson. "Cathode Rays." Royal Institution of Great Britain, Friday, April 30, 1897.
Chapitre 24
Jean Perrin. "Œuvres scientifiques." CNRS, 1950.

PART FIVE

Daisy Prescott: "On Guglielmo Marconi" (unpublished).
Lee De Forest. *Father of Radio*. Chicago: Wilcox and Follett, 1950, p. 4 et 31.
Mark Popovski. "URSS, la science manipulée." Ed. Mazarine, 1979.

CONCLUSION

Joseph Needham. "The grand Titration." London: George Allen and Unwin Ltd, 1969, p. 16 and 329.

BIBLIOGRAPHY

A

Ackerman Otto and Beck Edward: *Electronic Oscillograph* (Westinghouse Engineer, November 1944).

Aigrain P. and Nozières P. (ed): *De la thermodynamique à la géophysique: Hommage au Pr. Rocard* (Edition du C.N.R.S., 1977).

Aisberg E.: *La radio et la télévision... mais c'est très simple* (Ed. Radio, 1972).

d'Albe Edmund: *The Life of Sir William Crookes* (Fisher Unwin, London, 1923).

Alekseev N.F and Malyarov D.E.: *The Generation of Powerful Oscillations with a Magnetron in the Centimeter Band* (Zhurbal Tekhnichenkoi Fiziki, 1940).

Angello Dr. Stephen J.: *An Age of Innovation* (Mc Graw Hill, 1981), *Semiconductors - New Vigor in an Old Field* (Westinghouse Engineer July, 1950).

Arnaud Jean-François: *Dictionnaire de l'électronique* (Larousse, 1966).

Ashburn Edward V.: "Laser Literature: *A Permuted* Bibliography", 1958-1966 (Western Periodicals Co. North Hollywood).

Ashkin Arthur: "Pression de radiation et lumière laser" (Pour la Science. Issue out of print).

B

Baker W.J.: *A History of the Marconi Company* (Methuen and Co. London, 1979).

Balibar F.: "Des électrons pour l'étude des surfaces" (La Recherche n° 3 July-August 1970).

Balibar F.: "Microscope électronique: la visualisation des atomes" (La Recherche n° 31 February 1973).

Barbe D.F.: *Introduction to Very Large Scale Integration VLSI* (Springer-Verlag, 1980).

Bardeen John: *The Early Days of the Transistor* (Urbana Illinois, 1979).
Semiconductor Research Leading to the Point Contact Transistor (Stockholm, 1957).

Barthélémy Claude: *Evolution industrielle des applications des lasers* (Laboratoires de Marcoussis).

Barton David K.: "Historical Perspective on Radar" (Microwave Journal, August 1980).

Beauclair W. de: *Rechnen mit Maschinen* (Fried-Vieweg und Sohn, Braunschweig).

Bekker Cajus: *Augen Durch Nacht und Nebel* (Wilhelm Heyne Verlag, Munich, 1978).

Bello Francis: "The Year of the Transistor" (Fortune 47, 128, 133, 162, 164, 166, 168, March 1953).

Bernstein Jeremy: *The Analytical Engines* (New York, Random House, 1964).

Bloch F.: *Les électrons dans les métaux* (Actualité Scientifique et Industrielle, Hermann, 1934).

Blondel André: *Travaux scientifiques* (Gauthier-Villars, 1911).

Boettinger H.M.: *The Telephone Book* (Bell Telephone Laboratory, 1977).

Bohle G. and Hofmeister E.: *Electronic Semiconductor Components* (Siemens).

Bois Charles G. du: "A half-Century of Western Electric Achievement" (Western Electric News, 1919).

Bondelier René: *L'ordinateur à l'hôpital* (Masson, 1970).

Boutry G.A: *La connaissance et la puissance* (Albin Michel, 1974).

Braillard Pierre: *L'électricité* (Rencontre Lausanne).

Braun Ernest and MacDonald Stuart. *Revolution in Miniature* (Cambridge University Press, 1978).

Bredow Hans: *Im Banne der Äther Wellen* (Mundus Verlag, Stuttgart, 1954).

Bremenson Claude and Penicaud Etienne: "Télécommunication par satellite" (Commutation et transmission, n° 1 September 1979).

Briggs Asa: *The History of Broadcasting in the United Kingdom* (Oxford University Press, 1965).
I   The Birth of Broadcasting
II   The Golden Age of Wireless
III The War of Words
IV   Sound and Vision

Brillouin Léon: *Conductibilité électrique et thermique des métaux* (Actualité Scientifique et Industrielle, Hermann, 1934).

Brillouin Léon: *Les électrons dans les métaux du point de vue ondulatoire* (Actualité Scientifique et Industrielle, Hermann, 1934).

Brown Anthony: *Bodyguard of Lies* (Bantam, 1976).

Bruch Walter: *Die Fernseh-Story* (Telekosmos Verlag, Stuttgart, 1969).

Brüche Ernst and Schwerzer Otto: *Begründung der geometrischen Elektronenoptik* (AEG Telefunken, 1970).

Brüche Ernst and Mahl Hans: *Elektronenmikroscop und Elektronenmikroskopie* (AEG Telefunken, 1957).

Bruma Marc: "Applications scientifiques et techniques des lasers" (Palais de la Découverte, 1966 ; - paper).

Burc G. and the *Fortune* editors: "Le Monde à l'heure des calculateurs" (Fortune, 1967).

Bush Vannevar: *Pieces of the Action* (William Morrow and Co., New York: 1970).

Laennec cahiers: "L'information au service de la médecine" (December 1969, March 1970).

C

Campbell L. and Garnett N.: *The Life of James Clerk Maxwell* (London, 1882).

Capra Fritjof: *Le tao de la physique* (Tchou, 1979).

Carnap Rudolf: *Les fondements philosophiques de la physique* (A. Colin, 1973).

Carneal Georgette: *A Conqueror of Space* (Liveright, New York, 1931).

Carpenter B.E. and Daran R.W.: *The Other Turing Machine* (N.P.L., 1975).

BIBLIOGRAPHY

Casimir H.B.G. and Gradstein S. ed.: *An Anthology of Philips Research* (Centre Publishing Cy. Eindhoven).

Casimir H.B.G.: "Thoughts on Integrated Circuits and Microtechnology": Paper read before the Electronics Conference of the Hannover Fair, 1968. (Reviewed in Radio Mentor Electronics 34, 366, 1968).
Les cellules solaires (Radiotechnique, 1979).

CERN: "La conférence internationale sur les accélérateurs" (Courrier du CERN, September 1980).

Champeix Robert: *Simple histoire de la T.S.F., de la radiodiffusion et de la télévision* (L'Indispensable).

Chanson Capitaine: "L'optique électronique et ses applications au microscope électronique" (Communications à la Société des Radioélectriciens le 26 mai 1945 published in l'Onde Electrique).

Clark Ronald: *Einstein: The Life and Times* (Stock, 1980).

Clark Ronald W.: *Tizard* (Methuen and Co., 1965).

Clarke John: *La cryoélectronique* (La Recherche n° 38, October 1973).

Coltmann John W.: *Resonant Cavity Magnetron* (Westinghouse Engineer, November 1946).

Conférence internationale sur les télémanipulateurs pour handicapés physiques (I.R.I.A., 1978).

Connaissance de l'électronique (Editions du Tambourinaire, 1955).

Cook J.S.: "Les câbles optiques" (La Recherche n° 45, May 1976).

Couffignal Louis: *La cybernétique* (P.U.F. Que sais-je? 1978).

Crowther J.G.: *James Clerk Maxwell* (Hermann, 1948).

CR: "Today and Tomorrow" (Radio Mentor Electronics 44 n° 9, 334, 1978).

CR: "Intermetall develops faster" (Radio Mentor Electronics 44 n° 11, 452, 1978).

CRM: "Microprocessors for New Jobs" (Radio Mentor Electronics 33, n° 11, 452, 1978).

Crowley-Milling M.: (CERN Publications).

Crozon Michel: *New Projects in Particle Physics* (Endeavour Pergamon Press, 1979).

D

Daumas Maurice: *Histoire Générale des Techniques* (Vol. V P.U.F.).

David Pierre: *Le Radar* (P.U.F. Que sais-je? 1969).

Davies D.W.: "A.M. Turing's Original Proposal for the Development of an Electronic Computer" (N.P.L. 1972).

David Regis: *L'électronique* (P.U.F. Que sais-je? 1964).

Deloraine Maurice: *Des ondes et des hommes* (Flammarion, 1974).

Deloraine Maurice and Reeves Alec: "The Twenty-fifth Anniversary of Pulse Code Modulation" (I.E.E. Spectrum, May 1965).

Devons Prof. S.: *Rutherford's Laboratory* (Cavendish Laboratory).

Dickinson Dale F.: "Les lasers cosmiques" (Pour la Science, n° 10).

Draeger, 1963: *Le laboratoire d'électronique et de physique appliquée.*

Duarme Pierre and Rouquerel Max: *Les ordinateurs électroniques* (P.U.F. Que sais-je? 1961).

Dummer G.W.A.: *Electronic Inventions and Discoveries* (Pergamon Press, 1978).

Dupas Claire: "Avec l'holographie électronique: des clichés à l'intérieur des atomes" (La Recherche n° 49, October 1974).

Dupouy Gaston: *La mécanique ondulatoire et ses applications* (Académie des Sciences, 1967).

Dupouy Gaston and Frantz Perrier: "Microscope électronique fonctionnant sous une tension d'un million de volts" (Journal de microscopie, Vol. 1 n° 3 - 4, 1962).

Dupouy Gaston: "Le microscope électronique 1,5 MV" (Brit. J. Appl. Phys. 1969, séries 2, vol. 2).

Dupouy Gaston and Perrier Frantz: "Le microscope électronique à 1,5 millions de volts du laboratoire électronique du C.N.R.S. de Toulouse" (Onde Electrique, June 1967).

Dupouy Gaston, Perrier Frantz and Durrieu Louis: "Microscope électronique de trois millions de volts" (Journal de microscopie, vol. 9, n° 5 1970).

Dupouy Gaston: "Performance and Applications of the Toulouse 3 million volt Electron Microscope" (Journal of Microscopy, January-March 1973).

Dupouy Gaston: "La mécanique ondulatoire et ses applications" (Académie des Sciences, 1967).

E

Eames Charles and Ray: *A Computer Perspective* (Harvard University Press, Cambridge Massachussets, 1973).

"L'Echo des Recherches - CNET" (n° 100, May 1980).

Eden E.R.C. and Welch B.M.: *GaAs Digital Integrated Circuits for Ultra High Speed (LSI/VLSI Very Large Scale Integration)* (Springer-Verlag, 1980).

The Edison Era (Elfun G.E. Hall of History Publication, 1979).

Edgerton Harold E.: *Electronic Flash Strobe* (M.I.T. Press, 1979).

Edgerton Harold E. and Killian James R. Jr: *Moments of Vision* (M.I.T. Press, 1979).

Encyclopédie internationale des sciences et techniques (Presses de la Cité, 1971).

"Die Entwicklung des Rundfunkempfängers" (Telefunken).

L'épopée électrique (Régidée, 1973).

Erickson John: "Radio Location and the Air Defence Problem: the Design and Development of Soviet Radar 1934 - 1940" (1972).

Ernest J.: "Recherches et réalisations françaises dans le domaine du laser" (L'Onde Electrique, June 1967).

BIBLIOGRAPHY

Evans Walter C.: "Electronics - Prodigy of Electrical Science" (Westinghouse Engineer, January 1950).

Eve A.S., C.B.E.D., LLD, F.R.S.: *The Life and Letters of the Hon. Lord Rutherford* (Cambridge University Press).

F

Feynman Richard: *La nature de la physique* (Seuil, 1980).

Fiftieth Anniversary Golden Yearbook (R.C.A., 1959).

Fifty Years of Japanese Broadcasting (N.H.K., 1977).

Five Years at the Radiation Laboratory (M.I.T., 1946).

Fleming Sir Ambrose: (article in *Nature*, June, 1945).

Fleming John A.: *Memories of a Scientific Life* (London: Marshall, Morgan and Scott, 1934).

Forest Lee De: *Father of Radio* (Milcox and Follett, Chicago, 1950).

Forrester Jay M.: *Collected Papers* (M.I.T. Press, 1975).

Forte S.T.: "The Technology of Integrated Circuits" (Radio Mentor 32, 282, 1965).

Franc Robert: *Eugène Ducretet* (Tambourinaire, 1964).

Friedrich V.: "Conditions for and Examples of the Use of Microcomputers," Paper read at the Colloquium of the Austrian Tribological Society (Vienna, June 1979) (Siemens Publication).

Fritzsche H.: "Les semiconducteurs amorphes" (La Recherche n° 6, November 1970).

Fünfzig Jahre Telefunken (May 1953).

Fünfundzwanzig Jahre Telefunken (1928).

G

Garum Viktor: "Tonfrequenz: Verständertechnik" (A.E.G. Telefunken).

Garbrecht K.: *Microprocessors and Microcomputers. Large Scale Integrated Semiconductor Components. Festkorperprobleme XVII* (Vieweg-Verlag, 1977).

Geddes Keith: *Guglielmo Marconi 1874 - 1937* (Science Museum, London 1974).

Die Geschichte des Magnettons (Telefunken, 1973).

Gille Bertrand: *Histoire des techniques* (Gallimard, 1978).

Girardeau Emile: *Souvenirs de longue vie* (Berger-Levrault, 1968).

"Le grand débat de la mécanique quantique" (Unpublished letters between Max Born, Albert Einstein and Wolfgang Pauli,) (La Recherche n° 20, Feburary 1972).

Goldblat Josef: "La course aux armements stratégiques" (La Recherche n° 37, September 1973).

Good I.J.: "Early Work on Computers at Bletchley" (N.P.L. 1976).

Gosling W., Townsend W.G. and Watson J.: *Field Effect Electronics* (London, Butterworths, 1971).

Granier Jean: *L'électron* (P.U.F. Que sais-je? 1958).

Grivet Pierre: *Optique électronique* (Tambourinaire, 1958).

Guiho Gérard and Jouannaud J.P.: "Intelligence artificielle et reconnaissance des formes" (La Recherche n° 43, March 1974).

Guillien Robert: *L'électronique médicale* (P.U.F. Que sais-je? 1974).

Guillien Robert: *La télévision en couleur* (P.U.F. Que sais-je? 1978).

H

Hag-Hazen J.R.: *Fifty Years of Electronic Components* (Philips Elcoma Division, 1971).

Handel Samuel: *Histoire de l'électronique* (Marabout, 1970).

Harrel Mary Ann: *Those Inventive Americans* (National Geographic Society, 1971).

Hartcup Guy: *The Challenge of War* (Newton Abbott, 1970).

Heisenberg Werner: *La nature dans la physique contemporaine* (Gallimard, Idées 1962).

Hemardinguen P.: *Le superhétérodyne et la superréaction* (Chiron, 1926).

Henning W.: "Microelectronic Sensors of the Siemens AG" (Paper read at the Colloquium of the Austrian Tribological Society, Vienna, June 1979) (Siemens Publication).

Herbert Jean-Louis: "Histoire et perspectives du câble sous-marin" (I.E.E.E., 1980).

Hermann Peter Konrad: "Electronische Messtechnik" (A.E.G.Telefunken 1960).

Hertz Heinrich: *Erinnerungen* (Akademische Verlaggesellschaft, Leipzig, 1928).

Hertz Heinrich: *Schriften Vermischen Inhalts* (Leipzig, 1895).

Historique Thomson (In-house, Thomson-CSF).

A History of Engineering and Science in the Bell System (Bell Telephone Laboratory 1978).

Hogan D.C. Lester: "Reflections on the Past and Thoughts about the Future of Semiconductor Technology" (Fairchild Interface, 1977).

Hommages à Barthélémy, by Strelkoff and Le Duc (Compagnie des Compteurs, 1964).

Howeth Linwood S.: *History of Communications: Electronics in the United States Navy* (Washington, D. C.: Government Press Office, 1963).

Houssini Jean-Pierre: "Les télécommunications spatiales" (La Recherche n° 42, February 1974).

I

Iardella Albert B.: "Western Electric and the Bell System: A Survey of Service" (1964).

Ilopoulos Jean: "Les particules charmées" (La Recherche, May 1979).

"Information médicale et hospitalière" (Symposium de Toulouse, 1968).

J

Jammer Max: "Le paradoxe d'Einstein - Podopolsky - Rosen" (La Recherche, May 1980).

Johansen Robert, Vallée Jacques and Spangler Kathleen: *Electronic Meetings* (Addison-Wesley, 1979).

Johnson Brian: *The Secret War* (London: Hamish Hamilton, 1978).

Jones R.V.: *Most Secret War* (British Broadcasting Corp, 1978).

K

Kampner Stanley: *Television Encyclopedia* (New York Fairchild, 1948).

Kao K.C. and Hockham G.A.: "Optical Communications" (Proc. I.E.E., vol. 113, n° 7, July 1966).

Kitt Hauptmann V.: *Die Entwicklung des Sovjetischen Funkmesstechnik* (Militärtechnik Deutschen Militärverlag, East Berlin, 1970).

Kurylo F.: *Ferdinand Braun* (Heinz Moos Verlag, Münich, 1965).

L

Laffineur M.: "Radio-Astronomie" (L'Onde Electrique, June 1955).

Launois Daniel: *L'Electronique quantique* (P.U.F. Que sais-je? 1968).

Laures Pierre: "Les lasers, principes et applications" (Documentation Air Espace, n° 112, September 1968).

Law Spencer: *The New Brahmins* (New York:Morrow and Co., 1968).

Leaders in Electronics (McGraw Hill, 1979).

Le Duc Jean: *Au royaume du son et de l'image* (Hachette 1965).

Lee Colin K. and Blackburn Hendley N.: *Stories of Westinghouse Research*. Westinghouse

Lefebvre Antoine: "L'espionnage électronique" (Science et Vie, May 1979).

Lenard Philipp: "Uber Lichtemission und deren Erregung" (Heidelberg, 1909).

Lequeux James: "L'astronomie en ondes millimétriques" (La Recherche, June 1980).

Leroy R.: "Les missiles anti-missiles" (La Recherche n° 3, July-August 1970).

Lessing Lawrence P.: *Man of High Fidelity: Edwin Howard Armstrong* (Bantam Books, New York, 1970).

Lobanov M.M.: *Les débuts de la radiolocation en Union Soviétique* (Radio soviétique, 1975).

Loomis Alfred: (Physics Today, November 1975).

Lorenz G.: "Effects of Microelectronics on Consumer Goods and Medical Systems" (Eurocon lecture, March 1980) (Stuttgart Valvo Publications).

Lorenzi J.H. and Le Boucher E.: *Mémoires volées* (Ramsay, 1979).

Lovell Bernard: *Electronics and their Application in Industry and Research* (London, Pilot Press, 1947).

Lovell Bernard and Clegg A.: *Radio Astronomy* (John Wiley and Sons, New York: 1952).

Lucas Pierre: "La commutation électronique" (La Recherche n° 17, November 1971).

Luff Peter L.: "Le téléphone électronique" (Pour la Science n° 7).

Lyons Nick: *The Sony Vision* (Crown, 1976).

M

Mabon Prescott C.: *Mission Communication* (Bell Telephone Laboratory, 1974).

The Magic Crystal... how the Transistor Revolutionized Electronics (Bell Laboratories 1972).

Maillet Henry: *Les applications industrielles des lasers* (Laboratoires de Marcoussis, 1970).

Marconi Degna: *My father Marconi*.

Matra Jean-Jacques: *Radiodiffusion et télévision* (P.U.F. Que sais-je? 1978).

Marton L.: *Early History of the Electron Microscope* (San Francisco Press, 1968).

Massain R.: *Physique et physiciens* (Magnard, 1966).

Maxwell James Clerk: *Traité d'électricité et de magnétisme* (Gauthier-Villars, 1885).

Maxwell James Clerk: *A treatise on Electricity and Magnetism* (Oxford University Press, 1873).

Medical Data Processing (1976).

Microelectronics (special edition McGraw Hill, 1979).

Millikan Robert Andrew: *Autobiography* (Prentice Hall Inc., New York, 1950).

Millikan Robert Andrew: *The Electron, its Isolation and Measurement and the Determination of Some of its Properties* (University of Chicago Press, 1924).

Millman Jacob: *Introduction to VLSI Systems* (Addison Wesley Publishing, 1980).

M.I.T. Series published by the Radiation Laboratory.

Moralee Dennis: "The first ten years at the Cavendish Laboratory."

Morel Pierre: "La météorologie de demain" (La Recherche n° 22, April 1972).

N

Needham Joseph: *La science chinoise et l'Occident* (Seuil 1969). *Grand Titration* (University of Toronto Press)

Neumann John Von: *The Computer and the Brain* (New Haven: Yale University Press, 1958).

Niehaus Werner: *Die Radar schlacht* (Motorbuch Verlag Stuttgart, 1977).

Nora Simon and Minc Alain: *L'informatisation de la société* (Le Seuil, 1978).

O

"L'onde électrique" (special edition July-August 1971).

On the Shoulders of Giants (Elfun G.E. Hall of History Publication 1979).

BIBLIOGRAPHY

Osborne A.: *Round-up — in the Whirlpool of Microelectronics* (1980).

P

Page Robert N.: *The Origin of Radar* (New York: Doubleday, 1962).

Paulu Burton: *Radio and Television Broadcasting in Eastern Europe* (University of Minnesota Press, 1974).

Pelegrin M.: *Machines à calculer électroniques* (1964).

Percy J.D.: *John Baird* (Television Society, 1950).

Perrin Jean: *Oeuvres scientifiques* (C.N.R.S. 1950).

Perrin Jean: *La science et l'espérance* (P.U.F. 1948).

"Pictorial Resources in the Washington D.C. Area" Compiled by Shirley L. Green with the assistance of Diane Hamilton for the Federal Library Committee (Library of Congress Washington D. C., 1976).

Ponte Maurice and Braillard Pierre: *L'électronique* (Le Seuil 1964).

Ponte Maurice and Braillard Pierre: *L'informatique* (Le Seuil, 1969).

Pol Balthasar van der: *Selected Scientific Papers* (North Holland Public Co., 1960).

Poole Lynn and Gray: *Electronics in Medicine* (McGraw Hill, 1964).

Popovski Mark: *L'U.R.S.S. la science manipulée* (Edition Mazarine, 1979).

Powley Edward: *B.B.C. Engineering*.

Poyen Jacques et Jeanne: *Le langage électronique* (P.U.F. Que sais-je? 1960).

Prescott Daisy: *Reminiscences* (unpublished souvenirs by Marconi's cousin, January 1910).

Presentation (Electronique Marcel Dassault, 1978).

Price Albert: *Instruments of Darkness* (MacDonald and Jane's, London 1977).

Price J.A.: *Electrons: Waves and Messages* (Hanover House, 1956).

Prince J.L.: *V.L.S.I. Device Fundamentals Very Large Scale Integration, V.L.S.I.* (Springer-Verlag, 1980).

"Proceedings I.E.E." (special edition, Two Centuries in Retrospect, September 1976).

"Proceedings of the Second National Conference and Exposition on Electronics in Medicine" (San Francisco, 1970).

R

Rabinowicz Ernest: "Les exoélectrons" (Pour la Science n° 1).

"Radar" (Bell Telephone Magazine 1945 - 1946).

Radar Series from Radiation Laboratories.

"Radar Stories and The Navy's History of Radar" (Electronics, June-July 1943).

"The Radio and Electronic Engineer" (special edition Golden Jubilee of E.R.E., October 1975).

La Radiotechnique (1919 - 1969).

Randell Brian: *On Alan Turing and the Origins of Digital Computers* (University of Newcastle-upon-Tyne, 1972).

Read Oliver and Welch Walter Z.: *From Tin Foil to stereo* (Howard W. Sams and Co., 1940).

Reichard W.: "Wits versus Chips" (Newspaper Seminar of the German Institute for Correspondence Studies at Tübingen University, 1980).

Reichardt Jasia: *Robots* (New York: Penguin Books, 1978).

Renard Bruno: *Le calcul électronique* (P.U.F. Que sais-je? 1969).

Reuber C.: "After 1970 - Integrated Circuits in Entertainment Electronics" (Radio Mentor Electronic 32, 496, 1966).

Reuber C.: "Microprocessor and Consumer Electronics" (Yearbook of Entertainment Electronics 27 - 66, 1978).

Reuter Frank: *Microprocessor and Consumer Electronics* (Westdeutscher Verlag, Opladen 1971).

Revue Internationale de Défense: Two special editions (1979):
- Les systèmes de défense aérienne
- La guerre électronique

Reyner J.H.: *The Encyclopedia of Radio and Television* (London: Oldhams, 1950).

Ricard Patrick: *Le livre des Inventions* (Hachette, 1979).

Richardson O.W.: *The Electron Theory of Matter* (Cambridge Press, 1914).

Robieux Jean: *Les perspectives ouvertes par l'évolution des recherches dans le domaine du laser* (Académie des Sciences, 1975).

Robinson Donald: *The 100 Most Important People in the World Today* (Putnam and Sons, 1970).

Rogers: *L'empire I.B.M.* (Laffont, 1970).

Rose Albert: "La photoconductivité" (Société française de physique et Société des radioélectriciens, 12 October 1953, Published in l'Onde Electrique).

Rosenberg Jerry: *The Computer Prophets* (New York: MacMillan, 1967).

Röwentrop Klaus: *Entwicklung der modernen Reglungstechnik* (Munich, Vienna: R. Oldenburg 1971).

Rumebe Gérard: "Le laser" (Revue du Palais de la Découverte, June 1978).

Runge W.I.: *Elektronik ist keine Hexerei* (Econ-Verlag, Düsseldorf and Vienna, 1966).

Ryles Sir Martin: "Radio Astronomy: The Cambridge Contribution" (Cavendish Laboratory, 1970).

S

Sampson Anthony: *The Sovereign State: the Secret History of I.T.T.* (London: Hodder and Stoughton, 1973).

Sanders Frederic: "Radar Development in Canada" (Proceeding I.R.E. 195 - 200, February 1947).

Sarnoff David: *Looking Ahead* (McGraw Hill, 1968).

Satellberg Kurt: *Von Elektron zur Elektronik* (Berlin Elitera-Verlag, 1971).

Sauvy Alfred: *La machine et le chômage* (1978, Dunod).

Scarlott Charles A.: *Television Today* (Westinghouse Engineer, July 1949).

Schlesinger A.: *Principles of Electronic Warfare* (Prentice Hall).

Schüller Eduard: *Das Magnetton* (A.E.G. Telefunken, 1973).

Schwanker Robert: "Der laser 1917 - 1978" (in "Kultur und Technik" review of the Deutsches Museum in Münich, March 1979).

Schwitters Roy F.: "Les particules élémentaires avec du charme" (Pour la Science n° 2).

Segalen Jean: "La guerre électronique" (La Recherche n° 46, June 1974).

Selme Pierre: *Le microscope électronique* (P.U.F. Que sais-je? 1970).

Shiers George: *Bibliography of the History of Electronics* (The Scarecrow Press Inc. Metuchen, New Jersey: 1972).

Shockley William: "Transistor Physics" (American Scientist 42 1954).

Six pays face à l'informatisation (Documentation française 1979).

Sixty Years of Hitachi (1910 - 1970).

A Solid State of Progress (Fairchild).

Sonneman Rolf: *Geschichte der Technik* (Aulis Verlag, Derebner, D.D.R. 1968).

The Steinmetz Era (Elfun G.E. Hall of History Publications 1979).

Southworth Brian and Boixader Georges: *La chasse aux particules* (Tribune editions C.E.R.N. 1978).

Streetly Martin: *Confound and Destroy* (London: MacDonald and Jane's, 1978).

Stumpers F.L.M.H.: "L'œuvre scientifique de Balthazar Van der Pol" (Revue technique Philips 1960/61, vol. 22, n° 2) "Some notes on the correspondence between Sir E. Appleton and Balthazar Van der Pol."

Süsskind Charles: *The Encyclopedia of Electronics* (Reinhold, New York 1974).

Süsskind Charles: "On the Origin of the Term Electronics. Coming of Age and Some More" (Proc. of the I.E.E.E., September 1976).

Süsskind Charles: "On the First Use of the Term Radio" (Proc. I.R.E., March 1962).

T

Tellegen Bernard: "Some Milestones in Electronics" (Wireless World, February 1979).

Terman F.E.: "From Wireless to Radio to Electronics" (Stanford University, 1964).

Terrien Jean: "La cellule photoélectrique" (P.U.F. Que sais-je? 1965).

Thomson George Paget: *J. J. Thomson and the Cavendish Laboratory in His Day* (Nelson, 1964).

Thomson J. J.: *Recollections and Reflections* (Bell and Sons, 1936).

Thomson J. J.: *Cathode Rays* (Royal Institution of Great Britain, 30 April 1897).

"Toshiba Review" (n° 100, November-December 1978).

Tricot Jean: "Histoire de l'informatique" (Science et Vie, April-May-June 1979).

Trinquier Jacques: "Les microscopes électroniques géants" (La Recherche n° 13, June 1971).

Trenkle Fritz: *Die Deutschen Funkmessverfahren bis 1945* (Motor Buch Verlag, Stuttgart, 1979).

Trenkle Fritz: *Die Deutschen Funk-Navigations und Funk-Führungsverfahren bis 1945* (Motor Buch verlag, Stuttgart, 1978).

Tulkay Edgar: *Astronomy Transformed* (John Wiley and Sons, 1976).

Turing Sarah: *Alan Turing* (Cambridge: Heffer, 1959).

Tyne Gerald F.J.: *Saga of the Vacuum Tube* (Indianapolis: Howard W. Sams and Co. Inc., 1977).

V

Varian: "Twenty-five Years" (1974).

Vasseur Albert: *De la T.SF. à l'électronique* (Editions Techniques et Scientifiques françaises 1975).

W

Warnecke R. and Guénard P.: *Les tubes électroniques à commande par modulation de vitesse* (Gauthier-Villars,1951).

Watson-Watt Sir Robert: *Le Radar* (Lecture given at the Palais de la Découverte, 1946).

Watson-Watt Sir Robert: *Three Steps to Victory* (London Odham Press, 1957).

Watts R.K.: *Advanced Lithography, Very Large Scale Integration V.L.S.I.* (Springer-Verlag, 1980).

Weiher Siegfried and Goetzeler Herbert: *Weg und Wirken der Siemens Werke im Mortsbritt der Elektrotechnick 1847 - 1972* (Siemens, 1972).

Werk und Wirken (Tekade, 1978).

"Westinghouse in World War II: Radio and X-Ray Divisions" (Westhinghouse, October 1950).

Westmijze V.K.: "Studies on Magnetic Recording" (Philips Research Reports, April-June-August-October 1953).

White Sarah: *Guide to Science and Technology in the USSR* (Francis Hodgson, 1964).

White William C.: *The Early History of Electronics in the General Electric Company* (unpublished 1953 — 1955).

BIBLIOGRAPHY

White W.C.: "Some Events in the Early History of the Oscillating Magnetron" (Journal of the Franklin Institute, vol. 254, n° 3, September 1952).

Wiegand Otto: *Passive Bauelemente Siemens, 1881 - 1974).*

Wiener Norbert: *God and Golem Inc.* (M.I.T. Press, 1964).

Wildes Karl L.: "The Digital Computer Whirlwind" (unpublished M.I.T.).

Wilding-White T.M.: *Jane's Pocket Book II - Space Exploration* (London: MacDonald and Jane's, 1976).

Wilkes Pr.: *A History of Computing* (New York: Academic Press, 1979).

Wilkes Pr.: "Early Computer Developments at Cambridge: The Edsac."

Williams Trevor: *A History of Technology* (Oxford: Clarendon Press, vol. 11, 1978).

Windell Dr. Peter: "John Baird" (New Scientist, 11 November 1976).

Woodbury David O.: *Battlefronts of Industry: Westinghouse World War II* (New York: John Wiley and Sons, 1951).

Y

Yaviv Amnon: "L'optique des guides de lumière" (Pour la Science n° 17).

Z

Ziel A. van der: *Solid State Physics* (Prentice Hall, 1966).

Zuse Konrad: *Der Computer Mein Lebens-Werk* (Munich, 1970).

Zworykin V,K,: "Some Prospects in the Field of Electronics" (Washington, D. C.: Government Printing Office, 1952).

Zworykin at 89 (*Communicate*, the R.C.A. Magazine, 1978).

Zworykin and Wilson: *Les cellules photoélectriques et leurs applications* (1934).

Zworykin V.K. and Morton G.A.: *Television* (New York: John Wiley and Sons, 1940).

A FEW MILESTONES

1831: The laws of electrical induction (Faraday)
1873: Maxwell's *Treatise on Electricity and Magnetism*
1879: Cathode tube and Crookes's tube
1883: The Edison Effect
1887: Hertz demonstrates Maxwell's theories
The first international telephone line between Paris and Brussels
1890: Branly's iron filing coherer
1891: Stoney invents the word "electron" to describe the elementary particle of electricity
1895: X-rays (Roentgen)
In December Jean Perrin delivers an address to the Academy of Science on electrical particles
1896: Becquerel discovers radioactivity
Braun's cathode ray oscilloscope
1897: The "discovery" of the electron by J. J. Thomson (April 30)
1900: Quantum theory
Fessenden invents the high-frequency generator. Poulsen's "telegrafon"
1901: First wireless transmission across the Atlantic between Newfoundland and Cornwall, achieved by Marconi
1903: The theory of thermionic emission (Richardson). Fleming's diode. Poulsen invents the arc generator
1905: Wehnelt's oscilloscope
Von Lieben's tube. Einstein explores the photoelectric effect
1906: The triode. Dunwood discovers the detecting capacities of crystals
1909: Millikan's experiments: the electron's charge and weight. KQW, the first radio station set up by C. D. Herrold in San José, California, with a 15 W arc transmitter. It was later sold to Columbia Broadcasting System and became KCBS.
1910: Rutherford discovers the atomic nucleus and the proton
1911: Kamerlingh-Onnes discovers superconductivity
1912: Langmuir's pliotron
1913: Niels Bohr establishes the fact that electrons move in stable orbits that correspond to different energy levels
1917: Lucien Levy takes out a patent for the superheterodyne
1918: Sonar
1920: November: the first radio station in the world to broadcast regularly, in Pittsburgh: KDKA (Westinghouse)
1922: The Taylor-Young report on radio signals reflected by a metal object. September: the "Comintern" station in Moscow. November: Radiola Foundation of the B.B.C.
1924: Hull's magnetron
1926: The Yagi antenna, invented in Japan by Hidetsugu Yagi and Shintaro Uda
1929: The Japanese magnetron (Yagi and Okabé)
1930: Vannevar Bush's analogic calculator
Lawrence's particle accelerator
1931: The electron microscope (Knoll and Ruska) (Rüdenberger's patent)
1932: Cockcroft and Walton invent the linear accelerator
1933: The iconoscope. Armstrong's patent for frequency modulation
1934: The C.S.F. patent for "an obstacle locating system and its applications"

1936: C.S.F.'s resonating cavity magnetron
Electronic television in several countries
The B.B.C. opens the Alexandra Palace studios. Publication of Alan Turing's paper "on computable numbers"
1938: A magnetic recording tape is used for the first time in the world by a radio station in Germany
1939: The klystron
Regular television broadcasts begin in the United States in April
1940: The tape recorder (Telefunken)
1944: Aiken's "Mark 1" at Harvard
1945: E.N.I.A.C. (Von Neumann's theory)
1947-48: The 33 rpm microgroove
The point contact transistor
1948: The hologram (Dennis Gabor)
Mark I (Manchester University)
1949: 45 rpm
E.D.S.A.C.
1950: Discovery of the ferrite bead as a memory
Junction transistor
1951: The first digital calculator (UNIVAC Sperry Rand)
The video tape recorder
The field effect transistor
1953: N.T.S.C. The American color television system
1954: The maser
1957: Sputnik I, the world's first satellite (October)
1958: The laser
1959: SECAM (The French color television system)
PAL (The German system)
The first man in space (Gagarin)
February: Jack Kilby applied for a patent for the integrated circuit. August: Hoerni's planar transistor
1960: "Echo": the first telecommunications satellite (American)
1962: The first satellite broadcast (Telstar) between the US and Europe
1965: Corporation for communication by satellite (COMSAT) for telephone and TV relays between the USA and Europe by "Early Bird"
1966: The bubble memory
1967: PAL (August) SECAM (October) broadcasts begin
1968: L.S.I.
1969: Apollo II lands on the moon (the first man on the moon)
1970: M.O.S. technology developed
1971: First calculator on a chip (Texas Instruments)
First microprocessor (Intel)

FOUNDING DATE FOR SOME OF THE COMPANIES MENTIONED

1847: Siemens und Halske (Germany)
1869: Western Electric (U.S.A.)
1875: Shibaura Seisakusho (merged with Toshiba in 1939) (Japan)
1886: Westinghouse (U.S.A.)
1891: Philips (Holland)
1892: Groupe Thomson (France)
General Electric Company (U.S.A.) (Thompson-Houston & Edison General)
1898: C.G.E. (France)
1899: Nippon Electric Company or N.E.C. (Japan)
1900: Braun-Siemens (Germany)
1903: Telefunken (Germany)
1910: S.F.R. (France)
Hitachi (Japan)
1911: Federal Telegraph Company of California (U.S.A.) Computing Tabulating Recording Company (became I.B.M. in 1924) (U.S.A.)
1917: Plessey (Great Britain)
1919: R.C.A. (U.S.A.)
Radiotechnique (France)
Compagnie Générale de Télégraphie sans Fil (C.S.F.) (France)
1920: I.T.T. (U.S.A.)
Fairchild Aerial Camera Company (became Fairchild Camera and Instrument Corporation in 1944) (U.S.A)
1922: Matsushita (Japan)
1924: Bell Laboratories (U.S.A.)
I.B.M. (see 1911)
1927: Gründig (Germany)
C.B.S. (Columbia Broadcasting System) (U.S.A.)
1928: Galvin Manufacturing Corporation (became Motorola in 1947) (U.S.A.)
1930: Geophysical Service Inc. (became Texas Instruments in 1951) (U.S.A)
1931: E.M.I. (Great Britain)
Compagnie H.W. Egli Bull (became Compagnie des Machines Bull in 1933) (France)
1935: Fujitsu (Japan)
1939: Hewlett-Packard (U.S.A.)
Toshiba (merger between Shibaura Seisakusho and Tokyo Denki) (Japan)
1944: Ampex (U.S.A.)
Fairchild (see 1920)
1946: Tokyo Tsushin Kogyo K.K. (became Sony in 1958) (Japan)
1947: Motorola (see 1925)
1948: Varian (U.S.A.)
1951: Texas Instruments (see 1930)
1955: Shockley Semiconductor Laboratory (U.S.A.)
1957: Fairchild Semiconductor (part of Fairchild)
1958: Sony (see 1946)
1963: Electronique Marcel Dassault (France)
1966: C.I.I. (France) or Compagnie Internationale pour l'Informatique
1968: Intel (U.S.A.)
Thomson-C.S.F. (merger) (France)
1975: C.I.I.-Honeywell Bull (France)

DATE DUE			
JA11'85			
JY30'85			
DE10'87			

DEMCO